DISCOVERING QUARKS

This book describes the development of our understanding of the strong interactions in particle physics, through its competing ideas and personalities, its false starts, blind alleys, and moments of glory – culminating with the author's discovery of quarks, real particles living in a deeper layer of reality. How were quarks discovered, what did physicists think they were, and what did they turn out to be? These questions are answered through a collection of personal remembrances. The focus is on the reality of quarks, and why that reality made them so difficult to accept. How Feynman and Gell-Mann practiced physics, with their contrasting styles and motivations, presented different obstacles to accepting this reality. And how was the author, as a graduate student, able to imagine their existence, and act on it? Science buffs, students, and experts alike will find much here to pique their interest and learn about quarks along the way.

GEORGE ZWEIG discovered quarks, increased our understanding of the inner ear, and invented algorithms for signal and image processing. He received a MacArthur Fellowship the first year it was awarded, the Caltech Distinguished Alumnus Award, and the J. J. Sakurai Prize of the American Physical Society. He is a member of the National Academy of Sciences.

DISCOVERING QUARKS

Remembering
Feynman, Gell-Mann, and Tollestrup

GEORGE ZWEIG

Massachusetts Institute of Technology

CAMBRIDGE
UNIVERSITY PRESS

Shaftesbury Road, Cambridge CB2 8EA, United Kingdom

One Liberty Plaza, 20th Floor, New York, NY 10006, USA

477 Williamstown Road, Port Melbourne, VIC 3207, Australia

314–321, 3rd Floor, Plot 3, Splendor Forum, Jasola District Centre,
New Delhi – 110025, India

103 Penang Road, #05–06/07, Visioncrest Commercial, Singapore 238467

Cambridge University Press is part of Cambridge University Press & Assessment,
a department of the University of Cambridge.

We share the University's mission to contribute to society through the pursuit of
education, learning and research at the highest international levels of excellence.

www.cambridge.org
Information on this title: www.cambridge.org/9781009473507

DOI: 10.1017/9781009473514

When citing this work, please include a reference to the DOI 10.1017/9781009473514

First published 2025

A catalogue record for this publication is available from the British Library.

A Cataloging-in-Publication data record for this book is available from the Library of Congress

ISBN 978-1-009-47350-7 Hardback

To Erica Chunree Jen, my wife, who died unexpectedly before this book was completed, and to Kathleen Kay Corkins, whose many acts of kindness help those in need.

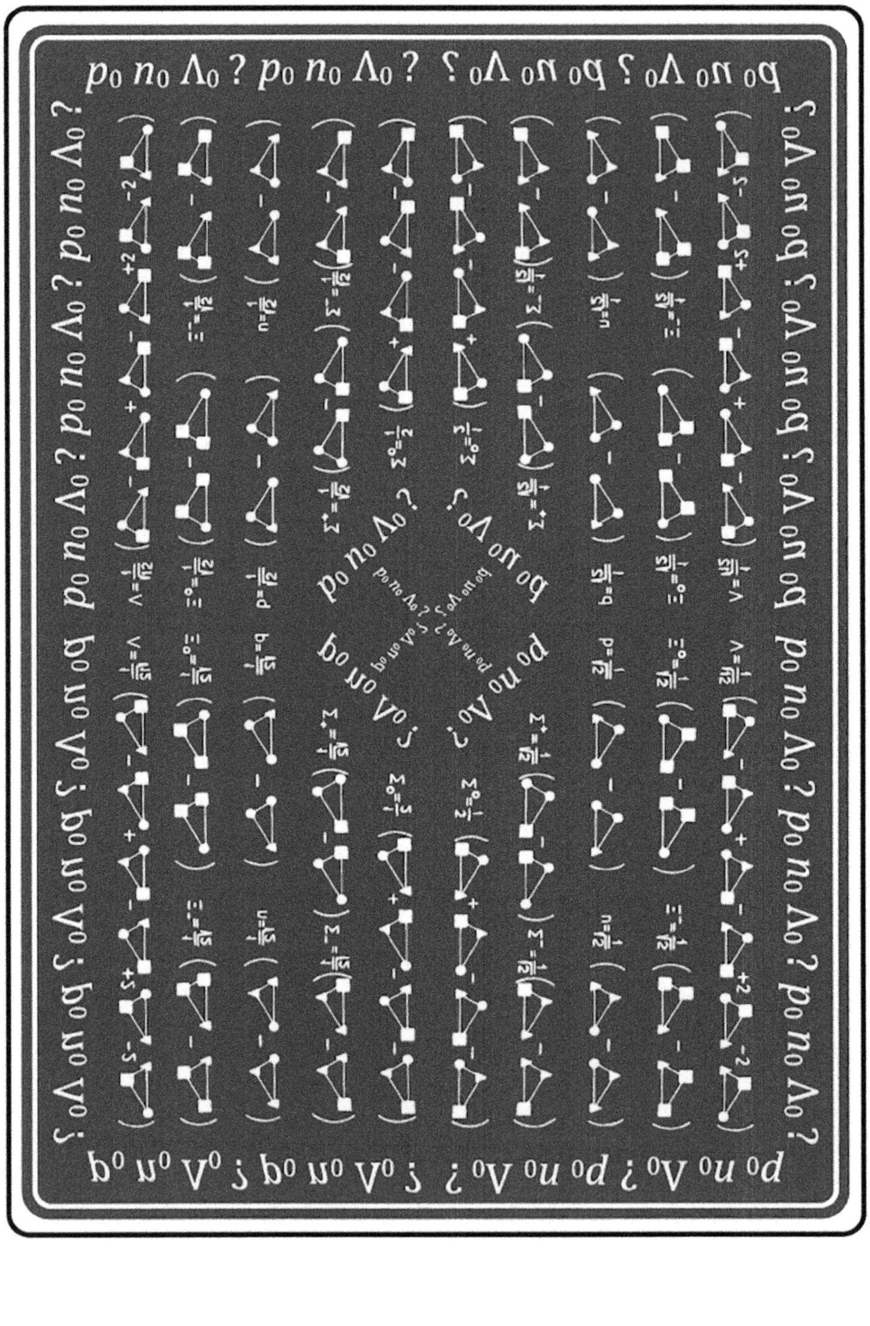
p0 n0 Λ0 ?

Contents

Preface

"If, in some cataclysm, all scientific knowledge were to be destroyed, and only one sentence passed on to the next generation of creatures, what statement would contain the most information in the fewest words? I believe it is the *atomic hypothesis* that *all things are made of atoms.*"

The Feynman Lectures on Physics, Volume I, 1963.

This book is for science buffs, students, and the curious, with material of interest to experts as well.

What is our world made of? The answer has changed with time. Not earth, water, air, and fire — not atoms — but electrons and quarks. That's it, at least for now.

How were quarks discovered, what did physicists think they were, and what did they turn out to be? The following chapters answer these questions with a collection of remembrances of physicists, the events they encountered, and their reactions to them. All these chapters have the discovery of quarks, to a greater or lesser extent, at their intersection. Since they were originally written at different times for different occasions, each examines that discovery through its own lens, from disparate angles, with differing levels of detail.

The focus of this book is the reality of quarks, and why that reality made them so difficult to accept. How Feynman and Gell-Mann practiced physics — their contrasting styles and sources of satisfaction — presented different obstacles to accepting quarks as real particles. And why was I, a graduate student, able to imagine their existence, and act on it?

Chapter 1, "The Road to Quarks," traces the period from 1896, with Henri Becquerel's accidental discovery of radioactivity, to 1935 with Hideki Yukawa's theory of the nuclear force.

Chapter 2 chronicles the explosion in the number of strongly interacting particles, and efforts to understand them. It ends with an introduction to the discovery of quarks (originally called "aces"), and the resistance to accepting them for what they are: real particles that live in a deeper layer of reality.

Chapters 1 and 2 describe the recurring chaos and confusion that existed during the time between the discoveries of radioactivity and quarks. Once discovered, the path to the acceptance of quarks as real particles was equally confounding.

Chapter 3 on Alvin Tollestrup, my experimental physics thesis advisor, describes the singular contributions he made to physics, and what was required to practice experimental particle physics at the highest level. What I learned from him affected me profoundly.

Chapter 4 on Richard Feynman, my theoretical physics thesis advisor, is a collection of vignettes that reveal aspects of behavior and thought that contributed to his mystique and unique accomplishments in physics.

Chapter 5 focuses on Murray Gell-Mann, who dominated particle physics for more than a decade starting in the mid-1950s. His perspective, style, and major contributions to physics, as I witnessed them, are described. Examples of a darker side — his pattern of inadequate attribution — that I only fully realized while writing this book, are also given. A comparison of Feynman with Gell-Mann concludes this chapter.

Chapter 6, "A Deeper Layer of Reality," describes my path to quarks, relating events beginning as a graduate student at Caltech in the spring of 1963, then a visitor at CERN, and finally a lecturer at a summer school in Erice, when my work on quarks was essentially completed. A way of judging the veracity of improbable theories is presented that, when applied to the quark model, pits the a priori likelihood that quarks exist against the difficulty of explaining the experimental data in a theory without them.

Chapter 7, "Epilogue," looks back on the discovery of quarks, identifies what of the original conception has survived, and what was missing. It also revives empirical relations that led to the discovery of quarks, long forgotten, that are relevant today. Concepts underlying quantum chromodynamics, the quantum field theory based on quarks and gluons, are also summarized.

Appendix A, "A Primer on Quarks," completes an unfinished manuscript Gell-Mann and I were writing with Francis Bello for the *Scientific American* in 1971. Here, as elsewhere, I follow the advice of Max Delbrück, ex-particle physicist and co-inventor of molecular biology, who advised me to assume "readers are infinitely ignorant, but infinitely smart."

Appendix B reproduces "The First CERN Report," distributed January 17, 1964, that first presented experimental evidence for the existence of quarks, but was never published.

The book concludes with the "Glossary: A Guide for the Perplexed," and an Index.

Viewed as a whole, this book tells the story of particle physics as it developed throughout most of the 20th century, with its competing ideas and personalities, its false starts, blind alleys, and moments of glory. It also will allow you to determine if you would have believed in quarks as real particles inside of protons and neutrons at the time of their discovery.

Much of the book is based on invited talks I gave to general physics audiences with students in attendance. Here those talks are revised, expanded, and integrated to make them more accessible and comprehensive, without losing their original spirit. The occasions of these talks were Richard Feynman's centennial celebration (2018), a conference in honor of Murray Gell-Mann's 80th birthday (2010) [227], the 50th anniversary of the discovery of quarks (2014) [228], the 50th anniversary of the deep inelastic scattering experiments at the Stanford Linear Accelerator that confirmed the existence of quarks as real particles (2019), and a memorial symposium for Alvin Tollestrup (2020). These chapters describe the work of phenomenal physicists, work that reflects different styles and temperaments. By necessity these chapters are fragmentary, "photographic flashes" from the past. But taken together they form a coherent picture of

the development of particle physics, while describing the associated human behavior on personal and institutional levels.

If you're primarily interested in Feynman, read Chapter 4, although Feynman makes cameo appearances in Chapters 2 and 5. For Gell-Mann, read Chapter 5. For quarks, after Chapter 2 and "A Primer on Quarks" in Appendix A, continue with Section 5.5 and Chapters 6 and 7. Supplemental material for an advanced high-school or undergraduate physics class is found in Chapters 1, 2, 6, and 7, and the Primer in Appendix A.

Experts should read Appendices 6.A and 6.B. This will help them understand if they, in January 1964 when the discovery of quarks was published, would have believed that quarks could form the foundation of a new quantum field theory for the strong interactions.

If you struggle with an unfamiliar technical term, skip it and focus on the qualitative ideas presented. Go with the flow. Consult the Glossary or Index if help is needed. As I hope you will see, it is possible to understand the substance of matter with a minimum of background information. Rarely can you go so deep so quickly. If you haven't been there before, give it a try. Stick with it; see how beautiful understanding can be.

Surprisingly, of my two thesis advisors, Alvin Tollestrup and Richard Feynman, Alvin influenced me more. Measurement provides information about the world, direct from observation, uncorrupted by history or thought. By watching Alvin I learned about measurements, not only how to make them, but how to evaluate the measurements of others, initially in physics, later in sensory physiology. I always began with measurements, looking for something that wasn't quite right — an anomaly — that told me where to start.

Richard Feynman, on the other hand, didn't start with measurements. It wasn't his style, and he didn't have to. As he explains [66]: "In general we look for a new law by the following process. First we guess it." Well that may have worked for Paul Dirac and Richard Feynman, but not for the rest of us. How to guess is not something that can be taught. Feynman's contributions were *sui generis*, in both their content and presentation. But he did provide a yardstick by which I could measure my work in theory, and the theoretical work of others. He helped form my professional superego.

I thought Gell-Mann — "Mr. Physics" — *the* leader in his field at the time, might become my thesis advisor after I switched from experimental to theoretical physics, but he went on sabbatical. He did, however, teach me an important lesson before he left. "Strange" particles β-decayed more slowly than their non-strange counterparts. I had an explanation for their slow decay, with a theoretical prediction! Excited, I walked into his office, explained my idea, and handed him two pieces of paper, each with a graph. One was a Xerox copy of a published experimental result, the other my prediction. Murray's head went back and forth from one graph to the other. Then he looked up at me and said, "In our field it is customary to put theory and experiment on the same piece of paper [to superimpose them]." I was mortified, although I'm sure that wasn't his intent. I learned a lot from Murray.

The context in which a revolutionary idea appears, and the reaction of the scientific community to it, is also the subject of this book. Two year before the discovery of quarks was published, Thomas Kuhn published *The Structure of Scientific Revolutions*. I read that book in 1965 and was amazed, even then, by the predictive power of Kuhn's ideas. Now that the long process leading to a theory of the strong interactions (quantum chromodynamics) has been completed, there is much more information for historians and sociologists to analyze and digest from both within, and outside, Kuhn's theoretical framework. Moreover, learning the history that led to the discovery of quarks, and analyzing the events in the decade-long process that led to their acceptance, may be of interest to practicing physicists. Understanding the history of a scientific revolution may help them realize when they are standing in the middle of one.

More than 60 years ago I started work in particle physics, but switched to neurobiology a decade later. Reprising the past, reliving it now with mixed emotions, makes sense as creation becomes more challenging, and remaining time more limited. In Fellini's movie *Amarcord* ("I Remember"), Titta — a surrogate for Fellini in his adolescence — is shown growing up among an uncommon cast of characters in the small seaside village of Borgo San Giuliano. Like Fellini, but at Caltech in Pasadena, I grew up surrounded by characters with remarkable eccentricities of their own, as you shall see.

Finally, why publish? Well, it's always fun to tell a good story, and discovering quarks is one of them, as are the remembrances of Feynman, Gell-Mann, and Tollestrup. These stories, my "Amarcord," are not well known; they derive from personal interactions, and are a blend of "witness literature" and intellectual history describing a unique time and place that was the center of theoretical physics in the middle of the 20th century.

There is a deeper reason for publishing, a *cri de coeur*. My life started off as one of almost a million, the cohort of Jewish children in Europe and the USSR at the start of World War II. This is the story of what happened to one of them. Imagine if the others had survived to tell their tales. Why tell mine? I owe it to them.

Chapter 1
The Road to Quarks

Henri Becquerel around the time of his 1903 Nobel Prize for the "discovery of spontaneous radioactivity." (Photo by Paul Nadar; Public domain via Wikimedia Commons.)

1.1 Prologue

The theory of strong interactions, quantum chromodynamics (QCD), developed in two phases: the first involved the discovery of quarks, the second specified the nature of their interactions. These phases arose from two different traditions, that of Rutherford and Bohr, and that of Einstein. The first was grounded in observation and the startling interpretation of what was observed; the second was created in a triumph of the imagination made possible by the power of quantum field theory, once its fields were known. The first phase, culminating in the discovery and acceptance of quarks as real particles, is the focus of this book.

1.2 Radioactivity

The story begins with the accidental discovery of spontaneous radioactivity by Henri Becquerel in 1896. Becquerel, like his father Edmond, grandfather Antoine, and later his son Jean, was a professor of physics at the Musée d'Histoire Naturelle in Paris. On the evening of January 20, 1896, at the weekly Monday meeting of the French Academy of Sciences, Becquerel heard Henri Poincaré describe Wilhelm Röntgen's recent discovery of X-rays. During his talk Poincaré wondered if X-rays were emitted by luminescent bodies (bodies that emit light not caused by heat).

Becquerel had long been interested in phosphorescence, the "magical" delayed emission of light of one color following a body's exposure to light of a different color. In his research he used phosphorescent uranium salts inherited from his father. Becquerel wondered if the light emitted in phosphorescence contained X-rays. He reported the results of his investigations to the French Academy of Sciences five weeks later on Monday, February 24, 1896 [11]:[1]

> One wraps a Lumière photographic plate with a bromide emulsion in two sheets of very thick black paper, such that the plate does not become clouded upon being exposed to the sun for a day. One places on the sheet of paper, on the outside, a slab of the phosphorescent substance, and one exposes the whole to the sun for several hours. When one then develops the photographic plate, one recognizes that

---

[1] Translation by Carmen Giunta: web.lemoyne.edu/giunta/becquerel .html

the silhouette of the phosphorescent substance appears in black on the negative. If one places between the phosphorescent substance and the paper a piece of money or a metal screen pierced with a cut-out design, one sees the image of these objects appear on the negative [Figure 1.2].

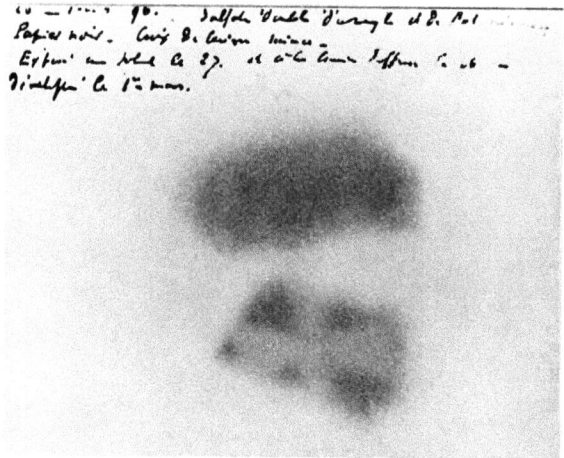

Figure 1.2. One of Becquerel's photographic plates exposed to radiation from potassium uranyl sulfate, a phosphorescent substance. Note the shadow of a copper medal, a Maltese Cross, placed between the plate and the lower piece of the radioactive uranium salt. (Public domain via Wikimedia Commons.)

Becquerel took this as evidence that the uranium salts had absorbed sunlight and later emitted a penetrating radiation like X-rays.

Seeking further confirmation he continued his experiments, but the weather did not cooperate; it became overcast for several days. When the French Academy of Sciences met the following Monday on March 2, Becquerel continued his story [12]:

> Since the sun was out only intermittently on these days [Wednesday and Thursday when the plates were wrapped], I kept the apparatuses prepared and returned the cases to the darkness of a bureau drawer, leaving in place the crusts of the uranium salt. Since the sun did not come out in the following days, I developed the photographic plates on the 1st of March, expecting to find the images very weak. Instead the silhouettes appeared with great intensity.

Becquerel then established that the penetrating particles differed from X-rays because, unlike X-rays, they could be deflected by electric and magnetic fields. The nuclear age had started.

Although Becquerel's grandfather, father, and son lived to the ages of 89, 71, and 75, Becquerel died at 55 in 1908. The cause of death is unknown, but he had developed serious burns on his skin, likely from handling radioactive substances.

Abel Niépce de Saint-Victor

Many years earlier someone else had made the same accidental discovery as Becquerel. In 1861 the French photographic inventor Abel Niépce de Saint-Victor observed that, even in complete darkness, uranium salts could expose photographic emulsions. He reported to the French Academy of Sciences [160]:[2]

> This persistent activity ... cannot be due to phosphorescence, for it would not last so long, according to the experiments of Mr. Edmond Becquerel; it is thus more likely that it is a radiation that is invisible to our eyes.

Seven years later Edmond Becquerel published the second volume of his "La Lumière: Ses causes et ses effets" [Light: Its causes and effects], mentioning Niépce de Saint-Victor's report. The phenomenon was inexplicable, but only Henri Becquerel followed up on his discovery, perhaps because he had a definite question to answer: "Did phosphorescent light contain X-rays?"

Lefty Gomez, an all-star baseball pitcher for the New York Yankees in the 1930s, is credited with saying "I'd rather be lucky than good." Niépce de Saint-Victor was lucky; Becquerel was both lucky and good.

1.3 Rutherford and His Group

Shortly after Becquerel's discovery, the nature of the radiation emitted in radioactivity was elucidated by Ernest Rutherford. It was sorted into three categories: α, β, and γ-rays. The α-rays (ionized helium atoms) and β-rays (energetic electrons) were distinguished by

their charge and penetration power, β-rays traveling further. The γ-rays (high-energy photons), which were neutral, differed from X-rays by traveling further in matter.

But what was radioactivity? In his 1921 Nobel Lecture, Frederick Soddy, one of Rutherford's early collaborators, said:

> The interpretation of radioactivity which was published in 1903 by Sir Ernest Rutherford and myself ascribed the phenomena to the spontaneous disintegration of the atoms of the radio-element, whereby a part of the original atom was violently ejected as a radiant particle, and the remainder formed a totally new kind of atom with a distinct chemical and physical character.

Rutherford received the Nobel Prize in Chemistry in 1908, five years after Becquerel and Marie and Pierre Curie received their Nobel Prize in Physics. Soddy did not get his Nobel Prize in Chemistry with Rutherford, but received one on his own 13 years later "for his contributions to our knowledge of the chemistry of radioactive substances, and his investigations into the origin and nature of isotopes." All told, 11 of Rutherford's students, collaborators, and members of his laboratory eventually received a Nobel Prize in either Physics or Chemistry. Rafi Muhammad Chaudhry, one of Rutherford's students, went on to pioneer experimental nuclear physics in Pakistan, and with his student Mustafa Yar Khan, founded Pakistan's nuclear weapons program.

A photo of Rutherford with his group at Manchester University, taken four years after his prize, is shown in Figure 1.3.

The Nucleus

Joseph John (J. J.) Thomson's discovery of the electron in 1897 indicated that the atom was divisible. This meant that the negative charges on electrons in a neutral atom must be canceled by positive charges. In Thomson's view [201]:

> The atoms of the elements consist of a number of negatively electrified corpuscles enclosed in a sphere of uniform positive electrification.

---

[2] Translation taken from: en.wikipedia.org/wiki/Abel_Niepce_de_Saint-Victor.

Figure 1.3. Rutherford's group in 1912. Rutherford is seated, center, second row from the bottom. Names mentioned in the text are in italics. Left to right, Front Row: R. Rossi, H.G.J. Moseley, J.N. Pring, H. Gerrard, and *E. Marsden*; Second Row: *H. Geiger*, W. Makower, A. Schuster, *E. Rutherford*, R. Beattie, H. Stansfield, and E.J. Evans; Third Row: C.G. Darwin, J.A. Gray, D.C.H. Florance, Miss Margaret White, May Leslie, H.R. Robinson, A.S. Russell, H. Schrader, and Y. Tuomikoski; Fourth Row: J.M. Nuttall, W. Kay, H.P. Walmsley (partially obscured), and *J. Chadwick*. The photo was donated by Lady Joyce Marsden from the papers of Sir Ernest Marsden. (Alexander Turnbull Library, Wellington, New Zealand.)

This so-called "plum pudding model" kept atoms neutral and stable by mixing a continuum of positive with punctate negative charges ("plums" in 19th-century English meant "raisins").

However, Thomson's model was incorrect. In 1909 Hans Geiger, a postdoctoral fellow from Germany, and Ernest Marsden, a 20-year-old undergraduate from New Zealand, showed that charge was concentrated within a very small nucleus within the atom. In a scattering experiment they bombarded a thin gold foil with α-particles, expecting them to pass right through with only small deflections. Instead they reported that [82]:

Conclusive evidence was found of the existence of a diffuse reflection of the α-particles. A small fraction of the α-particles falling upon a metal plate have their directions changed to such an extent that they emerge again at the side of incidence [they bounced back].

This innocuous sounding statement was revolutionary, apparently contradicting Maxwell's equations for electricity and magnetism and possibly Newton's laws of mechanics. Before his death in 1937, Rutherford recorded his reaction to their findings [187]:

It was quite the most incredible event that has ever happened to me in my life. It was almost as incredible as if you fired a 15-inch shell at a piece of tissue paper and it came back and hit you.

Its interpretation was provided by Rutherford two years later in 1911 [181]. In his 1920 Bakerian Lecture on the "Nuclear Constitution of Atoms," Rutherford recalls interpreting the Geiger-Marsden experiment, and discovering the nucleus [185]:

> The conception of the nuclear constitution of atoms arose initially from attempts to account for the scattering of α-particles through large angles in traversing thin sheets of matter. Taking into account the large mass and velocity of the α-particles, these large deflections were very remarkable, and indicated that very intense electric or magnetic fields exist within the atom. To account for these results, it was found necessary to assume that the atom consists of a charged massive nucleus of dimensions very small compared with the ordinarily accepted magnitude of the diameter of the atom. This positively charged nucleus contains most of the mass of the atom, and is surrounded at a distance by a distribution of negative electrons equal in number to the resultant positive charge on the nucleus. Under these conditions, a very intense electric field exists close to the nucleus, and the large deflexion of the α-particle in an encounter with a single atom happens when the particle passes close to the nucleus.[3]

This meant that *positive and negative charges were separated in an atom*, but how? Not statically, for electrons would fall towards the nucleus. Not dynamically with electrons circling the nucleus, for they would radiate energy like tiny spiraling antennae until they reached the nucleus. Rutherford had discovered the nucleus, an object that could not exist according to the known laws of physics, and with it, the separation of nuclear and atomic properties.

Rutherford gave credit where credit was due when he noted in his 1911 paper that Hantaro Nagaoka

seven years earlier had modeled the atom as a nucleus of positive charge surrounded by circulating rings of electrons [154]. The pattern of the discrete frequencies of light emitted from atoms, their "spectral lines," led him to this conclusion. Nagaoka had created a primitive precursor of the Bohr atom. With respect to scattering, his model was very similar to Rutherford's, except Rutherford's electrons were spherically distributed. As Rutherford explained [181]:

> It is of interest to note that Nagaoka has mathematically considered the properties of a "Saturnian" atom which he supposed to consist of a central attracting mass surrounded by rings of rotating electrons. He showed that such a system was stable if the attractive force was large. From the point of view considered in this paper, the chance of large deflexion would practically be unaltered, whether the atom is considered to be a disk or a sphere.

When Rutherford in 1919 presented the first experimental evidence that the central charge of atoms was divisible by knocking long-range protons out of nitrogen nuclei [184], another puzzle appeared: Why didn't the positively charged nucleus explode? What bound its protons together, overcoming their electrostatic repulsion? The existence of a "strong interaction or nuclear force" had been discovered.

The Range of the Nuclear Force

Rutherford and colleagues initiated a long series of experiments scattering α-particles off ever lighter nuclei. In 1927 Rutherford and James Chadwick conclude [186]:

> The study of the collisions of α-particles with hydrogen nuclei has shown that the force between the α-particle and the hydrogen nucleus obeys Coulomb's law for large distances of collision, but that it diverges very markedly from this law at close distances. The experiments of Chadwick and Bieler showed that for distances less than about 4×10^{-13} centimeters, the force between the two particles changed much more rapidly with distance than could be accounted for by an inverse square law of force. ...

---

[3] Rutherford concluded in his 1911 paper that the nuclear charge Z was "nearly proportional to the atomic weight," with a value of about 100 for gold (the charge is actually 79).

Possible explanations of the origin of these additional forces are discussed, and it is suggested tentatively that they may be due to magnetic fields in the nuclei.

Here is the first direct observation of the short-range nuclear force, but no new force was postulated!

The correct interpretation of novel phenomena can be impossible, even for giants like Rutherford and Chadwick, when insufficient information is available. They were conservative: Try with what you have. Don't wander off into a forest of speculations. There was no way they could have thought productively about a new force in 1927. The necessary experimental information, and theoretical concepts, were not yet available.

1.4   Nuclei Before Neutrons

Experiments had shown that electrons, protons, and the nuclei of ^{4}He and ^{3}He could be emitted from larger nuclei, leading Rutherford and many others to believe that these smaller particles were building blocks of all nuclei [185]. But there was a missing ingredient. Using Geiger's α-particle scattering data, Rutherford realized that the charge on the nucleus, measured in units of the magnitude of the electron charge, was roughly equal to half its atomic weight [181, 182]. For example, the atomic weight of helium is about 4, so its charge divided by its atomic weight is 1/2. The same calculation for gold gives 0.401. At most only half the mass of a nucleus came from its positive charge (its protons). What was responsible for the rest?

Electrons in the Nucleus? Rutherford's Doublet

To account for the neutral mass in the nucleus, Rutherford wondered in his 1920 Bakerian Lecture if there was an "atom within the nucleus" with a mass approximately equal to that of the proton, but which had no electric charge [185]:

> We also have strong reason for believing that the nuclei of atoms contain electrons as well as positively charged bodies, and that the positive charge on the nucleus represents the excess

positive charge. It is of interest to note the very different *rôle* played by the electrons in the outer and inner atom. In the former case, the electrons arrange themselves at a distance from the nucleus, controlled no doubt mainly by the charge on the nucleus and the interaction of their own fields. In the case of the nucleus, the electron forms a very close and powerful combination with the positively charged units and, as far as we know, there is a region just outside the nucleus where no electron is in stable equilibrium. While no doubt each of the external electrons acts as a point charge in considering the forces between it and the nucleus, this cannot be the case for the electron in the nucleus itself. It is to be anticipated that under the intense forces in the latter, the electrons are much deformed and the forces may be of a very different character from those to be expected from an undeformed electron, as in the outer atom. ... Such an atomic structure seems by no means impossible. On present views, the neutral hydrogen atom is regarded as a nucleus of unit charge with an electron attached at a distance, and the spectrum of hydrogen is ascribed to the movements of this distant electron. Under some conditions, however, it may be possible for an electron to combine much more closely with the H nucleus, forming a kind of neutral doublet.

The neutron had not yet been discovered. Rather than propose the existence of a new neutral particle that is subject to a nuclear force, Rutherford chose to combine an electron with a proton, binding them together so tightly that they could exist as a "neutral doublet" inside the nucleus. However, no known force was strong enough to create a doublet, or hold it together.

1.5   Neutrons

On June 1, 1932, Chadwick reported the observation of the neutron [31]:

> The properties of the penetrating radiation emitted from beryllium (and boron) when bombarded by the α-particles of polonium have been examined. It is concluded that the radiation consists, not of quanta [photons] as hitherto supposed, but of neutrons, particles of

mass 1, and charge 0. Evidence is given to show that the mass of the neutron is probably between 1.005 and 1.008. This suggests that the neutron consists of a proton and an electron in close combination, the binding energy being about 1 to 2×10^6 electron volts.

So Chadwick thinks he has discovered Rutherford's doublet (with its mass being slightly less than the proton mass)! It took a lot of time and effort for him and others to understand that he actually discovered a new particle — the neutron — and that the neutron was decaying through a new type of interaction in a process now called beta decay (β-decay). Understanding the constituents of the nucleus and what held it together was complicated by the existence of beta decay that made it look like the nucleus contained electrons. As for the neutron, unlike the proton and electron, now you see it, now you don't. No wonder they had a hard time imagining it.

In his Bakerian Lecture Rutherford indicated that his tightly bound doublet, appearing essentially uncharged, "should be able to move freely through matter," and "enter readily the structure of atoms," possibly uniting with the nucleus, unimpeded by its charge. The same was true of the neutron, making it very effective in producing changes within the nucleus, ultimately leading to the fission of uranium-235 by slow neutrons some 20 years later.

The Nucleon

Shortly after the neutron was discovered Heisenberg proposed that protons and neutrons were different states of the same particle, now called the nucleon $\mathcal{N}$ [117],

$$\mathcal{N} = (p, n).$$

In a 1970 interview he explains:[4]

> I had from the very beginning the idea that the neutron was a kind of a brother to the proton, and that there was a kind of symmetry between protons and neutrons. ... [T]he fact that in the nucleus [there] were an approximately equal

number of protons and neutrons ... pointed to a kind of symmetry between proton and neutron, not a complete symmetry, but rather good symmetry.

However, there were complications:

> And, on account of this symmetry, it was not so nice to say that the neutron is a compound particle consisting of proton and electron while the proton is an elementary particle. ... [B]ut still at that time we didn't know any way out yet.

Heisenberg tried to have it both ways, saying that a proton-neutron symmetry exists, while arguing that the "neutron is a compound particle consisting of proton and electron." He justified the latter by asserting that the composite nature of the neutron was responsible for the nuclear force.

It turned out that Heisenberg's intuition about an approximate symmetry between the proton and neutron was right on, and more important than he supposed. It would be generalized and applied to all strongly interacting particles, and be called "isotopic spin symmetry" or simply "isospin symmetry."

1.6 Heisenberg's Nuclear Physics

The H_2^+ Ion

To understand Heisenberg's thinking about the nuclear force, it is necessary to describe the origin of the *atomic* force between two protons created by an electron. Classically, two positively charged protons cannot be bound together by a single electron. The electron would become attached to one proton or the other, leaving the two protons unbound — the hydrogen ion H_2^+ cannot exist classically. Even in Bohr's model of the atom, H_2^+ is unstable [169]. In the reality of a quantum-mechanical world, the H_2^+ ion does exist because two protons can share an electron, exchanging it back and forth. This sharing creates an "exchange force" that varies with the protons' separation, and is responsible for their binding.

In a system with an electron and two widely separated protons, a measurement of the electron's position will find it primarily around one proton or

---

[4] Interview of Werner Heisenberg by Joan Bromberg on June 16, 1970. Niels Bohr Library & Archives, American Institute of Physics, College Park, MD USA. See https://repository.aip.org/heisenberg-werner-1970-june-16.

the other. As the two protons are brought closer, the electron will be found, with increasing probability, around the other proton. Eventually it will exist with equal probability around both protons, first near one, then near the other, thereby forming a stable H_2^+ ion. This behavior, which in humans would be called a "dissociative identity disorder," is a consequence of the electron's wave-like nature.[5]

Heisenberg's Nucleus

Following Chadwick's discovery of the neutron, Heisenberg attempted to understand the nuclear force in 1932 [117]. After stating that the neutron is an "independent fundamental component" of the nucleus, he proposed the existence of an ad hoc force between neutron and proton, created by a negative charge shuttling back and forth between them, an exchange force, where the act of sharing bound them together. What was a neutron became a proton, and then switched back again, eternally. He writes [117]:

> Suppose we bring the neutron and proton to a separation comparable to nuclear dimensions; then in analogy to the H_2^+ ion, the negative charge will undergo a "change-of-place" [Platzwechsel], whose frequency is determined by a function $J(r)/h$ of the distance r between the two parts [the neutron and proton, where h is Planck's constant]. The $J(r)$ corresponds to that "exchange" [Austausch], or more correctly the "place-change-integral" [Platzwechselintegral] of molecular theory.

Here was the fantastic prospect of neutrons and protons in a nucleus bound together in an ingenious flickering fashion through the constant exchange of negative particles.

Unfortunately the exchanged particle could not be a spin-1/2 electron, for suppose it were: Then according to the rules of quantum mechanics, the binding of the spin-1/2 electron with a spin-1/2 proton must result in a particle with a spin of either the sum or difference of the two spins, i.e., the spins will either point in the same direction and add, or point in the opposite direction and subtract. Therefore the interaction of a

spin-1/2 electron with a spin-1/2 proton will result in a particle of either spin 0 or 1, but not spin 1/2, which is the neutron's spin. Therefore the exchanged particle cannot be an electron.

Heisenberg recognized that his exchange force created a problem with the spins of the participating particles [117]:[6]

> For the following considerations it is assumed that neutrons follow the rules of Fermi statistics and have spin 1/2. This assumption will be necessary to explain the statistics of the nitrogen nucleus and corresponds to the empirical results on the nuclear moments. If one wanted to understand the neutron as composed of a proton and an electron, one would have to ascribe Bose statistics and zero spin to the electron. However, it does not seem appropriate to elaborate on such a picture. Rather, the neutron should be considered as an independent fundamental component, which, however, is assumed to split into proton and electron under suitable circumstances, whereby the laws of conservation of energy and momentum are probably no longer applicable.[1]

Here Heisenberg's footnote, denoted by [1] above, refers to Niels Bohr's May 1930 Faraday Lecture, published in 1932, which says [22]:

> At the present stage of atomic theory, however, we may say that we have no argument, either empirical or theoretical, for upholding the energy principle in the case of β-ray disintegrations, and are even led to complications and difficulties in trying to do so.

So following Bohr, Heisenberg was willing to throw "the conservation of energy and momentum" under the bus to create his theory of the nuclear force. It's not clear what "suitable circumstances" meant, and what replaced the successful rules for combining spins that came from quantum mechanics. In short, Heisenberg's theory of the nuclear force was a clever "fudge."

There was an additional problem in keeping an electron within a small nuclear volume. It was not consistent with relativistic quantum mechanics, as Enrico Fermi noted [56]:

---

[5] See Chapter 10-1 in Volume III of *The Feynman Lectures on Physics.*

[6] My translation.

The present relativistic theories of lightweight particles (electrons or neutrinos) are not capable of explaining, in a satisfactory manner, how such particles can be bound in orbits of nuclear dimensions.

These comments are not meant to be critical of Heisenberg or Bohr. As physicists, these two were as good as they get. Figuring out how the physical world works with incomplete information, and which cherished physical principles to ignore or abandon, if any, is an extraordinarily difficult task. Similar issues — violation of the spin-statistics theorem and binding energy problems — will arise some 30 years later with the discovery of quarks, but then the erroneous interpretation of experiments, due to the continuous creation of new technologies, will be an additional confounding problem.

Fermi's Theory of *β*-Decay

We now know that electrons are not a component of nuclear matter, although they are emitted from nuclei. They are created by the spontaneous disintegration of neutrons in neutron-rich nuclei. But initially there was a problem with this idea. The neutron's energy exceeded the combined energies of the electron and proton, leading Bohr, Heisenberg, and others to consider abandoning the conservation of energy.

Wolfgang Pauli was unwilling to do this. In December 1930 Pauli proposed that the electron emitted in neutron decay was accompanied by another neutral particle which would carry off the energy and momentum that appeared to be missing.[7] No one had predicted the existence of a new particle before, let alone one you couldn't see. It just wasn't done, bringing to mind the proverb: "The difference between a madman and a visionary is only apparent *a posteriori*."

---

[7] As reported in "This Month in Physics History," *APS News* **20** (7), p. 2, July 2011, "Pauli was reluctant to publish a paper on this unusual hypothesis, but he penned a letter to a group of prominent nuclear physicists gathering for a conference in Tübingen, Germany in December 1930 asking for input regarding means of detecting such a particle experimentally. 'I have done something very bad today by proposing a particle that cannot be detected; it is something no theorist should ever do,' he wrote, describing his idea as 'a desperate remedy.' When informed that the neutrino had been discovered 26 years later Pauli replied, 'Everything comes to him who knows how to wait.'"

Fermi took Pauli's idea seriously, developing a theory of *β*-decay requiring a new type of interaction, the predecessor of today's theory of the weak interactions. It explained why electrons were emitted from nuclei without being one of their constituents. He called Pauli's particle the "neutrino," which in Italian means "little neutral one."

Igor Tamm then considered the possibility that an electron-neutrino pair with integral spin, not the electron alone, would provide the negative charge needed in Heisenberg's exchange-theory of the nuclear force, without violating the conservation of angular momentum [200]. But Tamm concluded that "the exchange energy may be shown to be far too small," and the nuclear force problem remained. Proceeding further requires a little quantum mechanics and a bit of "field theory."

1.7   Quantum Mechanics

Shortly after receiving his doctorate in 1911 Bohr visited Rutherford's lab for several months, eventually settling there from 1914 to 1916. During his stay Bohr synthesized two disparate ideas, each violating a different aspect of classical physics. Rutherford's separation of positive and negative charge in the atom, and Planck's quantization of radiation, were combined into a heuristic dynamical model of the atom: Electrons with quantized energy existed around a nucleus in "stationary states," emitting or absorbing quanta of light when they jumped from one stationary state to another. This view of the atom resolved none of its contradictions with classical physics, but provided a conceptual framework within which the frequencies of light — the atom's colors or spectral lines — could be organized, and their patterns fruitfully contemplated.

Schrödinger's Wave Function

The contradictions with classical physics present in Bohr's atom, but not with its nucleus, were resolved with the introduction of quantum mechanics by Erwin Schrödinger, and independently by Heisenberg. In 1925, shortly after Louis de Broglie published his Ph.D. thesis attributing both corpuscular and wave-like properties to particles, Peter Debye asked Schrödinger to present de Broglie's thesis at an ETH seminar in

Zürich. After his presentation Debye is reported to have said "Erwin, when I went to school, when there was a wave, there was a wave equation. What is the wave equation?"[8] That's what got Schrödinger on the road to "Schrödinger's equation," but he had no idea of what was waving. The interpretation of the wave function was supplied by Max Born a few days after Schrödinger's paper was published: If $\psi(x, y, z, t)$ is the wave function of an electron in an atom, then $|\psi(x, y, z, t)|^2 \Delta x \Delta y \Delta z$ is the *probability* that the electron will be found at time t inside the volume $\Delta x \Delta y \Delta z$ centered at the point (x, y, z). Almost two years later Schrödinger still had not accepted Born's interpretation,[9] and Einstein was never comfortable with it, remarking: "God does not play dice with the Universe." Nevertheless, Born was right.

Heisenberg's Atomic Physics

Heisenberg's theory of the atom was formulated in terms of the frequencies of light it emitted. His theory consisted of both equations for these these frequencies, and *a philosophy of how theoretical physics should be practiced*. He emphasized that much of the original confusion in finding a theory for the atom would have been avoided if physicists had thought only in terms of observables, and the equations that connected them. In this 1925 paper titled "Quantum-theoretical re-interpretation of kinematic and mechanical relations" he writes [115]:

> It is well known that the formal rules that are used in quantum theory for calculating observable quantities such as the energy of the hydrogen atom may be seriously criticized on the grounds that they contain, as a basic element, relationships between quantities that are apparently unobservable in principle, e.g., position and period of revolution of the electron. Thus

these rules lack any evident physical foundation, unless one still wants to maintain the hope that the hitherto unobservable quantities may later come within the realm of experimental determination. ... Instead it seems more reasonable to try to establish a theoretical quantum mechanics, analogous to classical mechanics, but in which only relations between observable quantities occur.

Heisenberg's eminently reasonable insistence that physics should only involve observable quantities had far-reaching consequences, leading to the "bootstrap" theory of particle physics (p. 18), and also delaying the acceptance of quarks as the elementary constituents of matter for more than a decade. In the latter case, paraphrasing Heisenberg above, "it was necessary to maintain the hope that the hitherto only indirectly observable quantities (quarks) may later come within the realm of direct experimental determination," which they eventually did.

Heisenberg's Uncertainty Principle

Although it was possible, in principle, to measure a single property of an atom with arbitrary accuracy, measurements of certain pairs of variables, called "conjugate," or "complementary," inevitably interfered with each other if you tried to measure both of them simultaneously. In the quantum world conjugate variables cannot be measured with arbitrary precision. The more accurately one is measured, the less accurately its conjugate can be determined. Conjugate variables are "random variables" whose precision of measurement is related. For example:

- The position x and momentum p of a particle are conjugate random variables. If they were measured together again and again in identical systems (systems created the same way each time), their measured values would differ from one measurement to the next. In this sense there exists an uncertainty Δx in the position of a particle, and an uncertainty Δp in its momentum. The product $\Delta x \Delta p$ is bounded from below:

$$\Delta x \, \Delta p \geq \hbar/2,$$

where Δx and Δp are the standard deviations of the distributions of these variables, and $\hbar$ is Planck's

---

[8] George Uhlenbeck (co-discoverer of the electron's quantized spin, and my thermodynamics teacher), private communication, 1958. Peter Debye had studied under Arnold Sommerfeld, who later claimed that his most important discovery was Peter Debye. Four of Sommerfeld's students, Heisenberg, Pauli, Bethe, and Debye went on to win the Nobel Prize. Thomas Kuhn's AIP interview of Felix Bloch, who attended Schrödinger's talk, is also of interest: https://repository.aip.org/bloch-felix-1964-may-14.

[9] Schrödinger's "Four Lectures on Wave Mechanics," delivered at the Royal Institution, London, in March 1928

constant h divided by 2π (note the small upper "bar" in $\hbar$, not present in h).

• If E is the energy of a system, and T is the time it takes to measure it, then E and T are conjugate random variables. Their uncertainties are related in the same way as the uncertainties in position and momentum:

$$\Delta E\, \Delta T \geq \hbar/2. \qquad (1.1)$$

In order to measure energy more precisely (to make ΔE smaller), the measurement must take more time (ΔT must be increased).[10]

Zero-point energy A classical particle in a potential well with damping will roll back and forth, eventually settling at its bottom. In the quantum world the particle will always jump around. If it were to sit still at the bottom of the well, both its position and momentum would be precisely known, in violation of the uncertainty principle. The nonzero energy the particle has in its lowest quantum state is called its "zero-point energy."

1.8   Classical Fields

The Problem with Gravity

While it was possible to calculate many details of planetary motion that agreed with astronomical observation, the fact that Newton's laws required the gravitational force to act instantaneously over arbitrarily large distances was troublesome, especially to Isaac Newton. In a letter to Richard Bentley in 1692/93 he writes [131]:

> That gravity should be innate, inherent, and essential to matter, so that one body may act upon another at a distance through a vacuum without the mediation of anything else, by and through which their action and force may be conveyed from one to another, is to me so great an absurdity, that I believe no man who has in philosophical matters a competent faculty of thinking can ever fall into it. Gravity must be caused by an agent acting constantly according to certain laws; but whether this agent

be material or immaterial, I have left to the consideration of my readers.[11]

It took more than 125 years before Newton's problem was addressed in classical field theory, and finally solved in quantum field theory another 100 years later.

Fields of Force

The question of "action at a distance" arose again for Michael Faraday in his experiments with electric and magnetic forces in the early 1800s. He would have none of it. For Faraday, space was a "field of forces" filled with "lines of force." As James Clark Maxwell would show, charges, currents, and magnets created electric and magnetic fields that propagated with the speed of light, eventually filling all of space.

If you've ever held two magnets in your hands, moving them towards or away from each other, you have felt the effects of what Faraday called the "magnetic field," an invisible continuous medium that exists between magnets. Charged objects are surrounded by an electric field, and masses by a gravitational field. Knowing the electric or gravitational field at a point allows you to compute the force felt by a small charge or mass that's placed there. Since fields are invisible, forces seem to be acting at a distance.

Based on Faraday's work, Maxwell in 1861 published a theory for electric and magnetic waves that provided a unified mathematical description of both, connecting seemingly disparate phenomena, and abolishing action at a distance for electromagnetism.

1.9   Quantum Fields

When classical fields are quantized they become active. Transitory electron-positron pairs, quantum fluctuations of the quantum fields, are continually created, only to be destroyed before a violation of the conservation of energy can be detected. In a manner of speaking, the conservation of energy is not continuously enforced, as might be expected from

---

[10] A limiting case of Eq. 1.1 occurs in atomic physics where idealized excited atoms of definite energy are called "stationary states," i.e., they are assumed to live forever.

[11] Newton's extensive correspondence with Bentley is fascinating. At one point Newton says: "So, then, gravity may put the planets into motion, but, without the divine power, it could never put them into such a circulating motion as they have about the sun; and therefore, for this, as well as other reasons, I am compelled to ascribe the frame of this system to an intelligent Agent."

Heisenberg's uncertainty principle (Eq. 1.1, p. 11). To distinguish these quantum fluctuations from free electron-positron pairs, they are said to be "virtual."

Virtual pairs affect the world everywhere around them, before they disappear. The lighter they are, the longer they stick around, and the further they travel. This "magic trick" goes on and off forever, with every type of virtual particle-antiparticle pair, throughout all of space. Virtual particles *change the distribution of the real charges and currents that modified the quantum field responsible for their creation.*

The Lamb shift In particular, virtual electron-positron pairs within a hydrogen atom have observable consequences. These vacuum fluctuations in the space between the proton and electron affect the energy of the electron moving in its quantized orbit, and the frequency of light emitted or absorbed when the electron jumps from one orbit to another. Initial calculations of these energies with the Schrödinger equation (and even the relativistic Dirac equation), assuming the electromagnetic field was classical, predicted that electrons moving in two different orbits had the same energy. When the energy difference between them was accurately measured in 1947 by Robert Retherford and Willis Lamb, they found a small nonzero difference, called the "Lamb shift."[12] Once the classical field was quantized, the predicted difference matched the one observed. Virtual pairs exist!

Virtual Particles as Carriers of Force

The force between the electron and proton in the hydrogen atom is created by virtual photons shuttling back and forth between them. That's how the electron "knows" the proton is there, and should be attracted to it. The gravitational force between the Earth and Sun is created by the exchange of virtual particles called "gravitons." *No action at a distance here.* More generally, all forces are created by the exchange of one or more virtual particles, "carriers of force," the excitations of their quantum fields.

Forces have a distance over which they act that is determined by the mass of their carriers. Because

of the uncertainty principle, the lighter the carrier, the longer it can exist in its virtual state, and the further it can travel. Electromagnetic and gravitational forces, with their massless carriers, act over all distances, making it easy to experience their effects. The strong interactions responsible for the nuclear force have massive carriers that determining the radii of protons and neutrons. The weak interactions responsible for the slow decay of particles like the neutron have a range several hundred times less because their carriers are much heavier.

Yukawa's Nuclear Force

In 1935 Hideki Yukawa, a 27-year-old physicist at the Osaka Imperial University, proposed that the neutron and proton were bound together by exchanging a virtual particle with even spin he called a "heavy quantum," to distinguish it from the massless spin-1 photon of the electromagnetic field [217]. This heavy quantum, now called the "pion" or "π," created the short-range nuclear force that Rutherford [183], and Chadwick and Bieler [30], had observed.

According to Yukawa, a proton would emit a virtual π^+, becoming a neutron in the process. The π^+ would strike another nearby neutron, converting it to a proton. Similarly, a neutron would change into a proton and π^-, with the π^- turning a different proton into a neutron. If one could observe the nucleons in a nucleus, protons and neutrons in close proximity would appear to be flip-flopping identities, rarely choosing distant partners.

Using the measured range of the nuclear force, and assuming the virtual pions were traveling at somewhat less than the speed of light, Yukawa predicted that pions should be about 200 times heavier than the electron (the pion's mass turned out to be about 270 times heavier). Yukawa's force carriers were particles that had never been observed.

Scattering experiments showed that the force between two protons was approximately equal to the force between a proton and neutron [26, 202], i.e., the nuclear force was "charge-independent." Yukawa then added a third neutral force carrier, the π^0, to the $\pi^\pm$ [218]. Except for charge, and a very small mass difference between the $\pi^\pm$ and π^0, the three pions were indistinguishable (the π^+ and π^- have the same mass since they are each other's antiparticle).

---

[12] When created, the electron in a virtual pair tends to be closer to the proton, while the positron prefers the electron. The virtual pair shields the electron and proton from each other, decreasing their binding energy.

Just as Heisenberg had viewed the proton and neutron as different manifestations of the nucleon $\mathcal{N}$ when considering the nuclear interactions, the π^+, π^0, and π^- were eventually considered different states of the pion π, distinguishable by the electromagnetic, but not strong, interactions. Like the two nucleons, the three pions enjoyed "charge independence". The clustering in mass of strongly interacting particles, with similar properties except for charge, would become a common occurrence requiring explanation [Figure 2.2, p. 16].

For two years Yukawa's paper was overlooked in the West, until Seth Neddermeyer and Carl Anderson found a charged particle in cosmic rays whose mass was greater than the electron's, but smaller than the proton's [158]. Many physicists thought this particle was Yukawa's pion.

In 1943 Yukawa's former student and later collaborator Shoichi Sakata, and Takesi Inoue, suggested that the Neddermeyer and Anderson's particle was different from Yukawa's pion, the former being a decay product of the latter. Their paper was republished in English in 1946 [188]. Sakata and Inoue got it right: In 1947 César Lattes, Giuseppe Occhialini, and Cecil Powell observed tracks in their emulsion plates that corresponded to two different particles, one of which was the decay product of the other [140]. One particle corresponded to Yukawa's pion, the other to Neddermeyer and Anderson's particle — the "muon" μ. The muon did not interact strongly with nuclei; it behaved like a heavy electron. Like Chadwick, Powell had been one of Rutherford's students. The third pion, the π^0, was discovered three years later in 1950 [20].

History commonly records Chadwick discovering the neutron in 1932, and Anderson discovering the muon in 1936. This is chronological history, a genuflection to the past, practiced for efficiency. The history of ideas, intellectual history, tells a different story. The history of quarks as it is commonly related, and its intellectual history, also differ, for more reasons, and in other ways (Chapters 5, 6, and 7).

With these discoveries the cast of characters within the nucleus was complete, and Yukawa's theory of their strong interactions was widely accepted. When it came to nuclear forces, heavy hitters like Rutherford and Heisenberg struck out, while Yukawa, a young unknown in far-off Japan, hit the ball out of the ballpark.

1.10   Looking Ahead

This is just a glimpse of what confusion looked like in the 1930's when physics was relatively simple. Confusion also reigned in the 1950s and '60s, but by that time its scope and sophistication had mushroomed. There was Pauli, Dancoff, and Wentzel's "strong coupling theory"; Wightman's "axiomatic quantum field theory"; Heisenberg's "nonlinear spinor field theory"; Sakurai's "vector meson dominance model"; Schwinger's "source theory with dyons and magnetic charge"; Gell-Mann's "Global Symmetry," changed to "SU(3)"; Chew's — I'm not a field theory — "bootstrap"; Gell-Mann's — I'm not sure if field theory is right — "equal time current commutation relations"; and Sakata's — no theory here — "composite model." These were all different ways of thinking about the strong interactions, each with its own truth and possible ways forward. When quarks were first discovered, theorists wouldn't accept them for what they were, real elementary particles, different from strongly interacting particles, with their own laws of interaction. It took a decade before quarks became the fields of a quantum field theory that led to quantum chromodynamics and the "Standard Model."

Don't be surprised by the muddled state of physics for much of the 20th century. Confusion was no stranger to the field. Witness the history of the luminiferous (light-bearing) aether that originated in the 17th century with Robert Boyle and Christiaan Huygens. Aether rose to great prominence — confused with Faraday's fields — lingering on till the start of the 20th century.

Chapter 2
The Discovery of Quarks

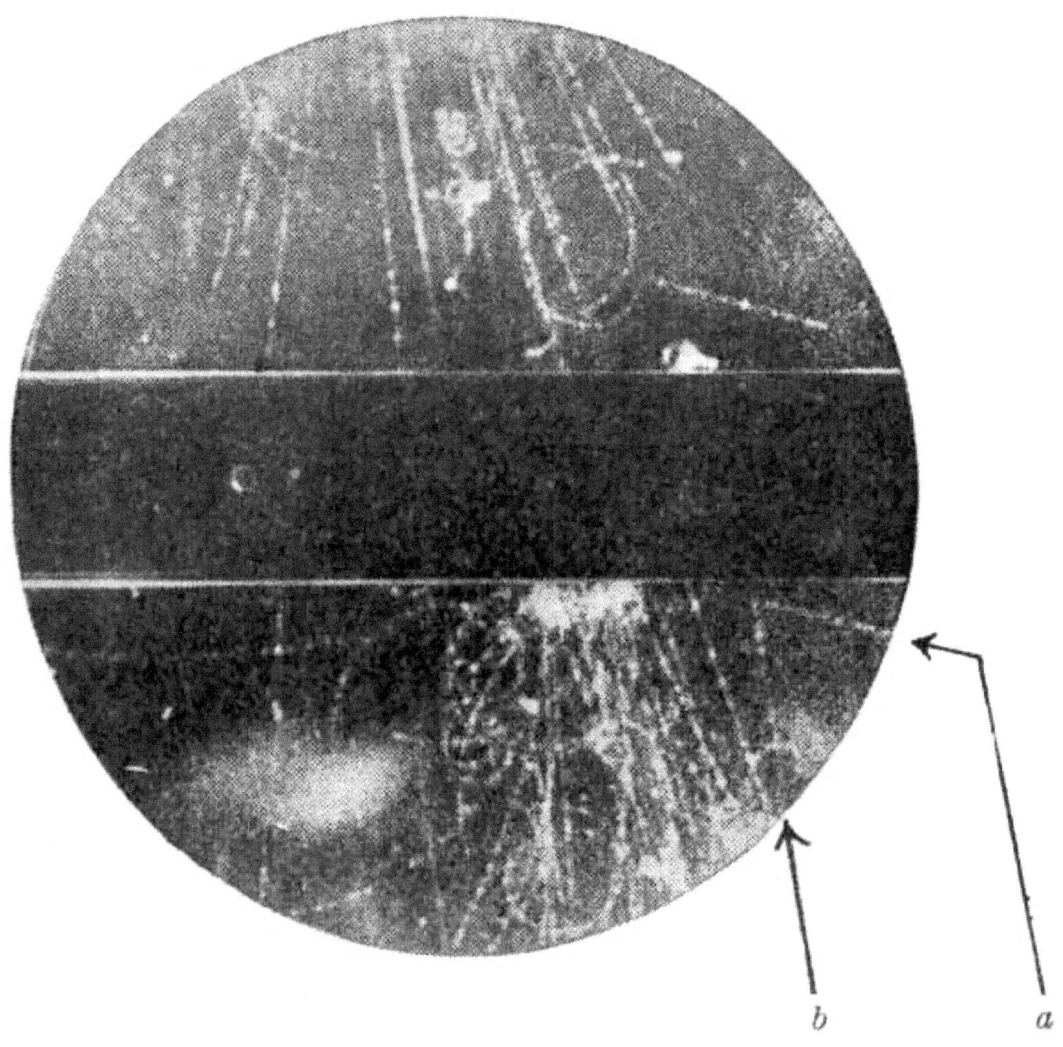

A 1947 cloud chamber "photograph showing an unusual fork (a [and] b) in the gas" [178]. The V-shaped event could only be explained by the decay of a neutral particle with a mass about 1,000 times greater than that of the electron. When the experiment was repeated at higher altitudes in the French Pyrénées — where the cosmic-ray flux is higher — it became apparent that the decaying particle was a new type of meson: a "kaon" or "K^0." (Nature Publishing Group.)

Of the celebrated English art critic and essayist John Ruskin, the 19th century French artist Rosa Bonheur — of even greater fame — is reputed to have said: "He is a gentleman, an educated gentleman; but he is a theorist. He sees nature with a little eye, *tout à fait comme un oiseau* [quite like a bird]" [107]. So it was with theorists, who dismissed the existence of quarks because of their inability to see the order before them.

2.1   Strangeness

In the same year the pion was discovered, George Rochester and Clifford Butler published two stereoscopic cloud chamber photographs of cosmic-ray events, providing the first evidence for the existence of the K^0 meson [178]. One of their photos appears at the beginning of this chapter. The V-shaped event Rochester and Butler observed in this photograph was a K^0 decaying into two charged pions,

$$K^0 \to \pi^+ + \pi^-. \tag{2.1}$$

Unlike the π meson, whose existence had been predicted, the kaon was entirely unexpected, not directly connected to the force inside the nucleus.

Three years later in 1950 Victor Hopper and Sukumar Biswas detected a new neutral particle, the Λ, in photographic emulsion flown in a balloon at 70,000 feet [126]. The Λ decayed into a proton and a negatively charged pion,

$$\Lambda \to p + \pi^-, \tag{2.2}$$

a reaction that reads: "lambda goes to proton plus pi minus."

The K^0 and Λ *were always produced together* by the strong interactions in reactions like

$$\pi^- + p \to K^0 + \Lambda. \tag{2.3}$$

Their "associated production" presented a puzzle. Like the K^0 and Λ, other "hadrons" (strongly interacting particles) like the Σ and Ξ, discovered shortly thereafter [Figure 2.2, p. 16], were also restricted in the way they were formed, interacted, and decayed. In 1954 Kazuhiko Nishijima realized that these restrictions could be explained by assigning a new property or "quantum number" to hadrons that was conserved

by the strong interactions. Two years later Nishijima's quantum number, "η," was renamed strangeness, "$\mathbb{S}$," and popularized in the West by Murray Gell-Mann (Section 5.8, p. 74) [85].

Strangeness $\mathbb{S}$ was an integer, with the strangenesses of particles and antiparticles having opposite signs. Nishijima assigned $\mathbb{S} = 0$ to p, n, and the three pions. He also assumed that the strangeness of a collection of hadrons was the sum of their individual strangenesses.

Nishijima's reasons for assigning the values $\mathbb{S} = 1$ and -1 to the K^0 and Λ could have gone something like this: In $\pi^- + p$ scattering the Λ was always created with a K^0 meson, but it could not be created with a π^0,

$$\pi^- + p \not\to \pi^0 + \Lambda,$$

even though charge was conserved in the reaction. And while the K^0 was created with a Λ, it could not be created with a neutron,

$$\pi^- + p \not\to K^0 + n.$$

Therefore the K^0 and Λ could not have $\mathbb{S} = 0$.

Since the K^0 and Λ were always created together from $\mathbb{S} = 0$ hadrons, their strangenesses must have opposite signs. By convention he gave the K^0 strangeness $\mathbb{S} = 1$, which meant that Λ had $\mathbb{S} = -1$.

While conserved by the strong and electromagnetic interactions, strangeness was not conserved by the weak interactions, as seen in the weak strangeness-changing decays of the Λ and K^0 in Eqs. 2.1 and 2.2.

Other strange mesons and baryons were found in short order:

- two more spin-0 mesons, the $(\bar{K}^0, K^-) = \bar{K}$ with strangeness $\mathbb{S} = -1$. The K and $\bar{K}$ are antiparticles.
- two sets of spin-1/2 baryons, the $\Sigma = (\Sigma^+, \Sigma^0, \Sigma^-)$ and $\Xi = (\Xi^0, \Xi^-)$. By considering the reactions in which they were, and were not, produced, they were assigned $\mathbb{S} = -1$ and $\mathbb{S} = -2$, respectively.

The members in each of the sets π, K, $\mathcal{N}$, Λ, Σ, and Ξ are close in mass [Figure 2.2], successively differ in charge by one, and have the same strangeness, behaving identically under the strong interactions. They are called "isospin multiplets." Why these multiplets existed, and not others, was not known.

Quantum Numbers

As noted above, strangeness is an example of a quantum number, a property of a particle or system of particles, that is conserved in one or more types of interactions. The sum Q of the electric charges in a system of particles is another more familiar example. It remains constant in time under all interactions, even if those particles are transformed into other particles, or change in number over time.

Some quantum numbers like strangeness $\mathbb{S}$ and baryon number B are only assigned nonzero values to hadrons. The $\mathcal{N}$, Λ, Σ, and Ξ have $B = 1$, their antiparticles $B = -1$. The deuteron has $B = 2$. All mesons have $B = 0$. The baryon number, like charge, is conserved by all interactions.

Charge, baryon number, and strangeness are "additive quantum numbers," i.e., their values for a multiparticle system are the sums of their values for the individual particles.

Other quantum numbers that will arise later, like "parity" P and "charge conjugation" C, are "multiplicative." For example, the parity of a multiparticle system without any angular momentum is the product of all the individual particle parities. Parity is conserved by the strong and electromagnetic, but not weak, interactions.

Quantum numbers are discussed in Appendix A: "A Primer on Quarks," p. 148. The story of parity violation begins in Section 4.2, p. 41.

2.2 The Elementary Particles in 1957

By 1957 the list of particles that were called elementary had grown to 19 [Figure 2.2]. Reworking a paper by Murray Gell-Mann and Arthur Rosenfeld [86], Gell-Mann and Edward Rosenbaum, an associate editor for *Scientific American*, wrote an article for the general public titled "Elementary Particles" [87]. These particles were grouped into two classes: point particles (three leptons and the photon at the top of Figure 2.2), and those that interacted strongly and were "extended in size" (the spin-1/2 baryons and spin-0 mesons listed below).

For future reference, five years later in 1962, the neutrino created in pion decay was shown to differ from the one created in neutron β-decay. Then there were four spin-1/2 leptons, the electron, muon, and

Point particles

Spin-1/2 leptons	
Particle	Mass
e^-	1
μ^-	206.7
ν	0

Spin-1 photon	
Particle	Mass
γ	0

Extended particles (strongly interacting)

Spin-1/2 baryons		
Multiplet	Particle	Mass (m_e)
Ξ	Ξ^0	?
	Ξ^-	2585
Σ	Σ^-	2341
	Σ^+	2325
	Σ^0	2324
Λ	Λ	2182
N	n	1838.6
	p	1836.1

Spin-0 mesons		
Multiplet	Particle	Mass
π	π^+	273.2
	π^-	273.2
	π^0	264.2
K	K^+	966.5
	K^0	965
$\bar{K}$	K^-	966.5
	$\bar{K}^0$	965

Figure 2.2. Particles called elementary in 1957. "Extended particles" were distinguished from "Point particles" by scattering electrons off of them. To the electron, extended particles looked like little balls of charge. These balls consisted of virtual charged particles held together by the strong interactions. Masses are given in multiples of the electron mass (0.511 MeV). (Figure from [87].)

their two associated neutrinos as observed in $n \rightarrow p + e^- + \bar{\nu}_e$ and $\pi^- \rightarrow \mu^- + \bar{\nu}_\mu$.

Too many elementary particles Willis Lamb, in the first paragraph of his 1955 Nobel Prize Lecture, joked that he had "heard it said that 'the finder of a new elementary particle used to be rewarded by a Nobel Prize, but such a discovery now ought to be punished by a \$10,000 fine.'" Leon Lederman recalled one of Fermi's famous quips: "If I could remember the names of these particles, I would have been a botanist" (and

Fermi died in 1954 before the particle explosion).[1] Yes, there were too many *elementary* particles.

Resonances

In the same issue of the *Annual Review of Nuclear Science* in which Gell-Mann and Rosenfeld listed the elementary particles, Sam Lindenbaum discussed a new phenomenon in strong-interaction physics: "resonances," short-lived particles found in two-particle scattering reactions (Appendix 2.C: "The First Pion-Nucleon Resonance," p. 32) [144].

> **Strange bedfellows** In 1957 elementary particles and resonances coexisted side by side, but unconnected, in journals and theoreticians' minds. Resonances (the "elephants in the room") were about to explode in number. Only later was it realized that the long-lived extended "elementary particles" like the neutron, and short-lived resonances, both of which interacted strongly, were cut from the same cloth. They just decayed with interactions of different strength.
>
> Resonances are both formed from, and decay through, the strong interactions. They disintegrate almost immediately after formation, typically living only a little longer than 10^{-23} seconds, the time it takes light to travel across their diameter, about 10^{-13} centimeters. The neutron β-decays through the weak interactions in about 14.6 minutes. The proton is stable, expected to live for at least 10^{34} years. Hadrons with such different lifetimes had to be considered together when trying to understand their origin.

2.3 Why So Many Hadrons?

Organizing hadrons with similar mass and identical quantum numbers, except for charge, was useful, but that organization didn't tell us anything about the structure of hadrons.

Composite Pions: Fermi and Yang

Fermi and Chen-Ning Yang published a paper in 1949 titled "Are Mesons Elementary Particles?," with the abstract [57]:

---

[1] L. M. Lederman, "Neutrino Physics," *Brookhaven Lecture Series Number 23*, BNL 787 T-300 (1963).

> The hypothesis that π-mesons may be composite particles formed by the association of a nucleon with an antinucleon is discussed. From an extremely crude discussion of the model it appears that such a meson in most respects would have properties similar to those of the meson of the Yukawa theory.

However, in the body of the paper they caution:

> Unfortunately we have not succeeded in working out a satisfactory relativistically invariant theory of nucleons among which such attractive forces act.

Honesty is the best policy. It's wonderful to see it practiced!

To form a pion like the π^+, the much heavier proton p and antineutron $\bar{n}$ would have to surrender almost all their rest mass to potential energy, which seems unlikely. It's not surprising that Fermi and Yang were unsuccessful.

Sakata's Composite Hadrons

Seven years later in 1956 Shoichi Sakata extended the ideas of Fermi and Yang by proposing that all mesons and baryons were composites of three fundamental baryons — the proton, neutron, and lambda — and their antiparticles, here collectively designated by S and $\bar{S}$ ("sakatons") [189]:

$$S = (p, n, \Lambda) \text{ and } \bar{S} = (\bar{p}, \bar{n}, \bar{\Lambda}). \qquad (2.4)$$

Composite mesons were created from $S\bar{S}$. For example,

$$\begin{aligned} \pi^+ &\sim p\bar{n}, \\ K^+ &\sim p\bar{\Lambda}, \\ K^0 &\sim n\bar{\Lambda}, \end{aligned}$$

where the symbol "$\sim$" reads "created from."

$S S \bar{S}$ created baryons. The charge, baryon number, and strangeness quantum numbers of composite mesons and baryons were the sum of their constituent quantum numbers.

Concerning strangeness, Sakata writes:

> The curious properties of the new [strange] particles could be reduced to those of [the] Λ,

just like the mysterious properties of the atomic nuclei were reduced to those of [the] neutron. Hence our theory contains less arbitrary elements than was the case for [the] original one of Nishijima and Gell-Mann.

While the spin-0 mesons were correctly accounted for, including the neutral η whose existence had not yet been established [128], the many additional spin-1/2 baryons that were predicted were not found, eventually killing Sakata's model.

Heisenberg's Nonlinear Spinor Field Theory

Heisenberg's response in 1957 to the proliferation of hadrons was different; he asked a more fundamental question: "What does it mean for a particle to be elementary?" [119]:

> Is there any criterion by which we can distinguish between an elementary particle and a compound system? Is it any more justified to introduce a meson field into the fundamental equations than, e.g., a hydrogen field or an oxygen field? The author believes that it is essential for any real progress in the theory of matter to recognize that such criteria do not exist.

He continued to argue that the very idea of an elementary hadron didn't make sense:

> A proton certainly looks like an elementary particle for energies < 100 MeV, but it may be considered as composed of a Λ particle and a K^+ particle in collisions of much higher energies. One might argue that the Λ particle and the K^+ particle are unstable while the proton is stable, that therefore the proton cannot be composed of Λ and a K^+. The fallacy of the argument is, however, seen at once from the case of the deuteron, which is stable but usually considered as composed of a proton and neutron, the latter being unstable. For an understanding of matter and of the atomic particles it is essential to realize that the question whether the proton is elementary or a compound system has no answer.

In a high energy collision between hadrons, energy may be converted into pions, lambdas, kaons, and many other types of particles. As Heisenberg saw it almost 20 years later in 1975 [120]:

> We have learned that energy becomes matter when it takes the form of elementary particles. The states called elementary particles are just as complicated as the states of atoms and molecules. Or to formulate it paradoxically: Every particle consists of all other particles. Therefore we cannot hope that elementary particle physics can ever be simpler than quantum chemistry.

Similar arguments had been made by Gell-Mann in 1966 to conclude that it didn't make sense to think of hadrons as being composed of quarks (Gell-Mann's quote, p. 71 in "Were Murray's Quarks Real?"). For Gell-Mann, as they would be for Heisenberg, constituent quarks, if they existed, would simply be hadrons, and the concept of an elementary hadron didn't make sense.

Hoping to create a fundamental theory for particle physics, Heisenberg proposed the existence of a single underlying spinor field (a spin-1/2 particle) that did not correspond to any hadron, but which was supposed to generate all observed hadrons on an equivalent basis (more on Heisenberg's thinking in Section 6.16, p. 111). While imaginative, his Spinor Field Theory turned out to be unrelated to reality.

Chew and Frautschi's Bootstrap: The Nuclear Democracy

When experimental particle physics came to consist primarily of two-particle collisions, the observables were scattering amplitudes, like those in the Geiger-Marsden experiment. During World War II Heisenberg argued that scattering amplitudes should be organized into a matrix he called the "S-matrix" (previously introduced by John Wheeler in 1937 [212]) and that fundamental laws governing particles and their strong interactions should be formulated in terms of this matrix and its properties [118]. Taken to its extreme some 15 years later, S-matrix theory replaced quantum field theory, which had dominated particle physics theory since the introduction of quantum mechanics (Section 1.9: "Quantum Fields," p. 11). While quantum field theory had been successfully used to develop a relativistic quantum theory of electricity and magnetism (quantum electrodynamics), its application to the strong interactions had not been successful (see Feynman's Section IV: "The Question of Dynamics" [68]).

The ever-growing list of hadrons made it clear to most physicists that none of them were elementary. Geoffrey Chew and Steven Frautschi, building on the work of Heisenberg and Stanley Mandelstam, also questioned the distinction between composite and elementary, advocating a "nuclear democracy" in which all hadrons were treated on an equal basis, collectively responsible for each other's existence. Chew and Frautschi sought to derive *all* information about the strong interactions from plausible assumptions about Heisenberg's *S*-matrix, with only the pion mass as an input.

In a paper titled "Principle of Equivalence for all Strongly Interacting Particles within the *S*-Matrix Framework," Chew and Frautschi motivated their "bootstrap" approach to hadron physics [33]:

> The notion, inherent in Lagrangian field theory, that certain particles are fundamental while others are complex, is becoming less and less palatable for baryons and mesons as the number of candidates for elementary status continues to increase. Sakata has proposed that only the neutron, proton, and Λ are elementary, but this choice is rather arbitrary, and strong-interaction consequences of the Sakata model merely reflect the established symmetries. Heisenberg some years ago proposed an underlying spinor field that corresponds to no particular particle but which is supposed to generate all the observed particles on an equivalent basis. The spirit of this approach satisfies Feynman's criterion that the correct theory should not allow a decision as to which particles are elementary, but it has proved difficult to find a convincing mathematical framework in which to fit the fundamental spinor field.

When hadrons scatter off one another the forces they experience arise from the exchange of virtual hadrons. When those forces are strong enough, the scattered hadrons may coalesce into other hadrons, which, in turn, can be scattered and exchanged. The bootstrap posits that the set of all hadrons, though scattering and exchange, create themselves in a self-consistent manner; no quantum field theory, with its many unknown parameters, is required.

It was a beautiful idea simply exemplified by Fredrik Zachariasen in his bootstrap of the ρ meson starting from two pions [Figure 2.3] [219]. The ρ was discovered in 1961 from its decay into two pions. The force generated between these pions, if they are allowed to scatter, by the exchange of a virtual ρ, turns out to be attractive, and strong enough to form a meson ρ' with the same quantum numbers as the ρ. The strength with which the exchanged ρ interacts with pions, i.e., the $\rho\pi\pi$ "coupling constant" $\gamma_{\rho\pi\pi}$, and the ρ mass M_ρ, determine the coupling constant and mass of the ρ'. By adjusting the two parameters for the ρ, the coupling and mass of the ρ' can be made equal to the coupling and mass of the ρ, i.e., the exchanged ρ and the ρ' become indistinguishable — the ρ will have "bootstrapped" itself into existence. The necessary values of the parameters were $\gamma_{\rho\pi\pi} = 2.8$ and $M_\rho = 350\,\text{MeV}$, as compared with the measured values

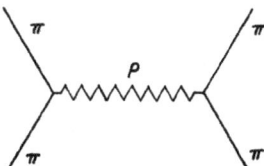

Figure 2.3. The first step in the ρ bootstrap. This Feynman diagram shows the interaction of two pions created by the exchange of a virtual ρ meson. That exchange creates a force strong enough to bind the two pions into a meson ρ' with the quantum numbers of the ρ. By properly adjusting the $\rho\pi\pi$ coupling constant and ρ mass, these two parameters can be equated to those of the ρ'. The particle created would have the same properties as the particle exchanged. (Figure from [219].)

of $\gamma_{\rho\pi\pi} = 1$ and $M_\rho = 750$ MeV.[2] Zachariasen hoped that a calculation including more hadrons, all coupled together, would give agreement.

So S-matrix theory in the form of the bootstrap became the dominant school of thought in particle physics for about a decade starting in the late 1950s. As described by Chew, Gell-Mann, and Rosenfeld in their 1964 *Scientific American* article, the bootstrap was the future [35]. Quarks were not mentioned.

While Gell-Mann believed that hadrons bootstrapped themselves into existence, forming the basis of the strong interactions, Feynman was less enthusiastic, as is evident in Tony Hey's recollection of Feynman [124]:

> Feynman enjoyed making a quick and amusing response. This trait was often in evidence in seminars given by visiting speakers. On one memorable occasion, the speaker started out by writing the title of his talk on the board: "Pomeron Bootstrap." Feynman shouted out, "Two absurdities," and the room dissolved into laughter.

Nevertheless, Feynman was sympathetic to some of the more limited goals of S-matrix theory, but never used it himself.

Heisenberg thought that abandoning field theory was going too far. He considered S-matrix theory correct and useful, but incomplete [119]:

> It is perhaps not exaggerated to say that the study of the S-matrix is a very useful method for deriving relevant results for collision processes by going around the fundamental problems [of quantum field theory]. But these problems must be solved some day and one will then have to look for a mathematical formalism that allows one to calculate the masses of the particles and the S-matrix at the same time. The S-matrix is an important but very complicated mathematical quantity that should be derived from the fundamental field equations; but it can scarcely serve for formulating these equations.

How true! Here Heisenberg was correct about the strong interactions.

As a student at Caltech, after reading Zachariasen's paper, I thought that although it *might* in principle be possible to bootstrap the hadrons, in practice it would be impossible. Fermi and Yang couldn't bind a nucleon-antinucleon pair into a pion; the bootstrap wasn't going to build it by scattering all hadrons off each other. I looked elsewhere for an understanding of the strong interactions.

2.4 Personal History

My early attraction to nuclear and experimental physics originated at the age of nine in 1947, two years after the atomic bomb had been dropped on Hiroshima. One of my favorite after-school radio programs was the "Lone Ranger," sponsored by Kix breakfast cereal. Quite unexpectedly during a commercial break, the announcer asked his little listeners to send away for an "Atomic Bomb Ring" [Figure 2.4].[3] I jumped at the chance, and my mother helped. After mailing in my name, address, 15 cents, and a Kix box top, I received the ring, took it into a dark closet full of musty winter coats, and waited until my eyes had adapted to the dark. Removing the red base from one end of the "bomb" and peering down its long axis, I was rewarded with brilliant punctate flashes of light as one α-particle after another, emitted from a tiny piece of radioactive polonium 210, barreled into a zinc sulfide screen (a spinthariscope invented by William Crookes in 1903).[4] This was the same kind of screen used by Geiger and Marsden in their 1909 α-particle scattering experiment. I was so enchanted by the magic of the ring that I ordered a second one, not realizing that the half-life of polonium 210 was only 138 days.

I also liked working with my hands, building countless contraptions. At my intermediate school in Detroit I chose to be in the Manual Arts Program

---

[2] The value $M_\rho = 950$ MeV given in Zachariasen's paper is incorrect. The correct value $M_\rho = 350$ MeV given here comes from his erratum [219].

[3] From orau.org/health-physics-museum/collection/spinthariscopes/lone-ranger-atom-bomb-ring-spinthariscope.html.

[4] Crookes also discovered thallium in 1861, invented the radiometer, and developed the Crookes tube that was later used by Röntgen and Thomson in their discoveries of the X-ray and electron. As a young teenager wandering the stacks of the main branch of the Detroit Public Library, I discovered a book written by Crookes chronicling seances held in his house in the 1870s, not knowing of his seminal contribution to the development of nuclear physics, and the creation of the atomic bomb ring.

Lone Ranger Atom Bomb Ring Spinthariscope (1947—early 1950s)

This ring spinthariscope was known as the Lone Ranger Atom Bomb Ring and advertised as a "seething scientific creation." The Lone Ranger was more closely associated with silver bullets than atomic bombs but that's what it was called. When the red base (which served as a "secret message compartment") was taken off, and after a suitable period of time for dark adaptation, you could look through a small plastic lens at scintillations caused by polonium alpha particles striking a zinc sulfide screen.

Distributed by Kix Cereals (15 cents plus a boxtop), the instructions stated: "You'll see brilliant flashes of light in the inky darkness inside the atom chamber. These frenzied vivid flashes are caused by the released energy of atoms. PERFECTLY SAFE - We guarantee you can wear the KIX Atomic "Bomb" Ring with complete safety. The atomic materials inside the ring are harmless."

The following advertisement was appearing in newspapers in early 1947.

Figure 2.4. The Atom Bomb Ring based on Crookes's 1903 spinthariscope. (Courtesy Oak Ridge Associated Universities Museum of Radiation and Radioactivity.)

learning woodworking, machining, and typesetting — which I foolishly choose over typing. At some point in High School my father insisted that I transfer to the College Preparatory Program. I did, and eventually got into the University of Michigan where I majored in mathematics. Upon entering graduate school at Caltech I switched to physics.

Caltech

After my first academic year I asked Bob Christy, my quantum mechanics professor, if I could do theoretical research with him over the summer. With disdain he replied, *"You know nothing.* Why don't you go over to the Synchrotron and learn experimental physics. If you do become a theorist, you won't have time to learn what experimental physics is all about." In retrospect this was great advice, although his tone of voice delivered a different message.

At the Synchrotron Alvin Tollestrup was testing his "fast electronics" that would be used to study the decay $K^+ \to \pi^+ + \pi^0 + \gamma$ at the Bevatron in Berkeley. I thought this K^+ could have other uses. By measuring the average spin direction of the μ in the decay $K^+ \to \mu^+ + \pi^0 + \nu_\mu$, time-reversal symmetry could be tested. I proposed looking for a violation of this symmetry, "piggybacking" on Alvin's experiment. My proposal was accepted, and I started my Ph.D. thesis with Alvin as my advisor.

Alvin's was the first "user group" experiment at the Bevatron, and probably at any national laboratory. It involved one faculty member (Alvin), two research fellows (Ricardo Gomez and Hans Kobrak), and two graduate students (Bob Macek and me). The experimental equipment was built and tested at Caltech over a two-year period, then trucked to the Bevatron where the experiment was run. I won't describe the next two years of 18-hour days of classes and experimental work. In the early spring of 1962, after 21 half-days of running at the Bevatron, an adventure chronicled in "The First User Group Experiment" p. 37, I returned to Caltech with a couple hundred thousand spark-chamber photographs. A preliminary scan showed that if a violation of time-reversal symmetry existed, it would be small, unlike the size of parity violation. I then faced an additional two years of tedious analysis, probably only establishing an upper bound to the violation. With this unpleasant prospect,

I embraced denial and went camping for a month in the Yucatán.

On returning in the fall of 1962, I switched to theory. At Alvin's suggestion, I had spoken with Murray Gell-Mann on several occasions about K decay, even showing him a theory I had developed that suppressed the β-decay of strange particles relative to those without strangeness, as had been observed. Screwing up my courage, I asked him if he would be my thesis advisor. He immediately said "No!," paused, and then told me he was going to the East Coast on sabbatical, but would "talk to Dick." I was too stunned to say anything. It wasn't until I returned from CERN in the fall of 1964 that Murray and I would speak again.

"Dick" was Richard Feynman. I would never have had the courage to approach him, but for Murray. I put off seeing Feynman for as long as I thought possible. When I finally went to see him, he was expecting me, and said in a deep formal voice, "Murray says you're OK, so you must be OK."

2.5 Fundamental Physics

The physics whose laws cannot be derived, even in principle, from any other laws is "fundamental physics." Those laws are analogous to the axioms of mathematical systems, except that physicists are not allowed to make them up.

In summarizing the state of particle physics at the 1964 Erice Summer School Lectures, Feynman writes [68]:

> Now let us look at the fundamental problems we have in physics. What do we not yet know? You can roughly divide these questions into two types, namely existential problems and dynamical problems. ... An existential problem appears in the form: "Why is such-and-such a thing there?"

With respect to dynamics, after listing a number of outstanding dynamical questions, Feynman continues:

> Far more important however, is the <u>absolute lack</u> of any dynamical theory for strong interactions. We have been in this frustrating situation since 1934 when Yukawa proposed his meson theory.

My concern was solving an existential problem: *Why do we have some hadrons, and not others?* The answer, given in two CERN Reports and the 1964 Erice Summer School Lectures [220, 221, 223], starts with the large suppression of ϕ decay into $\rho + \pi$.

2.6 An Overlooked Anomaly

The long-standing legal principle "de minimis non curat lex" states that "the law does not care about the smallest things." Not always so in science where the suppression of ϕ decay appeared to be a minor matter. By 1962, nine spin-1 mesons had been discovered. One of these, the ϕ, lived much longer than I expected, for no apparent reason. According to theory, *a particle with the properties of the ϕ should not exist.*

As information is acquired it must be integrated with what is already known, and somehow put in place. In this process I always looked for something that wasn't quite right, something that didn't fit, a crack in the edifice — a place to enter — hoping for something new. So it was that Thursday afternoon when I told Feynman about the ϕ meson [41]. It had been observed decaying into two kaons, not into the two particles I and others had expected, a rho and pion,

$$\phi \rightarrow K + \bar{K},$$
$$\phi \not\rightarrow \rho + \pi. \tag{2.5}$$

I calculated the decay rate for ϕ to $\rho\pi$. The measured upper limit was lower than my predicted value by more than *two orders of magnitude*, an unprecedented suppression for the strong interactions. Feynman thought there was something wrong with the experiment (p. 50), but I thought the experiment was clean, with enough $K\bar{K}$ events to make the absence of $\rho\pi$ events more than statistically significant.

An anomalously low decay rate of ϕ to $\rho\pi$ had already been noted in 1962 by Jun J. Sakurai [190], who was referenced in the ϕ-decay paper [41]. Sakurai's paper is carefully written with proper qualifications. Given the uncertainties, he summarized it with "All we wish to claim is that we should not be too much surprised if, despite its small Q value [released energy], the $K\bar{K}$ mode turns out to be almost as frequent as the $\rho\pi$ mode." In fact the decay into $K\bar{K}$ turned out to be much more frequent than the $\rho\pi$ decay!

After he had written this paper, Sakurai learned of the full suppression of ϕ decay several months before I did (he was in touch with Nick Samios, one of the authors of the ϕ-decay paper). Like Abel Niépce de Saint-Victor, who noticed that uranium salts could expose photographic emulsions (p. 3), but moved on, Sakurai did not follow up on the anomaly before him. Ironically, I was awarded the J. J. Sakurai Prize in 2015 for the proposal that "hadrons are composed of fractionally charged fundamental constituents," a conclusion that originated with the same data Sakurai had examined, but dismissed. Anomalies have always played an important role in physics (Appendix 2.A, p. 30). It's important to recognize them.

2.7 Constituents

A Predilection for Constituents

There were elementary particles for the weak interactions (leptons), so why not for the strong interactions? Elementary particles as constituents might not only account for the many new strongly interacting particles being discovered, but also for the absence of the still larger number of hadrons that were possible but not observed. The most familiar restriction on hadron quantum numbers was associated with their electric charge; baryons were found only with charges $Q = (2, 1, 0, -1)$. For mesons, charges were limited to $Q = (1, 0, -1)$. For strangeness $\mathbb{S}$ the restrictions were also stringent: $\mathbb{S} = (0, -1, -2, -3)$ for baryons and $\mathbb{S} = (1, 0, -1)$ for mesons, with the values of Q and $\mathbb{S}$ reversed in sign for antiparticles.

These constraints suggested that hadrons might be composed of a few elementary particles whose quantum numbers determined those of their hadrons. If true, "high energy physics" might become simpler, and could be called "elementary particle physics" once again. That, at least, was my hope when I switched from experimental to theoretical physics in the fall of 1962.

I couldn't stop thinking about the suppression of ϕ decay into $\rho + \pi$ until I realized that this anomaly could be explained by assuming that:

- Hadrons had constituents,
- When a hadron decayed through the strong interactions its constituents became constituents of its decay products.

If all of $\phi's$ constituents weren't also found in ρ or π, ϕ decay into $\rho + \pi$ could not occur.

But what were the constituents? Sakata's constituents were simple but arbitrary. Why should the first three long-lived baryons be the fundamental ones? But crucially, some baryons predicted by the Sakata model were not observed.

My Homage to Sakata

Like Fermi and Yang, as extended by Sakata, I assumed mesons were created from spin-1/2 particles, and their antiparticles; but these particles were point like and elementary, like the four known leptons. I called these constituents "aces," suspecting that there was a correspondence between aces and leptons, although only three aces were required at the time. Presumably a heavier fourth ace would be found in hadrons with higher mass. (The third ace was found in baryons heavier than those created from the first two, so why shouldn't an analogous situation hold for the fourth ace?) Aces were eventually called "constituent quarks," or simply "quarks." More on naming aces can be found on page 95.

The notation I used mirrored Sakata's sakatons (p. 17). The first three aces and their antiaces were designated by

$$\mathcal{A} = (p_0, n_0, \Lambda_0) \text{ and } \bar{\mathcal{A}} = (\bar{p}_0, \bar{n}_0, \bar{\Lambda}_0). \quad (2.6)$$

Each baryon divided its baryon number $B = 1$ equally among its three aces, so aces had a baryon number of $B = 1/3$. Like sakatons, aces were assigned strangeness $\mathbb{S} = (0, 0, -1)$. The charge, baryon number, and strangeness quantum numbers of a hadron were the sums of their corresponding constituent quantum numbers.

If baryons were created from three aces, it was easy to show that aces had charges $Q =$ (2/3, −1/3, −1/3), one-third less than the charges on (p, n, Λ), Appendix 2.D: "Ace Charges and Baryon Numbers," p. 33.

2.8   The Mesons

Mesons were created from ace-antiace pairs $(\mathcal{A}\bar{\mathcal{A}})$ called "deuces."

The Pseudoscalar Meson Octet

If the spins of $\mathcal{A}$ and $\bar{\mathcal{A}}$ pointed in opposite directions the meson had spin $S = 0$, and was called a "pseudoscalar meson." For example,

$$\pi^+ \sim p_0 \bar{n}_0,$$

$$K^+ \sim p_0 \bar{\Lambda}_0, \ K^0 \sim n_0 \bar{\Lambda}_0.$$

There were eight pseudoscalar mesons, the last of which, the η, had been discovered in 1961.

The Vector Meson Nonet

When the spins of $\mathcal{A}$ and $\bar{\mathcal{A}}$ aligned, the $\mathcal{A}\bar{\mathcal{A}}$ had a total spin of 1, and was called a "vector meson." There were nine vector mesons, all known by 1962. The ρ and K^* vector mesons had the same constituents as the π and K pseudoscalar mesons, but were significantly heavier.

Aces inside hadrons interacted; they had dynamics. Each deuce represented an ace and antiace connected by a spring [Figure 2.5]. A trey represented three aces at the vertices of a triangle whose sides were springs [Figure 6.15, p. 122], meaning that aces in baryons, like in mesons, interacted pairwise. Deuces and treys weren't pictures of hadrons; they were graphical representations of mathematical objects used for

Figure 2.5. Ace-antiace pairs bound by springs formed the vector mesons (spins not shown). Black symbols were aces, white their antiaces. Both p_0 and n_0 were represented by the same symbol — a circle — since their mass difference was small. Λ_0, appreciably heavier, is shown as a larger square. The more Λ_0 or $\bar{\Lambda}_0$ a meson contained, the heavier it was. Additional aces, when discovered, were to be represented by regular polygons with an increasing number of sides, and increasing area. Aces were never drawn alone since they hadn't been observed in isolation. (Figure 2d reproduced from the first CERN Report under a CC-BY license.)

the computation of hadron masses and couplings to each other.

A pictorial representation of the vector mesons is given in Figure 2.5.

Why Didn't the ϕ Decay as Expected?

As previously stated, I assumed that when a hadron decayed through the strong interactions, its constituents were conserved (Section 2.7, p. 23); the constituents in a decaying hadron flowed into its decay products. Since the ϕ was created from $\Lambda_0 \bar{\Lambda}_0$, and neither ρ nor π contained a Λ_0, the decay of ϕ into $\rho + \pi$ could not occur. Murray called this "Zweig's rule." He also enjoyed using Jonathan Rosner's term, the "twig rule" (the English word "twig" is derived from the German word "zweig," meaning "branch"). Figure 2.6 is a graphical representation of Zweig's rule showing the decay of $A\bar{A}$ into $A\bar{A}'$ and $A'\bar{A}$.

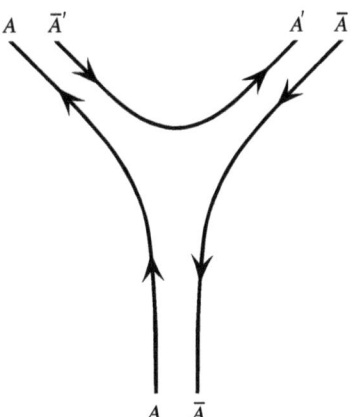

Figure 2.6. The decay of meson $A\bar{A}$ into mesons $A\bar{A}'$ and $A'\bar{A}$. Time runs upward in the diagram. Aces and antiaces are represented by arrows pointing up and down, respectively. As the ace and antiace (A and $\bar{A}$) in an unstable meson separate, a point is reached beyond which it is energetically favorable to create a virtual antiace-ace pair ($\bar{A}'A'$) from the vacuum, rather than to increase the A-$\bar{A}$ separation. The $\bar{A}'$ and A' also separate to bind pairwise with the A and $\bar{A}$ to form two new mesons ($A\bar{A}'$ and $A'\bar{A}$), the decay products of the initial meson. This is also an example of how Zweig's rule keeps aces confined. (Sampson Wilcox, Research Laboratory of Electronics at MIT.).

2.9 The Baryons

Baryons were created from ace triplets ($\mathcal{A}\mathcal{A}\mathcal{A}$) called "treys," where $\mathcal{A}$ designates a p_0, n_0, or Λ_0.

The Baryon Octet

The eight lowest mass baryons were:

$$
\begin{aligned}
p &\sim p_0 p_0 n_0, \quad n \sim p_0 n_0 n_0, \\
\Lambda &\sim p_0 n_0 \Lambda_0, \\
\Sigma^+ &\sim p_0 p_0 \Lambda_0, \quad \Sigma^0 \sim p_0 n_0 \Lambda_0, \quad \Sigma^- \sim n_0 n_0 \Lambda_0, \\
\Xi^0 &\sim p_0 \Lambda_0 \Lambda_0, \quad \Xi^- \sim n_0 \Lambda_0 \Lambda_0,
\end{aligned}
$$

(recall that "$\sim$" reads "created from"). Note that the charge, baryon number, and strangeness of the left- and right-hand sides of these relations match. Since these baryons had spin $S = 1/2$, as did their aces, the spins of two of the three aces canceled.

The Baryon Decuplet

If all three ace spins aligned, the resulting baryon had $S = 3/2$. All but one of the 10 spin-3/2 baryons that were predicted to exist had been discovered by the end of 1963.

2.10 Mass Relations

I conjectured that mass differences between hadrons primarily reflected the mass differences of their constituents, and to a lesser extent, the differences in their interaction energies (e.g., binding energies). The ace assignments for mesons required to suppress ϕ decay into $\rho + \pi$ gave mass formulae for vector mesons that worked surprisingly well [220]. For example, if the mass of a meson was the sum of its ace masses plus their interaction energy, and if interaction energies were approximately the same, then from Figure 2.5 it is clear that the ρ and ω masses are identical, and the K^* mass is the average of the ρ and ϕ masses,

$$
\begin{aligned}
M_\omega &\approx M_{\rho^0} \\
784 &\quad 750,
\end{aligned}
$$

$$
\begin{aligned}
M_{K^*} &\approx \frac{M_{\rho^0} + M_\phi}{2} \\
888 &\quad\quad 884.
\end{aligned}
\tag{2.7}
$$

The numbers below these equations are 1963 masses in MeV.[5] Both relations held to the known accuracy of the masses.[6]

Susumu Okubo also noticed these relations, as well as the suppression of ϕ decay [166]. He proposed a *mathematical* framework into which he could incorporate his observations. However, no interpretation of this framework existed. He honestly confessed:

> Unfortunately, the present author could not justify this "ansatz" on a more satisfactory mathematical ground, but it may not be impossible that such an assumption could hold if the interaction Hamiltonians are restrictive with some new kinds of symmetries.

As was commonly believed, such relations, if they were not the result of statistical fluctuations, must follow from a symmetry. Having them follow from the existence of elementary particles as constituents was unimaginable.

The first CERN Report *predicted* an additional mass formula, more accurate than either relation in Eq. 2.7, that related their errors:

$$\frac{M_\omega - M_{\rho^0}}{4} = \frac{M_\phi + M_{\rho^0}}{2} - M_{K^{*0}}$$
$$8.5 \pm ? \quad\quad - 4 \pm ? \text{ in 1963}$$
$$1.85 \pm 0.07 \quad 1.81 \pm 0.24 \text{ in 2023}, \quad (2.8)$$

where the 2023 values are from ref. [168]. Because of measurement errors, this relation was not informative in 1963, but agreement exists today! Since this relation still has not been derived from first principles (quantum chromodynamics) it has relevance now, reminding physicists that quantum chromodynamics, if properly formulated, might be simpler than it seems (Section 7.6: "Are Regularities Still Relevant?," p. 132).

Relations were also obtained for baryon masses in the second CERN Report:

$$M_{\Xi^-} - M_{\Xi^0} = (M_{\Sigma^-} - M_{\Sigma^+}) - (M_n - M_p)$$
$$6 \pm 1.3 \quad\quad 7.0 \pm 0.5 \text{ in 1963}$$
$$6.85 \pm 0.21 \quad 6.79 \pm 0.08 \text{ in 2023}. \quad (2.9)$$

---

[5] Mass M has units of energy E when the speed of light is set to 1 ($E = Mc^2$), as is done here.

[6] It wasn't clear in 1963 if meson relations involved masses or the squares of masses (right column, p. 69). For vector mesons, given the errors at the time, it made no difference. We now know that mass relations for vector mesons work better if they are linear in mass ("A Better Vector Meson Mass Relation," p. 100).

If hadrons were created from elementary particles, then hadron electromagnetic and weak interactions must result from the electromagnetic and weak interactions of these particles. When a neutron β-decayed, it really was one of its two n_0s that decayed [220, 221]; the n_0, not n, was governed by V–A interactions. The theory of strong and weak interactions were formulated in terms of aces, not hadrons, with predictions that worked ("The $\Delta\mathbb{S} = \Delta Q$ rule," p. 103).

Perhaps most remarkable, aces enabled the classification of all low-mass mesons and baryons, predicting which ones did, or did not, exist, and for those that did, correctly specifying all their quantum numbers, as described in the two CERN Reports, and a set of lectures at the 1964 Erice Summer School [220, 221, 223].

It was hard to imagine that aces were not real particles. They had spin, mass, and interacted with each other, albeit through some unknown force. They flowed from one hadron to the next in hadron decays. They vibrated and orbited around each other in quantized fashion. Confirmation of their discovery was provided by deep inelastic scattering experiments starting in 1969 [21, 25], experiments that were suggested in the second CERN Report (Section 6.9: "How to Look for Aces," p. 103).

2.11 An Analogy

Imagine playing a game with tokens (hadrons) having different combinations of properties (quantum numbers), the goal being to divide tokens into clusters, tokens in the same clusters having similar properties. Several properties for some tokens are unknown. To make matters worse, some tokens don't belong in the game (experimental error) — you don't know which ones — and other tokens are missing (haven't been discovered). Creating incomplete clusters, while leaving some tokens unassigned, is the best you can hope for.

But you have help. It's possible to connect tokens from different clusters based on their properties, connections corresponding to hadron interactions (couplings). This aids in clustering, since tokens in a cluster connect to other tokens that also cluster.

Now suppose you discover that several of the tokens — with known properties — that should connect, don't! What then?

Perhaps you should be playing a different game. In this game tokens have constituents (elementary particles or aces) whose properties determine those of their tokens. And the constituents in tokens, rather than tokens, connect, much like parts assembled in a Tinkertoy (a construction set for children). Connections are allowed only if all constituents in participating tokens can be connected.

With the game changed, the tasks are to:

- assign properties to the constituents.
- create rules for constructing tokens from constituents, so that the properties of tokens follow from those of their constituents.
- define rules for connecting constituents, so that their tokens can be properly connected (only the observed hadron couplings will be found).

You posit the constituents, their properties, and connection rules in order to *minimize the number of constituents, while maximizing the number of tokens that can be assigned to clusters*. To your amazement, only three constituents are required, and properties of tokens are simply sums or products of their constituent properties. Furthermore, tokens that were erroneously connected before, are now disconnected.

That's the way it was in particle physics in 1963 — bewildering — except people weren't playing games, and weren't postulating constituents. Many reputations were on the line. Dozens of hadrons, some confounded by statistical fluctuations, only masquerading as hadrons, were being organized into clusters ("group representations") and subclusters (isospin multiplets or "subgroup representations") according to their properties and rules of interaction. A symmetry proposed for clustering hadrons, Global Symmetry, failed. Then it was changed to SU(3). Much better, but even then a rule for hadron decay failed, the rule demanding $\phi \to \rho + \phi$, a decay that didn't occur. Miraculously, if hadrons had the right kind of constituents, theories which didn't work could be replaced by one built on only three constituents, but with fractional electric charge.

In 1963 no one knew what they were doing. There were too many hadrons, and they came with confusing combinations of quantum numbers. Why did particles with some combinations exist, while particles with different combinations didn't? Aces

changed all that. Life became much simpler, but a new question arose: What were the forces that held constituents together? It took a decade to convince theorists that constituents existed. Knowing that, they quickly found their forces, and built a quantum field theory around them — quantum chromodynamics (QCD).

The discovery of quarks, and what followed, is detailed in Chapters 6: "A Deeper Layer of Reality," and 7: "Epilogue."

2.12   The Reaction

The idea of constituent quarks was radical in 1964, and with some exceptions, not taken seriously — even ridiculed. It went against Heisenberg's admonition that one must always work with observables, and its bootstrap instantiation.

After returning from CERN in the early fall of 1964 I went into Murray's office and told him all about aces. Sometimes Murray would close his eyes when someone talked to him, but this time his eyes stayed open. After I finished at the blackboard he exclaimed from behind his desk "Oh, the *concrete* quark model. That's for blockheads!" Murray was referring both to me, and my graphical representation of aces as blocks connected by springs.

What about Heisenberg? In an interview in the early 1970s for broadcast as part of a CBC radio documentary series entitled "Physics and Beyond," he said [122]:

> Even if quarks should be found (and I do not believe that they will be), they will not be more elementary than other particles, since a quark could be considered as consisting of two quarks and one anti-quark, and so on. I think we have learned from experiments that by getting to smaller and smaller units, we do not come to fundamental units, or indivisible units, but we do come to a point where division has no meaning. This is a result of the experiments of the last twenty years, and I am afraid that some physicists simply ignore this experimental fact.

No, I'm afraid not. The experimental facts were not ignored. Heisenberg's mistake was thinking about quarks as if they acted like hadrons, composed of other hadrons, not elementary particles. While it is true that "a quark could be considered as consisting of two quarks and one anti-quark, and so on," this is not a problem since quarks are *elementary particles* that form the basis of a quantum field theory with interactions entirely different from those of hadrons. In this capacity quarks are *required* to become "two quarks and one anti-quark, and so on" for short periods of time.

Heisenberg's remarks have much in common with Murray's view that quarks could not be real particles (left column, p. 71). Murray used "proof by contradiction." He noted that hadrons are made from hadrons, *assumed that real quarks, if they existed, would act like hadrons*, and argued that "the idea that mesons and baryons are made primarily of quarks is difficult to believe." No wonder Murray said that aces were for "blockheads." He didn't understand what they were; he didn't entertain the possibility that there were two layers of reality, one created by the other, one directly observed, the other only indirectly seen in the spectrum of hadrons, their quantum numbers, and a decay anomaly.

Harry Lipkin, a staunch advocate of the constituent quark model, summarized its acceptance, or lack thereof, as of 1967:

> The quark model developed very differently in the Eastern and Western Hemispheres. In the East the model was taken seriously from the beginning and supported by top establishment figures like Bogolyubov, Sakharov, Zeldovich, Gribov, Thirring, Morpurgo and Dalitz. The Western approach was stated explicitly by M. L. Goldberger in introducing a colloquium speaker [presumably Lipkin] at Princeton in 1967. "A boy was standing on a street corner snapping his fingers and claiming that it kept the elephants away. When told that there had been no elephants around for many years, his response was 'You see! It works!' And now our speaker will talk about the quark model."[7]

Marvin (Murph) Goldberger, one of Murray's collaborators and a close friend, was *the* Western establishment par excellence, chairman of the Princeton Physics Department, future President of Caltech.

---

[7] I thank Jonathan Rosner for telling me this story, and Marek Karliner for finding the reference: [145].

Exceptional Times

Theorists have strict rules to follow that channel their thinking, wickets imposed by physical laws, enforced with mathematics. In the presence of anomalies, these rules must be relaxed, thoughts freed to wander. Incremental progress is no longer possible.

There are rules for how rules should be relaxed. While I was a graduate student Feynman told me about one such rule — Wheeler's principle of "radical conservatism" — which Wheeler attributed to Niels Bohr. In Feynman's retelling: "If you're going to violate principles, do so one principle at a time. That way the remaining number of principles you can use is maximized."

Constituent quarks violated too many principles, or preconceived notions, for most people's taste. They violated the spin-statistics theorem (as will be explained in "Were Aces Real?," p. 106), they had fractional charge, and they were not found in experiments that searched for them at accelerators and in cosmic rays [106]. There were just too many problems. As ideas go, aces were about as disruptive as they get.

Believers

The early adopters were Linus Pauling, Dick Dalitz, and Nikolay Bogolyubov, with Pauling coming first. Shortly after the two CERN Reports were circulated, I received galley proofs for the third edition of Linus Pauling's "College Chemistry," where Pauling presented aces to undergraduates [170]. He asked for corrections and comments. Essentially none were necessary. Pauling used the name "aces," not "quarks," with their symbols (p_0, n_0, Λ_0), as did Bogolyubov, but with λ_0 instead of Λ_0. Dalitz used the name "quarks" but labeled them p, n, λ.

Pauling recognized a good thing when he saw it. Aces didn't have anything directly to do with college chemistry, but I suspect he thought aces were so beautiful, so right, that he couldn't resist writing about them. That way he could start at the smallest scale and build up.

Dalitz's conversion was more natural. He came from nuclear physics where nuclei had nucleon constituents; describing nuclei in terms of their constituents was his daily bread and butter. He just changed constituents and what they created.

Surprisingly, the mathematician and theoretical physicist Academician Bogolyubov, winner of the Dirac Prize with contributions to quantum field theory, classical and quantum statistical mechanics, and dynamical systems, immediately believed in the reality of aces and instructed his students to do likewise, telling them to: "forget about Murray's quarks, think about aces." He was 55 at the time. I learned this from his students on a visit to Russia on the occasion of the 50th anniversary of the discovery of quarks.

Feynman accepted aces when he "found them for himself" (with two students) in 1971 in photo-production amplitudes, where no one else had looked [73]. Others were convinced by high-momentum-transfer scattering experiments at the Stanford Linear Accelerator (SLAC) [21, 25].

Most holdouts — but not Heisenberg — finally folded with the discovery of the J/ψ meson in November 1974. Zweig's rule, first formulated to suppress ϕ decay, was resurrected a decade later to suppress J/ψ decay. The $J/\psi \sim c\bar{c}$ was created from the fourth ace "c" and its antiace "$\bar{c}$," analogous to ϕ's creation from $\Lambda_0 \bar{\Lambda}_0$, but now the J/ψ decay was even more strongly suppressed, because low-mass mesons containing the c and $\bar{c}$, analogues of the K and $\bar{K}$ that appeared in ϕ decay, were too heavy to be the decay products of the J/ψ. As Yogi Berra was fond of saying, "Déjà vu all over again." The suppression of ϕ decay suggested the existence of aces; *the suppression of J/ψ decay, like the high-momentum-transfer scattering experiments at SLAC, confirmed their discovery.*

2.13 Invention or Discovery?

In the summer of 2013, some 50 years after the discovery of quarks, Murray and I attended a talk by Chris Llewellyn Smith at the Santa Fe Institute. In his introduction, Geoffrey West said "Murray and George invented quarks, which were later discovered at SLAC." I interrupted to suggest a different ending to Geoffrey's introduction: "Murray and George invented quarks, whose existence was later confirmed at SLAC." But even that wasn't right. Invention is defined to be "a product of the imagination," whereas discovery is "the act of finding or learning something for the first time *from observation*." Quarks might have been invented by Murray — a mathematical convenience — but for my part, aces — constituent quarks — were discovered in the Rutherford-Bohr tradition, objects buried in the data, obscured by the contradictions they implied. Bohr would have loved them.[8]

---

[8] Actually, Murray didn't invent quarks, Robert Serber did, and told Murray about them ("How Did Murray Come Up With Quarks?," p. 79). Only the name is Murray's.

Appendices for Chapter 2

2.A   Anomalies

Anomalies are important because the right kind cry out for explanation. In 1963, the ϕ meson lived much longer than expected, a striking anomaly if properly examined. To place it in context, compare it to previous anomalies:

1. In 1895, Wien and Lummer punched a small hole in the side of an otherwise completely closed oven, and measured the spectrum of the emitted light. The spectrum of this "blackbody radiation" was inconsistent with the prediction of classical physics. The discrepancy was serious — a blackbody at thermal equilibrium would emit radiation in the classical calculation at all frequencies: The higher the frequency, the more energy emitted. This was referred to as the Rayleigh-Jeans ultraviolet catastrophe. By quantizing the radiation emitted and absorbed by charged oscillators that Max Planck imagined existed in the interior wall of the oven, Planck computed a spectrum in agreement with experiment, initiating the quantum era.

2. As we have seen, wondering if the light emitted by phosphorescent uranium salts contained X-rays, Henri Becquerel, quite by accident in 1896, found that those salts emitted penetrating radiation on their own, without being exposed to light, leading him to the discovery of radioactivity.

3. The Michelson-Morley experiment in 1887 found no significant difference between the speed of light as the Earth traveled through a presumed aether, and its speed at right angles to the course of travel, killing the idea of an enveloping luminiferous aether, and leading to Einstein's Special Theory of Relativity in 1905.

4. The amount of heat required to increase a unit mass of gas by one degree in temperature, keeping either pressure or volume constant, is called its "specific heat." The specific heats of gases computed classically did not depend on temperature. While true at high temperatures, at low temperatures, as the temperature decreased, the specific heats decreased linearly towards zero. Understanding this behavior would require quantum mechanics.

5. The photoelectric effect was first observed by Edmond Becquerel in 1839. Subsequently it was rediscovered by Heinrich Hertz in 1887, immediately leading others to a series of experiments using different materials and wavelengths of light. Their results led Einstein in 1905 to propose that a beam of light is not a wave propagating through space, but a set of discrete packets of energy (quanta or photons), traveling with the speed of light, whose energy is proportional to the frequency of light. Evidently there was something wrong with Maxwell's beautiful theory of electricity and magnetism! Understanding how to correct it required the quantization of the electromagnetic field, which was initiated by Max Born, Werner Heisenberg, and Pascual Jordan in 1925, and refined by Dirac in 1927.

Anomalies are of two types:

1. Those for which no explanation has been established, but which might not contradict existing theory, e.g., the image of phosphorescent uranium salts created in Becquerel's first experiment, before he showed that exposure to sunlight was not necessary to create the image.
2. Those that are not only inexplicable, but also refute existing theory, like the five listed above.

Although both types should be investigated, the second is clearly more important because of the certainty that a serious problem exists. The long lifetime of the ϕ meson was an anomaly of the second type.

Anomalies can be missed by the best physicists. While Becquerel, Rutherford, and the Curies immediately recognized the importance of understanding radioactivity, Lord Kelvin did not. Radioactivity was not listed as a "third cloud" in his famous 1900 "two clouds" lecture (Section 6.2, p. 88). By overlooking the internal heat generated by radioactive decay within the Earth, he incorrectly estimated its age [53].[9] Kelvin became sensitive to this error, as is evident in Rutherford's 1936 talk at Cambridge University, recalling the discovery of the nucleus in 1909 [187]:

> I came into the room, which was half dark, and presently spotted Lord Kelvin in the audience

---

[9] See www.nobelprize.org/prizes/themes/how-the-sun-shines/. Kelvin's erroneous calculation of the Earth's age contradicted calculations in geology and evolution, e.g., Charles Darwin's 300 million year estimate of the time required for the observed erosion of a geological formation called the Weald.

and realized that I was in for trouble at the last part of my speech dealing with the age of the Earth, where my views conflicted with his. To my relief, Kelvin fell fast asleep, but as I came to the important point, I saw the old bird sit up, open an eye and cock a baleful glance at me! Then a sudden inspiration came, and I said Lord Kelvin had limited the age of the Earth, provided no new source [of heat] was discovered. That prophetic utterance refers to what we are now considering tonight, radium! Behold! the old boy beamed upon me.

Lord Kelvin's two clouds, cast by the aether and the specific heat of gases, were cleared quickly with the creation of the special theory of relativity and quantum mechanics, after five and 25 years. The cloud cast by radioactivity only lifted with the discovery of quarks and the development of quantum chromodynamics. It required legions of physicists working for almost eight decades.

Understand Everything You See

This maxim led Henri Becquerel to discover radioactivity, and Edmond Becquerel and Hertz to discover the photoelectric effect. Of course, they didn't understand the origin of these phenomena, but having seen something unusual they investigated — characterizing the phenomena — until it was clear that a discovery had been made. Many years later the same maxim drove me to find a reason for the ϕ's anomalously long lifetime, leading to the discovery of constituent quarks.[10]

2.B Hadrons and Their Halos

The vacuum contains vacuum fluctuations consisting of virtual particles (Section 1.9, p. 11). These fluctuations give hadrons an extended three-dimensional structure by surrounding them with virtual hadrons that interact with other hadrons, surrounded by virtual hadrons of their own. A proton moving through space-time will emit a positively charged virtual pion, becoming a neutron in the process. The neutron will then absorb the pion, changing back into the proton before any violation of energy conservation can be detected. Virtual particles like this pion, which are constantly emitted and reabsorbed, contribute to the proton's "sphere of influence."

Since the virtual pion is charged, its presence around the neutron can be detected in electron-proton scattering experiments. Photons emitted by electrons are absorbed by virtual pions, giving protons a size determined by the average pion-neutron separation. As we shall see, a proton is surrounded by a "halo" of charged pions whose spatial extent is determined by the pion mass (Eq. 2.10). Other heavier virtual particles contribute to the charge distribution inside the halo. This charge distribution was measured in low-energy electron-proton scattering experiments by R. W. McAllister and Robert Hofstadter in 1956 [151]. Its radius was

$$r_{\text{halo}} = 0.74 \pm 0.24 \times 10^{-13} \text{ cm.}$$

Hofstadter received the Nobel Prize in 1961 for this, and other, measurements of nuclear structure.

The halo's radius is approximately equal to the range of the nuclear force, which is determined by the mass M_π of its lightest carrier, the pion. This distance may be estimated by assuming it is proportional to whatever product of the constants $\hbar = h/2\pi$, M_π, and the speed of light c has the dimensions of length.[11] Using "$\propto$" to denote proportionality,

$$
\begin{aligned}
r_{\text{range}} \quad &\propto \quad \frac{\hbar}{M_\pi c} \\
&\propto \quad \frac{c\hbar}{M_\pi c^2} \\
&\propto \quad \frac{2 \times 10^{-11} \text{ MeV} \times \text{cm}}{140 \text{ MeV}} \\
&\propto \quad 1.4 \times 10^{-13} \text{ cm,} \qquad (2.10)
\end{aligned}
$$

a value twice that of halo's radius as measured by Hofstadter. Setting the proportionality constant equal to

---

[10] It used to be that computer programmers had to understand everything in the output of their code. Not so now with neural networks and artificial intelligence (AI), where ends justify means.

[11] Expressions in quantum mechanics containing Planck's constant h almost always have it divided by 2π. For example, spin is measured in units of $\hbar$, the electron having spin 1/2. Therefore it is conventional to work with $\hbar$, the "reduced Planck's constant," rather than h.

1/2 equates the range of the nuclear force to the halo's radius,

$$r_{range} = \frac{1}{2} \frac{c\hbar}{M_\pi c^2}$$

$$\approx \frac{1}{\sqrt{2}} \times 10^{-13} \text{ cm.}$$

2.C   The First Pion-Nucleon Resonance

Evidence for the existence of "hadron resonances" was first found late in 1951 by Fermi and several collaborators [6, 7, 8]. They studied the three reactions

$$\pi^+ + p \rightarrow \pi^+ + p,$$
$$\pi^- + p \rightarrow \pi^- + p, \text{ or}$$
$$\rightarrow \pi^0 + n,$$

at the newly constructed "cyclotron" at the University of Chicago, measuring the probability of interaction as a function of the incident pion energy. This probability is proportional to a quantity called the pion-proton

"total cross section." The concept of cross section, as its name suggests, is that of an effective area in a collision.

The cross-sectional area of the proton determined from the range of the nuclear force is a natural area to use for comparison,

$$\sigma_{range} = \pi r_{range}^2$$

$$\approx \frac{\pi}{2} 10^{-26} \text{cm}^2$$

$$\approx 15 \text{ millibarns,}$$

where 1 barn = 10^{-24} cm^2.

Remarkably, the total cross sections shown in Figure 2.7 vary dramatically with energy for low energies, with the $\pi^+ + p$ total cross section reaching a peak of 200 millibarns, much greater than the 15 millibarn cross section coming from the range of the nuclear force. Something unexpected is happening: Two short-lived particles, hadron resonances, have been created in $\pi^+ + p$ and $\pi^- + p$ scattering reactions.

In $\pi^+ + p$ scattering, as the kinetic energy of the π^+ increases from zero, the total cross section

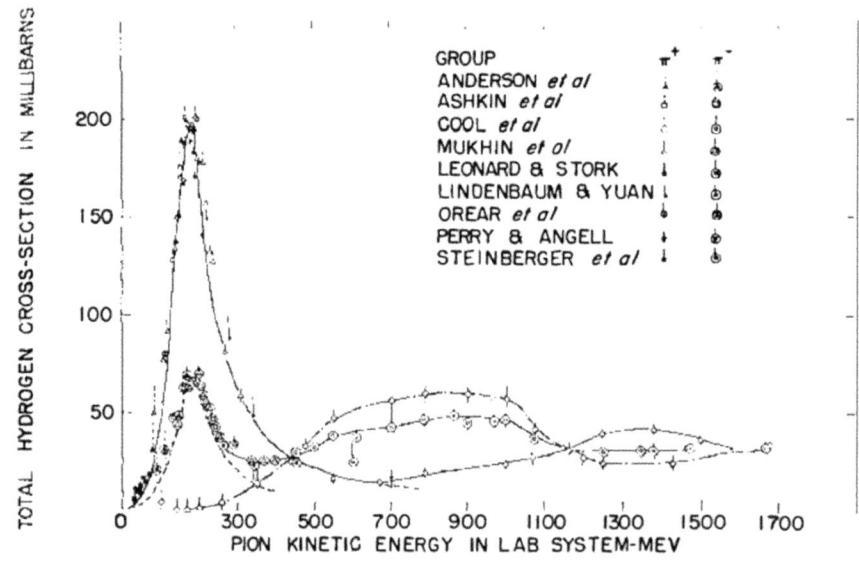

Figure 2.7. The $\pi^+ + p$ and $\pi^- + p$ total cross sections in 1957. These cross sections are proportional to the probability that a π^+ or π^-, incident on a proton, will interact with it. Focus your attention on the total cross sections below a pion kinetic energy of 300 MeV where two curves peak at about 150 MeV. The upper "resonant curve" was created by $\pi^+ + p$ scattering, as measured by several groups; the events on the lower resonant curve, at about one-third its height, came from $\pi^- + p$ reactions. These peaks were created by two different resonances, the Δ^{++} and Δ^0, with almost identical masses. The width of a peak, its "ΔE," is the interval of pion kinetic energies that formed the resonance. (Reproduced from [144].)

increases monotonically until it reaches a peak, and then begins to fall as the energy continues to rise to 300 MeV. The peak is created by a short-lived particle of charge $Q = 2$, the Δ^{++}, that was formed from, and then decayed into, a positive pion and proton. Its resonant curve reflects the distribution of pion kinetic energies that created a Δ^{++}.

Once created the Δ^{++} decayed with a variable lifetime τ having an exponential probability distribution. The average value T, and standard deviation ΔT, of the lifetime are $T = \tau$ and $\Delta T = \tau$, so the standard deviation equals the mean.

The lower peak, at essentially the same energy, was formed in $\pi^- + p$ scattering, and was due to a neutral resonance called the Δ^0. Both peaks were created in a 2-step process:

$$\pi^+ + p \quad \rightarrow \quad \Delta^{++} \rightarrow \pi^+ + p,$$
$$\pi^- + p \quad \rightarrow \quad \Delta^0 \rightarrow \pi^- + p, \text{ or}$$
$$\rightarrow \pi^0 + n. \qquad (2.11)$$

Scattering $\pi^\pm$ off neutrons created two other resonances at almost the same pion kinetic energy, with charges 1 and -1. The four "Delta" resonances are collectively called $\Delta = (\Delta^{++}, \Delta^+, \Delta^0, \Delta^-)$, an isospin multiplet (p. 15) with $I = 3/2$, spin $S = 3/2$, and strangeness $\mathbb{S} = 0$.

The shape $\sigma(E)$ of a resonant peak is a "Lorentzian" function that depends on the average Delta mass M and its range of values $\Delta M = \Delta E/c^2$ (p. 190 in [210]):

$$\sigma(E) \propto \frac{1}{(E - Mc^2)^2 + (\Delta E)^2},$$

where $\Delta E = \hbar/2\tau$ is the half width at half maximum of the cross section.

The product of the two uncertainties is

$$\begin{aligned} \Delta E \Delta T &= (\hbar/2\tau)\tau \\ &= \hbar/2, \end{aligned}$$

the limiting value of Heisenberg's uncertainty principle for energy and time (Eq. 1.1, p. 11).[12]

The width of a resonant peak is inversely proportional to the particle's lifetime, narrower peaks corresponding to longer-lived particles. An example of

a very narrow peak, indicating a very long lifetime for a hadron decaying through the strong interactions, is given by the mass distribution of the ϕ meson shown in Figure 6.2, p. 93. As will be explained ("The ϕ-decay Anomaly," p. 92), the suppression of this decay, immediately evident in its narrow mass distribution, led to the discovery of constituent quarks.

The shape of peaks in cross sections is familiar from classical mechanics where the idea of resonance originated. If a mass on a spring with damping is driven by a periodic force, as the frequency of the force increases from zero, and more energy is pumped into the system, the amplitude of oscillation increases until it reaches a peak at its "resonant frequency," and then falls as the frequency continues to rise. The width of the peak is determined by the damping. The smaller the damping, the narrower the peak, but the longer the lifetime of the oscillations created when the mass is struck impulsively. It's remarkable that the mathematics used to describe this simplest of mechanical systems is also used to calculate the shape of total cross sections, where in both systems the width of a peak is inversely proportional to a lifetime.

2.D Ace Charges and Baryon Numbers

Recall that

$$\begin{aligned} \pi^+ &\sim p_0 \bar{n}_0, \\ K^+ &\sim p_0 \bar{\Lambda}_0, \\ K^0 &\sim n_0 \bar{\Lambda}_0, \\ \Xi^- &\sim \Lambda_0 \Lambda_0 n_0, \qquad (2.12) \end{aligned}$$

where, as before, "$\sim$" reads "created from." Let the charge of a particle x be designated by Q_x. Balancing the charges on the left- and right-hand sides of the first three meson relations of Eq. 2.12 yields three equations for the π^+, K^+, and K^0 charges as sums of their constituent charges:

$$\begin{aligned} 1 &= Q_{p_0} + Q_{\bar{n}_0}, \\ 1 &= Q_{p_0} + Q_{\bar{\Lambda}_0}, \\ 0 &= Q_{n_0} + Q_{\bar{\Lambda}_0}. \end{aligned}$$

The first two equations imply that n_0 and Λ_0 have the same charge,

$$Q_{\Lambda_0} = Q_{n_0},$$

---

[12] ΔE cannot be expressed as the standard deviation of the energy distribution because that standard distribution is infinite.

like the equality of the lambda and neutron charge. The last two equations imply the p_0's charge is one more than n_0's,

$$Q_{p_0} - Q_{n_0} = 1.$$

like the charge difference between the proton and neutron.

If the charge on one ace can be determined, the charge on the other two will follow. That charge is found by considering the Ξ^- relation in Eq. 2.12. Since the charges on both the right and left hand sides of that relation must match,

$$2Q_{\Lambda_0} + Q_{n_0} = -1.$$

Since n_0 and Λ_0 have the same charge, this equation implies $Q_{n_0} = -1/3$; since p_0's charge is one greater than n_0's, $Q_{p_0} = 2/3$, and the charges on (p_0, n_0, Λ_0) are

$$Q = (2/3, -1/3, -1/3). \qquad (2.13)$$

The computation of ace baryon numbers is even simpler. Since $\pi^+ \sim p_0 \bar{n}_0$, and mesons have baryon number $B = 0$, p_0 and n_0 have the same baryon number (the baryon number of particles and their antiparticles have opposite sign). Similarly, $K^+ \sim p_0 \bar{\Lambda}_0$ implies that p_0 and Λ_0 also have the same baryon number. Therefore all three aces have identical baryon numbers. Since baryons have $B = 1$, and are created from three aces,

$$B = 1/3,$$

for all three aces.

Chapter 3
Alvin Virgil Tollestrup

Alvin Tollestrup in March 1963, six months after I left his group. (Caltech Archives.)

Visionaries aren't just born, they are forged through experience. So it was with Alvin.[1]

3.1 Caltech

Before We Met

Alvin Tollestrup came to Caltech as a graduate student shortly after the end of World War II. He had been an ensign in the Navy. Because of his interest in electronics he was sent to a preparatory radar school at Bowdoin College in Brunswick Maine, and then to radar school at MIT. Radar was new, as were pulse techniques and microwaves, and all this was classified. He learned the latest in electrical engineering not yet taught at universities. It was incredible training that made him useful to the nuclear physics group he joined at Caltech after the war, where he worked with William (Willy) Fowler, later to win a Nobel Prize for his measurements of nuclear reactions involved in the creation of the elements in stars.

Photomultipliers were used in radar as noise sources for jamming, and to look at zinc sulfide screens whose flashes of light from particle bombardment created a continuous current. When Alvin arrived Willy handed him a recent paper from Westinghouse that talked about using photomultipliers as particle detectors, and asked Alvin to build photomultiplier counters for nuclear physics experiments. He did, and that was his first contribution to Willy's group. It came directly from his previous training in the Navy — he had all the right tools to do it. He then measured the energy released in nuclear reactions to determine the mass of all the light elements, obtaining more accurate values than were available from mass spectrometry.

He was having fun as a graduate student when all that suddenly changed. Willy told him "You've got to graduate. We've got a job for you." Willy wanted Alvin to start working with three others on the design and construction of what would become the highest energy electron accelerator, used to create photons:

the Synchrotron. So Alvin stapled together a bunch of papers he'd published and graduated in the spring of 1950.

Building the Synchrotron Now Alvin had two responsibilities: to build part of the RF (radio-frequency) system to accelerate electrons, and to design and build a million-volt pulse transformer to inject electrons into the accelerator. It was baptism by fire. The pulse transformer was a crucial component that had to be ready on time. The transformer he built sat in a big tank of oil that acted to insulate, suppress corona discharge, and serve as a coolant. But when turned on it would invariably short. It was a difficult period, but he eventually identified the problem and fixed it — as he would some 25 years later with shorts in the superconducting magnets he was building for what would become the highest energy particle accelerator: the Tevatron.

The synchrotron was up and running in September 1952, less than three years after its design was initiated. Alvin then studied pion photoproduction off hydrogen and deuterium [204], confirming the existence of the first hadron resonance, the Δ, that Fermi and collaborators had just discovered.

Finding the $\pi \to e + \nu$ decay In 1957 Alvin took a sabbatical at CERN where its accelerator was just being completed. Parity violation had been discovered a year earlier, but how was it violated? There were competing theories, championed by giants. Sudershan's and Feynman's V–A theory predicted the existence of the decay $\pi \to e + \nu$. But Herb Anderson and César Lattes in 1957 failed to find the decay at the predicted rate. Two years earlier S. Lokanathan and Hans Jakob "Jack" Steinberger [148] failed to see the decay because their experiment was not sensitive enough. Alvin, with his knowledge of how photomultipliers worked, discovered a flaw in one of the experiments, and with four collaborators found the decay [55], removing the last obstacle to the acceptance of the V–A theory of the weak interactions. As a testimony to the difficulty of measuring this decay, both Anderson and Steinberger were outstanding experimentalists, students of Fermi. Steinberger later shared the Nobel Prize for demonstrating that the electron and muon each have their own neutrinos. As I

---

[1] This chapter is based on a talk given at a memorial symposium for Alvin Tollestrup; https://cerncourier.com/a/remembering-alvin-tollestrup-1924-2020/. It is used with CERN's permission.

was later to discover, there were things that I would never be able to learn from Alvin. His intuition for electronics was beyond my grasp, a gift from the Gods.

The First User Group Experiment

After returning to Caltech Alvin realized that the future of experimental particle physics was not with academics running experiments on campus accelerators, like the Synchrotron, but with "user groups" running experiments at national laboratories with more expensive higher-energy accelerators. Alvin managed to get approval to run an experiment at the Bevatron in Berkeley to measure a rare decay mode of the K^+ meson. This was the first time an outsider was allowed access to Berkeley's machine, much to the consternation of Luis Alvarez and other University of California (UC) faculty.

When I joined Alvin's group, he, as Willy Fowler had done 15 years earlier when Alvin joined Willy's group, handed me a recent paper. It was titled "A New Type of Particle Detector: The Discharge Chamber," written by two Japanese physicists [79]. Alvin asked me to design, build, and test a new type of charged particle detector that would reveal a particle's trajectory by the trail of sparks it created. The discharge chamber was not what he wanted, but a place to start. His postdoc, Ricardo Gomez, watched over me as I built Caltech's first "spark chamber." In retrospect it was remarkable that Alvin was willing to risk the success of his experiment on the creation of a new technology, and willing to give the project to a first-year graduate student.

In addition to building equipment for my own experiment, Alvin also asked me to design a transport system of magnetic lenses that would capture as many Ks as possible at the "thin window" of the accelerator, and guide them to our "hut" on the accelerator floor, where K decays would be observed. I did my calculations using punch cards on an IBM 709 at UCLA, Alvin checked them by tracing rays at his drafting table. When the beam design was completed and the chain of magnets were in place on the accelerator floor, Alvin threaded a single wire through all of them from the thin window to our hut. I had no idea what he was doing, or why. I'm not usually the quiet type, but around Alvin the Zen master, I didn't say much or ask many questions. After turning the magnets on and

running current through the wire, the wire snapped to attention tracing the path a K would follow from where it left the accelerator to where its decay could be observed. The wire floated through the magnet centers far from their walls, tracing an unobstructed path. Calculations — how much current was required in the wire — followed by testing were Alvin's *modus operandi*.

A couple of months later in 1962, run-time arrived. All the equipment for the experiment that had been built and tested over a two year period at Caltech was shipped in a moving van to the Bevatron floor and assembled in our hut. We had 21 half-days to make our measurements. The circulating proton beam inside the accelerator was steered into a tungsten target behind a thin window. The collision of protons with tungsten nuclei would produce Ks, some of which would be captured by a magnet just outside the thin window, and then focused into a beam and steered, by other magnets, to our hut. Inside the hut we waited for the scintillation counters to start clicking wildly as Ks streamed through them. But there was hardly a click. No K beam, and the 21 half-day clock had started to tick. In complete silence Alvin set out to find the beam, slowly moving a scintillation counter from one magnet to the next until he reached the thin window. Hardly any Ks were coming through. After explaining the situation to the machine operator in the control room, Alvin asked the operator to shut the machine down and remove the thin window to expose the target — an unprecedented request that meant losing the vacuum the proton beam required. Time restoring the vacuum would be lost to us, Luis Alvarez, and others. There was a long silence while the operator mentally processed Alvin's request. Several phone calls later the operator complied. With a pair of long tongs Alvin pressed a small square of dental film against the radioactive target. When developed it showed a faintly illuminated edge at the top of the target. The Bevatron surveyors had placed the target one inch below its proper position, a big mistake.

No panic, no finger pointing, just measurement and appropriate action. That was Alvin's style, always diplomatic with management, never asking for something without sufficient cause, and persistent. Unfortunately, we were unfairly charged a full day's running time, which Alvin didn't contest. Not everyone at UC Berkeley was happy with outside users coming in to use "their machine," and Alvin did not want to antagonize them any more than was necessary.

Alvin was my first thesis advisor. When he taught me how to think about my measurements, he also taught me how to analyze and judge the measurements of others. This was essential in understanding which of the many "discoveries" of hadrons in the early 1960s were believable. Without his influence, I never would have discovered quarks, whose existence was later definitively confirmed in high-momentum-transfer scattering experiments at SLAC.

3.2   The Fermilab Years

More than a dozen years later, true to his belief that users of accelerators should improve them, Alvin left Caltech in 1975 for a sabbatical at Fermilab where he would create the first large-scale application of superconductivity. He never returned.

Physics at Fermilab was limited by the energy of the protons it produced. If superconducting magnets could be built, its copper magnets could be replaced, energy costs significantly reduced, higher magnetic fields produced, and the energy of protons doubled to almost 1 TeV. Furthermore, protons and antiprotons could eventually be accelerated in the same ring, traveling in opposite directions, colliding at nodes around the ring where experiments could be performed. All this without digging a new tunnel.

Designing Superconducting Magnets

I visited Alvin shortly after he arrived at Fermilab and found him at a drafting table once again tracing rays, this time through lenses representing superconducting magnets. Looking up he told me of the magnetostrictive forces trying to tear each magnet apart, and the enormous energy stored within each one, as much energy as a one-tonne vehicle traveling more than 100 km/hr, all within a bath of liquid helium bombarded by stray high energy protons (it's magnetostriction that causes transformers to hum). If a superconducting magnet is not quenched properly (returned to its normal state), this energy will suddenly be released and serious damage will occur. There was also the possibility of a domino effect, one magnet quenching the next. Alvin had learned about magnetostriction from the

legendary William Smythe whose problem-oriented course on electromagnetism at Caltech was used for decades to weed out would-be theory students. Alvin was deeply indebted to Smythe and his book "Static and Dynamic Electricity" [196].

With a number of ingenious inventions — always experimenting, but making only one change at a time — Alvin combined an understanding coming from physics with the practicalities of engineering, to make essential contributions to the successful design, testing, and commissioning of Fermilab's superconducting magnets. With the superconducting magnets working, and the energy of the proton beam doubled, the new accelerator, the Tevatron, began operation in 1983. Alvin immediately began working on converting the Tevatron to a proton-antiproton collider, which was completed in 1987.

The Tevatron was the world's most powerful particle collider for 25 years until CERN's Large Hadron Collider (LHC) was completed. Both the bottom and top quarks, as well as the tau neutrino, were discovered at the Tevatron. Alvin's critical contributions to the creation of the Tevatron and its Collider were recognized in 1989 with a National Medal of Technology and Innovation. Alvin was also co-spokesperson for the Collider Detector at Fermilab (CDF) collaboration from its founding in 1983 to 1992.

Designing fail-safe superconducting magnets that could be mass produced was extremely difficult. Physicists at Brookhaven National Laboratory working on their next generation accelerator — Isabelle — failed repeatedly, even though they received substantially more government support and funding. When they did succeed it was too late, and Isabelle was canceled in 1983 [44]. Twenty-five years after the birth of the Tevatron, 10 days after CERN's LHC was first switched on in 2008, an electrical fault occurred in the LHC at a connection between adjacent superconducting magnets. Four megajoules of energy were released into an electrical arc in the first second, and many magnets quenched, one after another, with massive damage. The 27 kilometer Collider was closed for several months. No such problems at Fermilab.

Recognition The virtuosity required to create new accelerators sometimes exceeds that necessary to run the resulting prize-winning experiments. Alvin once told me that the Bevatron's first director, Edward Lofgren, never got the credit he deserved. The Bevatron was designed and built to find the antiproton, and sure enough Emilio Segré and Owen Chamberlain found it in 1955 after the Bevatron was completed a year earlier. They were recognized for their discovery with a Nobel Prize in 1959. But the work Lofgren and others did to create the machine for them was of a higher order than that required to run their straightforward experiment. Alvin also didn't get the recognition he deserved. His modesty only exacerbated the problem.

Designing a muon collider Alvin remained a visionary all his life. His last big project started in the early 1990s when he began work on a "muon collider." By then fundamental physics was no longer focused on hadrons and their scattering, but on their constituents, scattering them off each other at ever higher energies. Proton colliders create collisions between proton constituents, but each of those constituents carries only a fraction of the overall collision energy. Because muons are elementary particles, like quarks, every muon-muon or muon-antimuon collision will involve the muons' entire energy, making muon colliders important for fundamental high energy physics, if technical problems can be overcome.[2] They also would help us determine if a muon really is a point particle, just a heavy electron, or light tau.

3.3 Life Is on the Wire

Alvin did not suffer fools gladly, but outside of work he created a community of collaborators, an extended family. He fed and entertained us. His pitchers of martinis and platters of whole hams are memorable.

As a child my parents took me to a traveling circus where we saw a tight-rope walker, Karl Wallenda, who had an incredible high-wire act. Wallenda is quoted as saying, "Life is on the wire. The rest is waiting." Alvin showed us how to have fun while waiting, and shared a long and phenomenal life with us, sometimes off — but mostly on — the high wire.[3]

---

[2] Muon Collider Forum Report, 2023 arxiv.org/abs/2209.01318.
[3] Tributes to Alvin may be found at events.fnal.gov/honoring-alvin-tollestrup/tributes/.

Chapter 4
Richard Phillips Feynman

Richard Feynman in 1974 as I remember him. (Floyd Clark, Caltech Archives.)

4.1 Prologue

Be suspicious of autobiographies because of their inevitable conflicts of interest. This is what I learned from my uncle, a psychoanalyst, while I was growing up. So you and I have a problem. Anything original I have to say about Feynman comes from my interactions with him, and therefore relating them carries the risks of autobiography. The writer Cormac McCarthy once asked me if he should write his autobiography. "Only if you tell the truth," I replied.

Discussing Feynman will be difficult, not as difficult as talking about my father, but almost. With my advice to Cormac as a guide, and with my uncle watching, I will try my best to do justice to a man who meant so much to so many of us.

4.2 Before Meeting Feynman: Parity Violation

In his career Richard Feynman would confront two remarkable discoveries related to the weak and strong interactions. The discovery of parity violation in the weak interactions occurred in the mid 50s before I met him. The discovery of constituent quarks began while Feynman was my thesis advisor in 1962–63. To better understand Feynman's reaction to the discovery of constituent quarks, it is helpful to examine his response to the discovery of parity violation, and his subsequent explanation of this phenomenon.

Parity, a Symmetry of Nature?

When you look at a photograph can you tell if the camera was pointed at a mirror, or faced the world directly? If you can't tell, then parity is conserved — the equations governing the world are the same as those of its reflection. If parity is conserved in an interaction, then changing the signs of all spatial variables in the equations of the corresponding quantum field theory will leave its equations unchanged. These equations contain symbols representing particles, and the signs of *some* of those symbols also have to change when spatial variables flip sign, to leave the equations unchanged. Particles whose symbols remain the same are said to have positive parity $P = 1$, the others parity $P = -1$. The parity of the pion is $P_\pi = -1$, compared to that of the nucleon whose parity was set to

$P_\mathcal{N} = 1$ by convention. This negative parity of the pion was measured by William Chinowsky and Jack Steinberger in 1954 [36]. It involved the deuteron reaction $\pi^- + d \rightarrow n + n$, that was governed by the strong interactions, whose equations are parity conserving.

Applying the parity operation twice has no effect ($P^2 = 1$); applying it three times has the same effect as applying it only once ($P^3 = P$). Parity is a "multiplicative quantum number," as opposed to an "additive quantum number" like charge. The parity of a system of particles (without any angular momentum) is the product of their individual parities.

Parity Violation

Although parity was conserved in the strong and electromagnetic interactions, was it also conserved in the weak interactions responsible for β-decay, and the slow decay of strange particles like the K and Λ? The conservation of parity in all interactions was universally believed.

There was enormous confusion at the Sixth Annual Rochester Conference in April 1956, which Feynman was attending. Two new particles called theta (θ^+) and tau (τ^+), with the same mass, had been discovered, one decaying into two pions, the other into three,

$$\theta^+ \rightarrow \pi^+ + \pi^0,$$
$$\tau^+ \rightarrow \pi^+ + \pi^+ + \pi^-.$$

Both particles had long lifetimes, and therefore decayed through the weak interactions. Since none of the pions in the two decays had any angular momentum,[1] the parities of the two and three pion states were just the products of their pions' parities, and therefore differed. If parity was conserved in the weak interactions, the parity of θ^+ and τ^+ would differ, and therefore they would be different particles.

Why did two particles whose decay products had different parity have identical mass and charge? Theorists were confounded, but a young experimentalist, Martin Block, had a simple answer. Feynman, who was rooming with Block, relates [70]:

---

[1] The uniform distribution in a "Dalitz plot" of the three pions in τ^+ decay was used to show that no internal angular momenta existed among the pions [Figure 4 of [45]]. A good discussion of the Dalitz plot as applied to the tau-theta puzzle is given in https://www.slac.stanford.edu/slac/sass/talks/BrianL.pdf.

So Martin Block said to me when we were going to bed, after the discussion of the experimental situation on this problem — he says, "All you guys worrying all the time about this tau-theta puzzle. ... You know, from an experimental point of view, it's very easy. It's just the same particle. It's only that your principle of conservation of parity is cockeyed." He said to me, "What would be wrong with assuming that the conservation of parity is wrong?" So I said, "Well, let's see ... . I don't see anything wrong with it. ... That's a very good question, and you should ask the guys tomorrow." So he said, "No, you ask them for me. They won't even listen to me."

Next day when J. Robert Oppenheimer, hoping for clarification of the tau-theta puzzle, said "We need to hear some new, wilder ideas ... ." Feynman rose to the occasion [75]:

> I got up and I said, I'm asking this question for Martin Block: "What would be the consequences if the parity rule was wrong [parity was not conserved]?"
> Murray Gell-Mann often teased me about this, saying I didn't have the nerve to ask the question for myself. But that's not the reason. I thought that it very well might be an important idea.
> Lee [Tsung-Dao Lee] answered something complicated, and as usual I didn't understand very well.

The discussion ended with Oppenheimer saying: "Well, I think it's time to close our minds again."

Feynman continues:

> At the end of the meeting Block asked me what he [Lee] said, and I said I did not know, but as far as I could tell, it was still open — there was still a possibility. I didn't think it was likely, but I thought it was possible. ... There were all kinds of experiments, all kinds of interesting discoveries about parity. But the data were so confusing that nobody could put things together.

That summer (1956) Lee and Chen-Ning Yang took the possibility of parity violation seriously, stated that there was no existing experimental evidence for parity conservation, and proposed a β-decay experiment to test it [141].[2] By the end of the year Chien-Shiung Wu, together with the Low Temperature Group of the US National Bureau of Standards, performed the experiment that Lee and Yang suggested and showed that parity was *not* conserved. The tau and theta were the same particle, later renamed the K^+. Simultaneously, violation of parity was established by two other groups in muon decay [78, 83].

The conservation of parity had been deeply ingrained in the best of physicists. Rudolf Peierls, in his biographical memoir of Pauli, describes the contents of a letter Pauli wrote to Victor Weisskopf after having offered to bet substantial odds on parity conservation [171]:

> He [Pauli] was lucky that the bet had not been taken, because he would have lost some money, which he could not afford, whereas as it was he had only lost some of his reputation, which he thought he could afford.

How should a world with parity violation be described? Lee and Yang proposed a theory at the next Rochester Conference in April 1957. As Feynman relates [75]:

> I was still always behind, and Lee was giving his paper on the violation of parity. He and Yang had come to the conclusion that parity was violated [because of Wu's experiment], and now he was giving the theory for it.
> During the conference I was staying with my sister in Syracuse. I brought the paper home and said to her, "I can't understand these things that Lee and Yang are saying. It's all so complicated."
> "No," she said, "what you mean is not that you can't understand it, but that *you didn't invent it* [my italics]. You didn't figure it out your own way, from hearing the clue."

Feynman continues with his sister's questioning [70]:

> "Did you go down through it step by step?" I said, "No, I tried to figure the same thing out

---

2 Lee and Yang were wrong on one point: Evidence for parity violation did exist, and had for many years ("Turn Every Page," p. 137).

myself." She said, "Listen — for once, it will do you good, my young friend — " (my sister's a physicist) "you sit down with the paper, like a student, and instead of guessing what the heck it is, you sit down and you do it step by step. Ok." So I sat down and did it step by step, and it was very simple. It described the particular way that the parity might be violated, that the neutrino only spun one way. When I looked at the forms of the thing, I realized that there was another way of expressing it. It may be that the electron, muon and so on, are also coupled with a certain component of the Dirac equation. In playing with path integrals, I was forced into using a second order form for the Dirac equation, which was not obviously parity conserving. There was a wave function in the thing. It just looked parity conserving. And if I used that wave function, that was just the combination of what's called, $1 + \gamma_5$ times the other wave function, which was appearing in front of the neutrino. But I thought, why not put it also in front of the electron? So I put it in front of the electron and the neutrino and the mu and so on, at that time, and concluded that I would have to have vector and axial vector coupling, and that would get the same result that those people got for the mu spectrum, but with the opposite sign on the spin. And electron disintegration in every beta decay would always have to have the same polarization, whereas there were some clues that had different directions of polarization for different elements disintegrating.

So this was how Feynman did physics. An experiment shows current theory is wrong — parity is not conserved. After some prodding by his sister, he sits down, and with some fiddling, replaces Lee and Yang's theory of parity violation with the correct one. His process of invention, the reasoning behind his guess, is remarkably idiosyncratic.

The Theory of Parity Violation

The experimental situation couldn't have been more confusing. To quote Steven Weinberg, who was a postdoc at Columbia University where Lee worked:[3]

Beta decay interactions were known to be a mixture of scalar (S), tensor (T) and perhaps pseudoscalar (P) four-fermion interactions. This was the result of a series of wrong experiments, each one of which as soon as it was discovered to be wrong was replaced by another wrong experiment.

At the start of 1957 when β-decay experiments were still in error, Lee and Yang considered combining five possible weak interaction theories, called S, P, V, A and T, into a theory for β-decay [142]. Looking to experiment, they chose "A" and "T." Feynman's simpler theory required "V" and "A" in the combination of V–A ("$1 + \gamma_5$," i.e. left-handed neutrinos), but in disagreement with experiment. Uncertain what to do, Feynman left for Brazil that summer. On his return he relates [70]:

I came in from the vacation and I said to the boys [Felix Boehm, Aaldert Wapstra, and Hans Jensen at Caltech], "What's new with the data?" ... And they told me there were all kinds of experiments, everything's inconsistent, everything's all mixed up. They told me about their own experiments, other people's experiments, and so on. I didn't believe other people's experiments because I didn't see their equipment (because I know how to check, evaluate experiments). I'd seen what those fellows had done, and how they measured, and I knew everything inside out. So I knew it was right, whereas I didn't know the others were right. So I only paid attention to what they said, and not to what the others said. ... I had the advantage over some people, at least, in that I had a little selector that would select some experiments at least that were right. And I threw away everything I hadn't looked at.

This quote shows Feynman's confidence — he didn't have to examine *all* the experimental evidence once he found a convincing experiment that agreed with his theory — and also his disinclination to understand nature by synthesizing truth from milliards of contradictory observations, a talent later necessary for making progress in understanding the strong interactions. For the strong interactions, as Robert Caro observed while excavating information from the massive archives of Lyndon Johnson for his multi-book

---

[3] From "The Making of the Standard Model," an edited text of a talk given at CERN on September 16, 2003.

biography, it would be necessary to "turn every page" (Chapter 7, Footnote 10, p. 137).

The many experiments that led physicists to rule out V–A were overturned, one by one, with only the absence of the pion's β-decay to an electron and neutrino remaining [77]. Finally an experiment with enough sensitivity to detect this decay was performed at CERN by Alvin Tollestrup and his collaborators (Chapter 3, p. 36), and the V–A theory prevailed.[4] It quantitatively described how a piece of nature works, starting with a qualitative idea: All neutrinos spin in the left-handed direction. If you see a picture of a neutrino spinning in the other direction, that picture was taken through a mirror.

It turned out that the V–A theory was first discovered by E. C. George Sudarshan, an Indian graduate student working with his thesis advisor Robert Marshak at the University of Rochester. They proposed a theory of β-decay, first circulated as a preprint, like the one that Feynman had invented, but which he put on the shelf because it contradicted several experiments. Their theory was developed through a careful analysis of all existing experiments. It was not theoretically motivated like Feynman's formulation. Sudarshan described his discovery to Gell-Mann at a lunch arranged by Marshak. Sudarshan indicated which experimental results were wrong and why, and proposed a universal V–A theory of the weak interactions.

The rest of the parity violation story is fascinating, but not relevant to this chapter, shedding more light on Gell-Mann than Feynman. It continues in "Murray and Parity Violation," p. 76.

Of all his work, Feynman was proudest of his theory of parity violation in β-decay with its equation for a massless left-handed neutrino. He explains why [70]:

> When, say, Dirac got the equation [for the electron] he knows something about nature that nobody else knows. And it is a miracle that it's possible, by doing experiments over here, to predict what's going to happen over there. It

is not as much a miracle to predict something if you know the laws about it. In other words, it's enough of a miracle that there are laws at all, but *what's really a miracle is to be able to find the law.* It's another kind of miracle. You see, knowing a law to figure out that such and such is going to do something, and then have nature do it — OK, that's pretty good. But to look at other aspects and *to guess, and to know that there's a pattern under there, and to tell nature that in this experiment she's going to do that — not by deduction, strictly speaking, from what's known but by guessing from what's known — it seems a wonderful thing to me* [my italics]. And I always wanted to do that. Now, my work in electrodynamics was really using other people's formulas. My electrodynamics is not inequivalent to the electrodynamics of Pauli, Dirac, and so on, in 1929, with some technical improvements and methods of analysis and so on. It's fundamentally the same thing. Also, even the diagrams and so on only help people make calculations, and therefore makes predictions, but with a basic theory which is essentially not my own. The work with helium I got a great deal of pleasure out of also, but it still wasn't exactly that same category, because in the work on helium I had the Schrödinger equation which I thought was going to give the helium. The puzzle here is, how can that equation ever lead to that phenomenon? But that's still not exactly the same. But here, for a moment, that night, a couple of nights, I have a knowledge of a law and I can make predictions analogous to, but nowhere near as important or as vital and marvelous as, the Dirac equation or Maxwell equation.

This completes the description of events that occurred before I arrived at Caltech. Besides teaching us what to value, they provide background helpful in understanding Feynman's reaction to quarks.

4.3 First Impressions

Plenty of Room at the Bottom

There was lots of excitement as I entered the Bridge Laboratory of Physics just after my first Christmas at Caltech in 1959. Feynman had just given an after-dinner talk titled "Plenty of Room at the Bottom"

---

[4] Alvin's experiment found 40 $\pi \to e + \nu$ events. At the same time a bubble chamber experiment found six in 65,000 photographs, also demonstrating the existence of pion β-decay to an electron [129].

to the West Coast Section of the American Physical Society. He pledged to pay [63]:

> $1,000 to the first guy who can take the information on the page of a book and put it on an area 1/25,000 smaller in linear scale in such manner that it can be read by an electron microscope.

Feynman explained that a reduction of 1/25,000 made "enough room on the head of a pin to put all of the Encyclopedia Britannica." Then he added:

> And I want to offer another prize ... of another $1,000 to the first guy who makes ... a rotating electric motor which ... is only a 1/64 inch cube.

The prizes represented a significant fraction of Feynman's annual salary. No one claimed the first prize.

Later Feynman told me that he regretted making the second challenge. One crackpot after another came to his office and he had to listen to them all. Then Bill McLellan, a 1950 Caltech graduate in mechanical engineering, came in with a big wooden box, and Feynman said [10], "Oh, here's another one of them," but when he saw its contents remarked, "Uh-oh, nobody else brought a microscope."

"So I set it up," McLellan recalled [10], "and he played with it for a while." Finally Feynman admitted McLellan had done it. "He wrote me a cheque, and in the letter he said it met the specifications."

Although not explicitly stated, there was no theoretical reason that machines could not be built one molecule at a time. Feynman's enduring message is that if what we want to do is not forbidden by the laws of physics, it is possible, and we should try.

Biology

Six months later Feynman took a sabbatical to work in molecular biology in Max Delbrück's lab. Delbrück, trained as a theoretical physicist but unfortunate enough to get his Ph.D. in 1930 just after the discovery of quantum mechanics,[5] changed fields, and with Salvador Luria and Alfred Hershey, went on to

---

[5] Delbrück got his Ph.D. under Max Born at the University of Göttingen. In physics he is best known for calculating the scattering of light by light, a process Hans Bethe named "Delbrück scattering."

invent molecular biology. In 1947 Delbrück came to Caltech. He periodically visited the Physics Department to recruit physicists, and eventually recruited Feynman. When I contemplated a switch to biology a decade later, also encouraged by Delbrück, Feynman told me that he found biology interesting, but was "wasting his talents" by working in the field.

What was Feynman thinking? To typical high energy theorists who, once trained, continue to work in their narrow specialty until retirement, Feynman's trajectory after his work in quantum electrodynamics in the late 1940s may seem puzzling. He solved problems in superconductivity [61], solid state physics (the polaron problem [60]), and then contributed to molecular biology [52]. Feynman provided an explanation for his temporary abandonment of "fundamental physics" [71]. In the 1950s and most of the 1960s he thought there wasn't enough information available to understand the strong interactions. Feynman told me he liked to solve problems in at least two different ways to check his work. But in strong-interaction physics there wasn't even one way to start. His intuition was right. Real progress in understanding only became possible with the discovery of quarks in January 1964, and later in 1969 when new information about the constituents of strongly interacting particles was becoming available from "deep-inelastic" scattering experiments at SLAC.

So, disengaged from research in the high energy physics group that was centered around Steve Frautschi, Murray Gell-Mann, and Fred Zachariasen, he worked by himself picking his own problems, pushing for breakthroughs where he thought they might occur. He believed that explaining the strange behavior of superfluid liquid helium might require modifications to quantum mechanics. However he showed how conventional quantum mechanics provided an explanation of liquid helium's puzzling behavior.

Quantum Electrodynamics (QED)

My second introduction to Feynman needs some background. In 1947 Lamb and Retherford found that the energy difference between two electron orbits in hydrogen was not zero, unlike that predicted by quantum mechanics and classical field theory (p. 12).

Although the mismatch in energy, the Lamb shift, was small, when theorists had previously tried to compute it using quantum field theory, they found that it was infinite!

Lamb announced the value of the energy difference at a conference focused on Quantum Electrodynamics on Shelter Island, June 2-4, 1947. Now that theorists knew the answer, they began to redo their calculations. Immediately after the conference, Hans Bethe, who had previously figured out what makes the Sun shine, calculated a finite value on his train ride from the conference to the General Electric Research Lab in Schenectady, New York.[6] The resulting *Physical Review* paper was two pages long with only 12 equations [15]. Although Bethe hadn't included relativistic effects, Lamb and Bethe's numbers were in good agreement.

Then Feynman and Julian Schwinger (Big Julie) figured out how to repeat the calculation in quantum electrodynamics, a relativistic quantum field theory. Feynman emphasized the particle aspect of QED, Schwinger its quantized fields. The advantages of Feynman's method, that rested on a visualization of the calculation with his "Feynman diagrams," were simplicity and ease of application, while those of Schwinger's were generality and completeness built on familiar ideas in quantum field theory (p. 10 in [58]).

Schwinger and Feynman presented their results the following year at a conference Oppenheimer had organized in Pocono Summit, Pennsylvania. Now Bohr, Dirac, Fermi, and Wigner were attending. While Schwinger's highly mathematical presentation was well received, Feynman's was met with confusion, skepticism, and even hostility. Bohr objected to the idea that positrons were electrons traveling backward in time, and to Feynman's diagrams. Bohr argued that particle trajectories in Feynman diagrams were at variance with Heisenberg's uncertainty principle, not appreciating that these diagrams were just mnemonics for performing calculations. It was only later after Freeman Dyson showed the mathematical equivalence of Schwinger and Feynman's approaches that Feynman's simpler formulation was universally adopted. I suspect this initial lack of acceptance helped poison Feynman's already negative attitude towards formal theories and their "priests."

During my second Christmas vacation at Caltech I was determined to learn QED and compute the Lamb shift, so I started to read two of the classic papers, one by Feynman, intuitive and down-to-earth, the other by Schwinger, elegant but so formal that it was difficult to digest [59, 193]. One was titled "The Theory of Positrons," the other "Quantum Electrodynamics. I. A Covariant Formulation." You don't have to guess who wrote which paper. It was for their work in QED, largely defined by these two papers, that Feynman and Schwinger, together with Sin-Itiro Tomonaga, shared a Nobel Prize in 1965.

Early in his paper Feynman states that since a positron is an electron going backwards in time, both an electron and a positron can be represented by a single continuous line zigzagging through space-time. He writes:

> It is as though a bombardier flying low over a road suddenly sees three roads and it is only when two of them come together and disappear again that he realizes that he has simply passed over a long switchback in a single road.

I had a hard time understanding how the electron-positron pair could appear, and then disappear, without violating both energy and momentum conservation, but the idea was fantastic.[7] As Feynman describes in his Nobel Lecture:

> I received a telephone call one day at the graduate college at Princeton from Professor Wheeler, in which he said, "Feynman, I know why all electrons have the same charge and the same mass." Why? "Because, they are all the same electron!"[8]

It's like Feynman to give credit where credit is due. It's also interesting that Wheeler called him Feynman, not Richard or Dick. Feynman never seemed like a Richard or Dick to me. Like Wheeler, I called him Feynman. Feynman called me Zweig.

In later years Feynman covered the entire side of his VW van with giant Feynman diagrams, a little over the top, but that's Feynman.

[6] www.webofstories.com/play/hans.bethe/104.

[7] This was before I knew about virtual particles.

[8] If Wheeler were right, there should be as many positrons as electrons in our universe, which apparently is not true.

Figure 4.2. Rudolf Mössbauer at 32, the year he received the Nobel Prize. (Public domain via Wikimedia Commons.)

Mössbauer's Colloquium

Once again there was a lot of excitement as I entered Bridge Laboratory after summer vacation in September, 1961. I heard that Rudolf Mössbauer [Figure 4.2], a Senior Research Fellow in Felix Boehm's Nuclear Physics Group housed below my office in Bridge (who looked like a Hollywood movie star) had just been promoted to Full Professor. Evidently as a courtesy to the Caltech administration, the Physics Nobel Prize Committee informed the Caltech Trustees that Mössbauer was about to receive the Nobel Prize in Physics for his 1957 discovery of the "Mössbauer Effect." Most of us had never heard of Rudy or his Mössbauer Effect.

Shortly after Mössbauer returned from Stockholm he gave the Thursday physics colloquium. A young and inexperienced Mössbauer began to explain his work. As usual, Feynman sat in the front row center section of the auditorium on the side closest to the door. Although he was an experimentalist, Mössbauer presumably was technically strong, having been an assistant lecturer at the Institute of Mathematics at the Technical University in Munich. As Mössbauer started to explain the Mössbauer Effect, Feynman asked a question trying to tease out the essence of Mössbauer's discovery. Unsatisfied with the answer, Feynman continued his questioning, showing ever greater frustration with Mössbauer's responses. After several exchanges, in utter exasperation, Feynman declared in disgust, "Evidently Mössbauer doesn't understand the Mössbauer Effect," and answered the questions for him. I was so embarrassed for Mössbauer that I can still hear the sarcasm in Feynman's voice, with emphasis on "Evidently." Sometimes Feynman was brutally direct.

4.4   Quantum Gravity

Describing Feynman's work to those not trained in particle physics is about as difficult as describing da Vinci's drawings to the blind. However, an overview of Feynman's work on gravity can give the uninitiated a glimpse of the Colossus.

In the fall of '62, a year after returning from his brief interlude in molecular biology, Feynman was all revved up as he told me about a quantum gravity course he was about to teach. Teaching would provide the motivation he needed to create a quantum theory of gravity, combining quantum mechanics with Einstein's General Theory of Relativity.

In the first lecture Feynman asked, "What if Einstein had never been born and general relativity had never been created? How would high energy theorists understand gravity?" He argued that since the electromagnetic force between charged particles was created through the exchange of a particle (a photon), they would assume that particle exchange was also responsible for the gravitational force. Feynman called this exchanged particle the graviton, previously mention in "Virtual Particles as Carriers of Force," p 12.

Then he considered its properties. Gravity was very long-ranged, so the graviton must be very light. He assumed it was massless. Gravity was blind to a particle's charge, so it must be electrically neutral. What about its spin? The electrical force between particles of the same charge was repulsive, and the photon had spin one. Since the gravitational force between identical particles was attractive, he argued that the graviton's spin must be even. The simplest possibility was spin 0, but special relativity puts energy and mass on the same footing. Therefore "two tea pots, as they got hotter, felt a stronger attraction," not possible with a spin-0 graviton. Feynman concluded that spin two for the graviton was the simplest possibility, and coupled the graviton to the 2-indexed stress-energy tensor, since mass (= energy/c^2) was affected by gravity, and energy's generalization was the stress-energy tensor.

The next time I went to class, Feynman wrote down the Feynman rules for "quantum gravitational dynamics" (QGD) in the upper left-hand corner of a large slate blackboard. Here was a modern Moses giving us the commandments for gravity, the force that sticks us to the surface of the Earth.

I had seen the Feynman rules for QED used to calculate the scattering of positrons off electrons to obtain the energy levels of positronium. Now Feynman used QGD to calculate the scattering of Mercury off the Sun, as if these objects were elementary particles, obtaining an advance of the perihelion of Mercury of 43 seconds of arc per century, a tiny number that agreed with observation. Forget Einstein's curvature of spacetime, the geodesic paths of planetary motions, the Ricci tensor, and the Christoffel symbols. Just compute the rotation of Mercury's perihelion directly, like in an ordinary scattering problem! Magic without magic, all in a 50-minute hour.

I had taken a course in general relativity from one of its masters, Howard Percy Robertson of the "Robertson-Walker metric." Robertson had the distinction of rejecting a paper by Einstein submitted to the *Physical Review* on June 1, 1936. Einstein incorrectly claimed that an accelerating mass would *not* emit gravitational radiation. Robertson's review was longer than Einstein's paper; Robertson's evaluation was correct.[9]

The course I took from Robertson started in September, but it wasn't until the following year that he showed us how Einstein calculated the advance of

the perihelion. He needed all that time to teach us what Feynman didn't need. Feynman's lecture was the most brilliant display of intellectual prowess I would ever witness, and I knew it at the time.

Feynman on Tape

The job of transcribing Feynman's lectures into class notes fell to Bill Wagner, a brilliant graduate student whose office was next to mine. Bill dragged me into his office to play back the recording he taped that day in class. Unlike Schwinger who spoke in eloquent publishable sentences, Feynman used colloquial English with a heavy Queens accent, his elocution and syntax promising infinitely less intelligence than it delivered. When you heard him talk, you were swept away, thinking you understood everything. But the reality was quite different. Bill was beside himself, crying, "What should I do with this?" You might remember that it took two Caltech full professors, Robert Leighton and Matthew Sands, working together for several years to edit *The Feynman Lectures on Physics*.

Disappointment

The rest of the course was tough going, and didn't end on a satisfactory note. Of the 27 lectures delivered, notes for only 16 were published.[10]

The Feynman rules for QGD allowed one to perturbatively calculate quantum gravitational effects to any order, but at each order a measurement was required to determine a new unknown "renormalization" constant. Feynman was unable to eliminate this expanding complication, which still exists today. In addition, nonperturbative quantum effects that are encountered in the physics of black holes remain incalculable.

I sensed that Feynman's inability to create a satisfactory renormalizable quantum theory of gravity had a profound effect on him. He lost his edge. He threw more of his energies into drawing and other activities unrelated to science. Feynman was at an impasse. It took more than five years for his return [72]. The long road to recovery and what we

---

[9] Einstein was furious that the editors had shown the paper to a reviewer, rather than publishing it directly [136]. After making corrections in response to a conversation he had with Robertson (he did not know that Robertson was the reviewer), rather than resubmitting, Einstein submitted the paper to the Journal of the Franklin Institute. Einstein never again submitted a paper to the *Physical Review*.

[10] The english in the lectures transcribed by Wagner was edited by Fernando Morinigo. The published notes were further edited by Brian Hatfield [76]. Notes were so extensively edited that Feynman's voice is largely lost, and some notes no longer accurately reflect what was said in lecture.

can learn from it, recorded in his own words, thread throughout this chapter.

4.5 Physics with Feynman

When I entered Caltech as a physics graduate student I wanted to be a theorist, but wasn't sure I could hack it. Theory, with its content and practitioners, was very intimidating. I also came to understand that an entering graduate student could not do meaningful theoretical research, so at the start of my first summer at Caltech I talked to experimentalists, searching for an experiment to join.

In 1956 Lee and Yang asked if space-reflection symmetry held, so four years later I asked if "time-reversal symmetry" held. If shown a movie in reverse of an egg cracked open onto a sizzling frying pan, could you tell if the movie was run backwards? This sounds like a really stupid question, but at the time no known law of physics would be violated by an egg leaping up from the pan into a pair of closing shells. All that existing laws could say was that this was an extremely unlikely event, but possible in principle. Time-reversal symmetry was thought to hold.

Does time-reversal symmetry hold? Like space-reflection symmetry, the existence of time-reversal symmetry must be established experimentally, but it hadn't been. By building additional equipment for an experiment that I could add to the one that Alvin and his collaborators would run at the Bevatron in Berkeley, I could determine if time-reversal symmetry held in the decay $K^+ \rightarrow \pi^0 + \mu^+ + \nu_\mu$ ($K_{\mu 3}$ decay). If the average μ^+ spin pointed out of the plane of decay, time-reversal symmetry would be violated. So I joined their experiment, designed the K^+ beam, and built equipment to measure the direction in which the μ^+ spin was pointing. I also took theoretical physics classes to hedge my bets.

After more than two years of intense experimental work, when I couldn't find any evidence for symmetry violation, I switched to theory with Feynman as my thesis advisor.[11]

---

[11] It turned out that Jim Cronin and Val Fitch in 1964 found a very small violation of CP symmetry in the decay of the K^0 [38]. Assuming that CPT is conserved, as is widely believed, time-reversal symmetry is also violated. As luck would have it, I had a K^+ beam. To my knowledge time-reversal symmetry has never been tested in $K_{\mu 3}$ decay. Besides directly testing this symmetry, it would also test CPT conservation, if time-reversal symmetry held.

Feynman told me that when Wheeler became his thesis advisor, Wheeler told him to come in once a week to talk physics for an hour. Their first meeting began when Wheeler clicked his stop watch. At their second meeting, when Wheeler pulled out his stop watch, Feynman pulled out his. That's how their eventual collaboration began.

Feynman told me to come to his office every Thursday afternoon at 1:30. We would talk physics until 4:15, and then adjourn for tea before the 4:45 physics colloquium. Each week I prepared frantically for our Thursday meeting, picking some subject that I thought would interest him, never touching any subject more than once, not even my thesis. My job was to make sure that Feynman was never bored. Fortunately Feynman was not a passive listener, so time flew. Unlike many professors, Feynman was not interested in his students' calculations (in fact he was not interested in anyone's calculations). Later he told me that if he really cared about a calculation, he did it himself in a fraction of the time it would have taken any student.

This routine continued for the entire 1962–63 academic year. During that period Feynman also taught his graduate course on Quantum Gravity, and a new physics class to sophomores whose lectures were eventually published as the second volume of his famous *The Feynman Lectures on Physics*. He was a busy man, but still had time to meet each Thursday afternoon. On three of these occasions I showed him what I was working on (which now follow).

Weak Interactions of Hadrons — Advice Not Taken

Gell-Mann and Ne'eman had independently proposed a scheme for organizing strongly interacting particles of similar mass and quantum numbers into clusters, where only certain numbers of particles could appear in each cluster. These numbers were 1, 8, 10, 27, ... I asked: How do these strongly-interacting particles β-decay?

The description of weak interactions involved "weak currents" that I assumed would form one or more irreducible representations of SU(3). I told Feynman that the currents came in a group of 8 (an octet of currents), but also in another group of 27. The 27 dimensional representation was necessary to accommodate a single event, the "Barkus event," a Σ^+ β-decay supported by exceptionally strong experimental evidence.[12] With gusto Feynman declared, "Forget the Barkus event. Throw away the 27 and keep the 8!" He then launched into a tirade about how unreliable experiments were, and explained that at the time he proposed the V–A theory for the weak interactions the experiments were against him, and those experiments were wrong.

But I couldn't ignore the Barkus event because Barkus was an excellent experimentalist and his experiment was clean. The probability that his event was a statistical fluctuation was only 2 or 3×10^{-5}. Eventually it became clear that the Barkus event was indeed a very unlikely statistical fluctuation. I should have followed Feynman's advice, but since I was chained to observation I had to include the 27 currents. Much to my chagrin, almost a year later Cabibbo published his now famous paper proposing an octet of currents to govern the weak interactions [29]. He finessed the issue of the possible existence of other currents by not mentioning them, something that should have occurred to me.[13]

The Clue — Advice Not Taken

I was very excited on the afternoon of Thursday April 25, 1963 when I showed Feynman a paper in *Physical Review Letters* titled "Existence and Properties of the ϕ Meson" [41]. Other papers had presented evidence for the existence of the ϕ [14], but this paper found no evidence where evidence was most expected. It was the nonexistence of a decay mode that absorbed my attention:

$$\phi \not\rightarrow \rho + \pi. \qquad \text{(2.5 revisited)}$$

I had learned from Feynman that in the strong interactions everything that can possibly happen does,

and with the maximum strength allowed by unitarity. Suppression implied symmetry, but no symmetry was present. Therefore the suppression had to be dynamical, but no dynamical mechanism was available.

My view that "suppression implied symmetry" was later expressed by Heisenberg in 1976 [121]:

> For a ϕ meson, it is its symmetry properties that determine whether it can disintegrate with the emission of a pion into a ρ meson.

Heisenberg doesn't comment on the fact that after 12 years no symmetry had been found for the suppression of ϕ decay!

Feynman didn't think that ϕ decay could be suppressed, arguing that "unitarity mixes all states with the same quantum numbers, so the suppression of one reaction, ϕ decaying to $\rho + \pi$, is impossible." Feynman didn't object to my calculations, just to the measurement, *but without looking at how it was performed* ("The ϕ-decay anomaly," p. 92).

So unitarity cut both ways. Believing the experiment was correct, I thought unitarity meant that there was a serious anomaly that required explanation. For Feynman, unitarity implied that the experiment was in error.

Because observation is often faulty, Feynman's personal experience, going back to the parity-violating days, was at odds with his deep-seated belief that truth was only found in observation. Paraphrasing the words of the infant psychoanalyst Melanie Klein, experimentalists offered the "bad breast," and Feynman rejected it, living with all the consequences that followed.

I had no confidence in my ability to invent theories; I was wed to experiment. So as on a previous Thursday afternoon when Feynman told me to forget the Barkus event, I once again disregarded his advice, but this time knowing that he had been right the first time. As with parity violation, the ultimate problem would be deciding which of many ancillary experiments were incorrect.

My Thesis

One afternoon I announced that I would tell Feynman about my thesis. "Very good. Very good," he responded. I can still hear his voice, low-pitched,

---

[12] In Σ^+ β-decay ($\Sigma^+ \rightarrow n + \mu^+ + \nu_\mu$) the change in strangeness and charge have opposite signs ($\Delta\mathbb{S} = -\Delta Q$), a property requiring the inclusion of 27 currents. Nine other Σ^+ decays were found with $\Delta\mathbb{S} = \Delta Q$, which could be explained with only eight currents.

[13] More than 30 years later I ran into Cabibbo and asked him if he hadn't worried about the Barkus event before he published his paper proposing that the weak interactions were governed by eight currents. "Yes," he said, "but I had an important advantage over you. I was at CERN when they observed an additional 10,000 Σ^+ decays, none requiring 27 currents, so I decided to publish."

formal and brisk. In all our time together over more than 20 years, Feynman would become irritated with me only three times. This would be the first time: He had no additions or corrections to my presentation, but when I showed him the written version of the thesis he strenuously objected. The thesis consisted of two parts, each concerned with a different aspect of high energy physics. Each part would eventually be published separately. The introduction was perfunctory. Feynman said that a thesis should be written so that students and faculty in other departments could understand what it was about, and why it was important. A thesis was not supposed to be a "stapling job."[14] Back to the drawing board; I wrote a new introduction and improved the text. Feynman now approved, but when the thesis was submitted to the Caltech administration it was rejected because of improper pagination. By the time I resubmitted it was too late to graduate that year. I started work at CERN later that fall as a graduate student, the authorities non the wiser.

4.6 Feynman and Aces

We both lectured at the 1964 Erice Summer School. Six months had passed since the distribution of the 25 and 80-page CERN Reports where I introduced aces [220, 221], and the publication of Gell-Mann's two-page paper in *Physics Letters* where he introduced quarks [93]. Feynman talked about the implications of SU(3) symmetry for the weak interactions [67]. His lectures were unusual; they mostly described other people's work, rather than his own.

I spoke about aces that existed inside strongly interacting particles and the broken SU(6) symmetry their spin, and other quantum numbers, suggested [223]. When I talked to Feynman again at the summer school about the suppression of ϕ decay, he said, with some exasperation, "suppression doesn't make any sense." This was the second time Feynman expressed his irritation with me.

Feynman was not sympathetic to the idea that hadrons had elementary constituents. The fact that I treated aces as real particles, constituents of hadrons, went against his intuition that the constituents were other hadrons. Feynman still thought the nut of high

energy physics was too hard to crack, a view he held till late spring of 1968.

Feynman also gave the "Closing Lecture" to the summer school, a remarkable summary of the state of particle physics. He asked [68]: "Where do we go from here?" He addressed his question, but didn't answer it:

> What should we measure in order to perceive the correct dynamical theory of strongly interacting particles? I think that nothing new is required. One day when the dynamical theory of these strong interactions is discovered, we will ask ourselves: "Why did somebody not think of it way back in 1964?" We have all the evidence and enough data on strong interactions for an Einstein to find their law. I do not believe that there is a fundamental clue missing.

It turned out that Feynman was right; no new clue was required. The clue was in full sight: the suppression of ϕ decay, and its interpretation in terms of ace dynamics, as described at the summer school where Feynman was lecturing!

The description of ϕ decay that was given above in Section 2.7, p. 23, and the regularities in hadron masses and weak decays to be described in Chapter 6, indicated what Feynman and others were asked to accept. It wasn't that aces were the right answer for the wrong reasons, or the right answer for no reason. Aces were the right answer for the right reasons, but believing that required studying many experimental papers, and overcoming a natural aversion to revolutionary change. In addition, my constituent quark model used a novel graphical calculus with Tinkertoy constructions that made the CERN Reports hard to understand ("The Origin of Graphical Calculus" and "Couplings Computed Graphically," pp. 108 and 121). So in 1964, faced with two possibilities — aces or measurement error — *and not having the interest in systematizing and judging the voluminous data available*, Feynman made the conservative choice.

The existence of aces eventually became apparent with large momentum transfer scattering experiments at SLAC starting in 1969, and capped by the discovery of the J/ψ meson in November 1974. Nevertheless, when the J/ψ was discovered Feynman insisted [147]:

---

[14] You can tell when a theorists falsely claims that Feynman was their thesis advisor if their thesis consists of their papers. I know of one such case.

That this "crazy Zweig rule" could not give such a large suppression [of J/ψ decay] ... . There must be some new symmetry principle with a new conserved quantum number.

Feynman no longer believed experiments were in error. Like Heisenberg (Section 6.16, p. 111), he thought that a symmetry must be responsible for the suppressions observed in ϕ and J/ψ decay.

4.7 What Constitutes the Craft of Theory?

Style is a child of personality. While theory predicts and predictions are tested, the style with which theory is created is not tested, and remains contentious. Nothing is more important than style. Style is where everything starts, and limits what emerges. The rest is rational and relatively automatic. For humans the imponderable is how to think, for machines it is what to think about, at least for now.

There are conflicting views of how theory should be practiced, and where emphasis should be placed. Different views are appropriate for different times and different talents. In a 1964 lecture at Cornell, Feynman says [66]:

> In general we look for a new law by the following process. First we guess it. Then we compute the consequences of the guess to see what would be implied if this law that we guessed is right. Then we compare the result of the computation to nature, with experiment or experience, compare it directly with observation, to see if it works. If it disagrees with experiment it is wrong. In that simple statement is the key to science. It does not make any difference how beautiful your guess is. It does not make any difference how smart you are, who made the guess, or what his name is — if it disagrees with experiment it is wrong.

Here Feynman tells us how *he* worked. What he meant can be understood by examining the process that led him to the V–A theory of the weak interactions, as he describes at the beginning of this chapter (long quote starting on page 42).

Dirac's equation for the electron was motivated by the requirement that its motion be governed both by special relativity and quantum mechanics. Given that, he created a relativistically invariant quantum mechanical equation for the electron. Analogously, in attempting to quantize gravity Feynman asked, "given gravity, special relativity, and quantum field theory, how should gravitational effects be calculated?" Then he wrote down the "Feynman rules for gravity," a perturbation theory for quantum gravity that sprang full formed from his head like Athena from the head of Zeus.

Others look closely to experiment for guidance. Recall the measurement of blackbody radiation forcing Planck to its quantization; the backscattering of alpha particles from gold guiding Rutherford to a model of the atom; and regularities in the atom's spectral lines leading Bohr to quantize electron orbits. These examples have common elements: observation leading to discovery, followed by prediction, and further observation for confirmation.

In the early 1960s the problem was to recognize the problem: to interpret an anomaly (the long-lived ϕ), as well as to systematize a large body of data made possible by the industrial-scale application of bubble chambers. The former was reminiscent of Rutherford's challenge to interpret the puzzling pattern of α-particle back-scattering off gold, the latter of Bohr's challenge to organize and understand the pattern of atomic spectral lines. Neither of these challenges matched Feynman's style. Feynman was interested in inventing theories, and by his own observation, it was too soon for that ("What was Feynman thinking?," p. 45). Quarks had to be discovered before a theory of strong interactions could be created.

Reformulations of Theory

Theory from experiment, one way or another, is the basis of science, but theories also can be reformulated in useful ways. After Newton developed his theory of classical mechanics, it was reformulated by Joseph-Louis Lagrange about a 100 years later, and again 50 years after that by William Hamilton.

When the limits of classical mechanics were reached, and had to be extended, its different formulations led to different ways of thinking about quantum mechanics. Thus Lagrange's idea that particles follow a path of "least action" in classical physics led to Dirac's Lagrangian-based formulation of quantum mechanics, and the principle of least action in quantum mechanics [50]. This, in turn, resulted in Feynman's "sum over all paths" formulation of quantum mechanics where all possible paths of a particle are traveled simultaneously with equal amplitude, but different phase [69]. Starting with Hamilton's reformulation was not as useful to Feynman because it was difficult to make it relativistically invariant, although it was used by Schrödinger in his more familiar nonrelativistic formulation of quantum mechanics.

Multiple ways of representing a theory can be useful because different predictions may be easier to obtain in different formulations. Also, if quantum mechanics fails in some new domain, perhaps one of its many formulations can be modified, and built upon [197]. Feynman emphasized the value of approaching old problems in new ways, even if there were no immediate benefits.

4.8 Dirac's Visit

Dirac was Feynman's hero. As just noted, he wed quantum mechanics to special relativity by inventing the Dirac equation, leading to the prediction that the electron had an antiparticle. He pulled his equation out of thin air, an inspired guess.[15] Understanding that every particle has an antiparticle had profound implications. The vacuum, previously passive, thought empty, suddenly teemed with life. All particles, paired with their antiparticles, could — and did — momentarily appear in the vacuum, disappearing before a violation of the conservation of energy could be detected.

When Dirac visited we took him to lunch at Caltech's faculty club, the Athenaeum. Millikan had the Athenaeum constructed in the Mediterranean Revival Style just before the Great Depression, and furnished it with oriental rugs and heavy ornate furniture to entertain his wealthy donors. Diners were required to wear coat and tie. Dirac, very thin, formally suited, and Gell-Mann, slightly portly, wearing a tweed sports coat and tie, were dressed for the occasion. Feynman, who rarely ate at the Athenaeum, was sartorially neat,

Figure 4.3. Dirac listening to Feynman in July 1962 at the International Conference on Relativistic Theories of Gravitation in Warsaw. (Marek Holzman, Caltech Photo Archives.)

but coat and tieless, with dark grey pressed pants, white shirt, and navy-grey Hush Puppy shoes. For such wannabe diners the management had a back room where guests could borrow coat and tie. To match his newfound jacket, Feynman picked a bright tie with a string of women in red bathing suits diving from a high board, head to toe. Roger Dashen, a brilliant young Caltech professor who just a few years earlier had been the state wrestling champion of Montana, was unable to find a jacket that fit. His jacket was bursting at the seams, lapels a half a foot apart, pushed out by his bulging chest. As for me, I dressed in the back for obscurity.

When finally seated we were quite a sight — the most proper Dirac, an acceptable Gell-Mann, and then the rest of us. Gell-Mann did most of the talking. Feynman was subdued. No antics here. I was disappointed

---

[15] Only later did Dirac fully understand the predictions that followed from his equation. Before the positron was discovered by Carl Anderson in 1932, Dirac speculated that the proton was the electron's antiparticle. Gell-Mann told me that when he asked Dirac why he didn't predict the existence of a new particle — the positron — Dirac replied: "Cowardice, pure cowardice." Times have changed.

that these two, who could have had so much to say about each other's work, especially about the problematic renormalization of QED, did not quiz each other. Perhaps they did in private. Feynman's sketch of Dirac in Figure 4.6, p. 62 was presumably drawn at that time.

Nobel Prize

Feynman received the Nobel prize in physics in 1965, with Schwinger and Tomonaga, "for their fundamental work in quantum electrodynamics, with deep-ploughing consequences for the physics of elementary particles." This was work Feynman finished in the late 1940s. Feynman made QED calculable with "Feynman diagrams," almost without thought, to ordinary physicists, much as Leibniz made calculus accessible to engineers. I didn't understand why awarding the prize took so many years. Almost everyone at Caltech was enormously happy. Afterwards Feynman told me that he went back to his high school to talk to students. While there he looked up his school records and was surprised to find the results of his IQ test. Beaming, he told me that it was 125, and explained, "Anyone can get a Nobel Prize with an IQ of 180, but I got it with 125!"

4.9   A Difficult Period

Dejection

In 1964 Feynman, Gell-Mann and I published a paper together [65]. The last particle physics paper Feynman had published was six years earlier — the β-decay paper with Gell-Mann — although he did write a solid-state physics paper with three other authors in 1962. Feynman describes how and why he became an author on our paper in an interview with Charles Weiner [70]:

> Feynman: "Murray was developing an idea, and I had a little, but not much, to do with it. But I did discuss it with him, and in fact had suggested some aspects of it. But I really did not completely understand what he was doing. But while he was discussing it, I would suggest, 'Look at this. It would be like this; if this is like that, then that's like this,' and make some suggestions, and we worked together, somewhat. And Zweig was doing independently

something related to it. I must say, I didn't really understand that paper very well. It was written in a very great hurry by Murray, and he asked me if I wanted to get my name on it — if it was right, you know, to put the name on it. I had discussed it with him, and much of the discussion influenced him, and so on. So my first reaction was: Well, no, I'm perfectly willing for you to use my influence. It's all right; I don't need my name on it. On the other hand, I was getting depressed by having not done anything for so many years. So I made what I would consider now an error. I don't mean that I think the paper's bad or good. That's not the question. I still don't even know. I don't even know whether what they now know is in there. I didn't understand it very well. I didn't check everything that was written. I had a principle that everything that I wrote, I should understand inside out; that there was just a little bit less written than what I knew; and that whatever I wrote would be right. I didn't like the papers that somebody would write; suggesting an idea which in three months they find is cockeyed. And there was just a possibility that was such a paper, because I didn't check everything — in and out, back and forth — like I did with the β-decay. But he [Murray] came to me when I was eating lunch and asked me if I wanted my name on it. They were getting it out right away, because it was a big hurry."

> Weiner: "It was sent to *Physical Review Letters*."

> Feynman: "Right — it's a big hurry. They're writing it this morning, and do I want my name on it? And as I ate lunch I decided: yeah, because I haven't done anything, and it would be nice. ... That was a terrible thing. That was stupid. Then I looked at the paper, and I asked a number of questions to see if it was likely to be right. But I should have done the work myself. And I decided, OK, to put my name on it. I don't know whether it's right or not, but I must say that I put my name on that, and I did so little in it that it isn't really my work."

> Weiner: "Let me ask you some ... "

> Feynman: "It isn't really my work, and I've felt uncomfortable about that, since it was through

weakness, a human weakness, that I got my name on that. I did a little, a small fraction of the work, and it wasn't deserving to have my name on it. It was dumb."

Feynman was not happy with himself.

Although Feynman became an author on our 1964 paper, for some unknown reason he removed his name from a paper with Gell-Mann and Lévy just one month before it was published four years earlier [64, 89]. That paper contains a famous footnote due to Feynman, but uncredited in the publication, in which what later came to be called the "Cabibbo angle" was introduced. Feynman told me about his idea, and advised me to use the Cabibbo angle in the octet of currents I showed him at one of our Thursday afternoon meetings (p. 49). This was several months before Cabibbo's publication. If Feynman thought he got too much credit in 1964, he certainly didn't get enough credit in 1960.

Particle Theory Moves On

While Feynman was disengaged from physics in the middle 1960s, profound changes were occurring in how to think about theoretical physics, eventually leading to the "Standard Model," a quantum field theory which provides a description of the strong interactions and the unification of the electromagnetic and weak interactions. Remarkably, many theorists also began working in areas where theory and experiment had no connection (e.g., string theory).

The interaction between theory and experiment is optimal when experiments explore uncharted territory, and theorists use experimental results to make testable predictions. If there are many theorists and few experiments, what is considered acceptable theory begins to change. Theory decouples from experiment. Such a change began during the period of Feynman's malaise. As Steven Weinberg so correctly observed [207]:

> The history of science is usually told in terms of experiments and theories and their interaction. But there is a deeper level to the story — a slow change in the attitudes that define what we take as plausible and implausible in scientific theories. Just as our theories are the product

of experience with many experiments, our attitudes are the product of experience with many theories.

Paraphrasing Reverend Thomas Bayes, the probability with which we believe a theory is worth exploring is based both on observation and prior knowledge, influenced by attitudes that change with time.

A slow change in attitudes occurred through the collective interaction of many particle theorists over many years, a community to which Feynman did not belong, and with which he had little patience. As Feynman often said, "physicists chase one fashionable idea after another, like a pack of hounds." His criticism of many of their ideas — grand unifying theories (GUTS), supersymmetry, and string theory — was justified [104]. However, he did miss important new directions that fundamental physics took. One example is provided by spontaneous symmetry breaking which was introduced to particle physics in 1960 by Yoichiro Nambu [156]. Spontaneous symmetry breaking eventually led to the electroweak theory and the Higgs field responsible for the masses of quarks, and three of the six leptons.

Unlike Einstein who in his later years ignored experiment in his quest for a Unified Theory — and so lived on a "tangent plane to reality" — Feynman continued to be grounded. Starting in 1968, when he needed to evaluate the constituent quark model because he had to teach it to his students, and when scattering experiments begin to probe the internal structure of baryons, Feynman again contributed to high energy physics in his own inimitable manner [72].

4.10   Recovery

As already noted, I believe that Feynman's confidence suffered when he was unable to fully quantize (renormalize) gravity in 1962–63, and this loss of confidence was later augmented by a sense of despondency, that is evident in Weiner's interview above. In a reminiscence, David Goodstein describes a turning point [108]:

> In the immediate aftermath of his Nobel Prize in 1965, Feynman suffered a brief period of dejection, during which he doubted his ability to continue to make useful, original contributions at the forefront of theoretical physics.

In the fall of 1966 Feynman accepted an invitation to give a public lecture the following February at the University of Chicago about his ideas on teaching. Goodstein, then a new physics faculty member at Caltech, became involved and shared a suite with him at the University's Quadrangle Club. Feynman had just finished reading a typed manuscript that Jim Watson had given him, and said to Goodstein [108]:

> "You've gotta read this book," he said. "Sure," I said, "I'll look forward to it." "No," he shot back, "I mean right now." And so, sitting in the living room of our suite, from one to five in the morning, with Feynman waiting impatiently for me to finish, I read the manuscript that would become "The Double Helix." At a certain point, I looked up and said, "Dick, this guy must be either very smart or very lucky. He constantly claims he knew less about what was going on than anyone else in the field, but he still made the crucial discovery." Feynman virtually dove across the room to show me the notepad on which he'd been anxiously doodling while I read. There he had written one word, which he had proceeded to illuminate with drawings, as if he were working on some elaborate medieval manuscript. The word was "Disregard!"
> "That's what I'd forgotten!" he shouted (in the middle of the night). "You have to worry about your own work and ignore what everyone else is doing." At first light, he called his wife, Gweneth, and said, "I think I've figured it out. Now I'll be able to work again!"

As soon as he had a chance, in the 1967–68 academic year, he taught a survey course in particle physics. Late in May of 1968 I bumped into him as we both walked to the "Greasy" (the Caltech cafeteria) for lunch. He was very excited about the course he had just finished teaching. After listing the many areas of research he had just covered, he stopped, turned to me and asked, "Did I miss anything Zweig?" Patiently, once again, I told him about aces. This time he was quiet, intent, and listening. After I finish, he hitched up his pants with both thumbs, looked me straight in the eye, and in his most official voice replied, "All right, I'll look into it!"[16]

A couple of weeks later Feynman went to Santa Barbara. Earlier that spring he had been thinking about high-energy hadron collisions by representing hadrons as discs flattened by the Lorentz contraction. Now Feynman imagined that these discs were really partitioned into smaller pieces he called partons [72]. According to his biographer Jagdish Mehra (p. 508 [152]):

> As recorded in his notebook on 19 June, Feynman began to think of each hadron as a collection of smaller parts, of unspecified quantum numbers, which he christened "partons." He thought of partons as arbitrary "bare," ideal particles — the quanta of some underlying field, their exact number in any hadron being indeterminate.

This was before he visited SLAC in the second week of August where he saw experimental results from a high-momentum-transfer inelastic scattering experiment that could be understood as resulting from the scattering of electrons off point particles — his partons — in the nucleon. Aces may have suggested breaking hadrons into pieces, but at that time aces and partons had nothing to do with one another, except that they both were point particles inside of nucleons, ready to become part of a quantum field theory of the strong interactions.

Feynman's work on partons gave a physical interpretation to earlier work by James Bjorken that anticipated point-like scattering in deep inelastic scattering [17, 18]. Ironically Bjorken's papers were, in turn, based on the paper I had written with Feynman and Gell-Mann [65], that Feynman was embarrassed about (Section 4.9: "A Difficult Period," p. 54).

A year later in the fall of 1969 Feynman and two graduate students, Finn Ravndal and Mark Kislinger, looked for evidence of constituent quarks in reactions involving a photon, an area largely unexplored, but natural given Caltech's photon-producing

---

[16] As I recently discovered after writing my recollection of our conversation, Feynman recalled the same incident [71]: "And at the end of the course I was always feeling confused as to whether I

took care of them — the students should learn everything so they can read the papers, right? And I was not feeling confident in the subject. You know I told you I was worried about giving it. So after I gave my course I had about two weeks left. I went all around and I said, listen this is what I talked about, is there anything else? So I went to Zweig and I told him. He said, well, you haven't told them much about the quark model. ... So I say, Oh, OK. So I sat down and started to evaluate it myself."

synchrotron.[17] Ravndal described the beginning of their collaboration [177]:

> One day early in the fall [of 1969] Feynman was suddenly standing in the door to my office, smiling broadly as he often did. He wanted to know everything about the quark model I had used and the calculations of strong and electro-magnetic decay amplitudes. It sounded like he never had heard about quarks before.

Feynman wanted to see if real quarks exist inside of hadrons. This work was much more phenomeno-logical than his work with partons and looked to experiment, not his mind, for answers. After assuming springs connected quarks, he, Ravndal, and Mark Kislinger (another graduate student) computed 75 matrix elements, compared their values with experiment, and sent their results in for publication at the end of 1970 [73].[18]

At that time, as I was walking down the corridor on the fourth floor of Lauritsen Laboratory where we both had offices, I noticed Feynman in the distance with an enormous grin, swaggering like a sailor, thumbs hooked inside his belt, fingers splayed apart. When he was almost in my face, no more than a foot away, his right hand shot out and he said, "Congratulations Zweig! You got it right." It's hard not to love a man like that!

A Tribute from the Master

In 1977 Feynman nominated Murray and me for the Nobel Prize in Physics [Figure 4.4].[19] When I learned

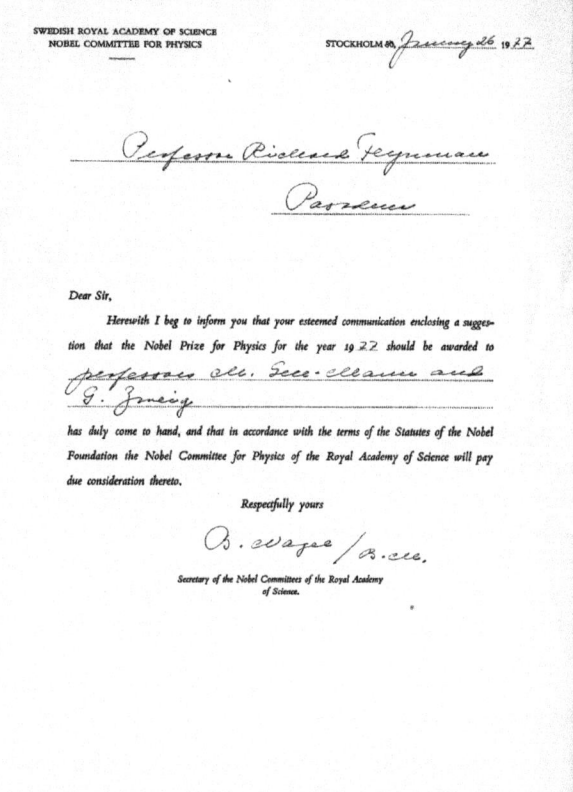

Figure 4.4. Acknowledgement of Feynman's nomination letter by the Nobel Committee for Physics. (Caltech Archives.)

about this I was very happy. Murray, on the other hand, might think that this was no big deal, at least for him. He already had a Nobel prize, and presumably had previous nominations for a second one. But to my knowledge, Feynman never nominated anyone for anything, so I think this was a real tribute both to Murray and me.

4.11   Feynman's Character

An Allergy to Philosophy

James Boswell in his biography of Dr. Samuel Johnson relates the following anecdote:

> Returning from church one day, Johnson encountered a former fellow-collegian at Oxford, an Oliver Edwards, whom he had not seen for almost half a century. In the ensuing

---

[17] Finn told me that when he first came to Caltech as a graduate student in 1968 he was eager to work on the constituent quark model, and was very surprised to find that Gell-Mann and I were not thinking about quarks. Luckily Feynman had decided to "look into it," and did so with Finn.

[18] Feynman's 1970 solution to the spin-statistics problem was to make quarks bosons! This might not be as crazy as it sounds. After all the spin-statistics theorem was only shown to hold for particles that were free (color singlets). The abstract to a paper written by Feynman, Kislinger, and Ravndal titled "The $\Delta I = 1/2$ Rule from the Symmetric Quark Model" states: "The $\Delta I = 1/2$ rule for the weak non-leptonic hyperon decays will result from quark currents interacting at a point, if the quarks obey Bose statistics." The paper was withdrawn before publication when Feynman learned that this idea had previously been proposed (Ravndal, private communication).

[19] I thank Jim Gleick for sending my wife a copy of the Nobel committee's letter to Feynman thanking him for his nomination.

conversation, Edwards humbly offered a confession, "You are a philosopher, Dr. Johnson. I have tried too in my time to be a philosopher; but, I don't know how, cheerfulness was always breaking in."

Feynman also had his issues with philosophy, but for different reasons. I suspect he agreed with Nietzsche who said that some philosophers, "muddy the waters so that they appear deep." For Feynman, philosophy missed the point and spoiled everything. The lethal craving for abstraction that was present in so many theorists was absent in Feynman. I suspect that Feynman's abhorrence of abstract theory, divorced from observation, and his contempt for philosophy, had a common origin.

It's curious that his son Carl got his undergraduate degree in philosophy and linguistics. However, his master's degree was in computer science, an area that interested his father. In the 1980s they both worked for Danny Hillis at the Thinking Machine Corporation.

Independence

It was rumored that John von Neumann, a Hungarian emigrant, was envious of another Hungarian emigrant, Theodore von Kármán, because von Kármán invented consulting.[20] No matter who invented consulting, Feynman used it to buy his freedom. At times Feynman's salary was considerably less than that of others on the physics facility, presumably because he was not supported by Caltech's Department of Energy (DOE) high energy physics contract. For many years Feynman consulted for the Hughes Aircraft Company in Malibu on subjects that had little to do with Hughes' mission.[21] Not indebted to DOE, Feynman worked on any problem that interested him. His view was consistent with that of my mother's, although neither

of them expressed it explicitly, "You work to please yourself, not others."

Truthfulness

Feynman talked to an enormous number of people, sometimes to learn, sometimes to perform. His behavior was extraordinary; no word describes it. If it didn't have sexual connotations, "promiscuous" would be fitting.[22] There was a remarkable consistency in what Feynman said, reflecting an aspect of his character that I associate with a story about Sam Rayburn, a congressman from Texas from 1913 to 1961, and Speaker of the House for almost half that time. Lyndon Johnson was his protégé. Rayburn's power and accomplishments were enormous, but he lived modestly in a Washington hotel. His estate was valued at only $30,000 at his death. An interviewer once asked, "You talk to so many people about so many things, how can you possibly keep your stories straight?" To which Rayburn replied: "I tell the truth."

What Came First?

Feynman was a hero to many people for many reasons. To some, as expressed to me by Jeffrey Mandula:

> His salient characteristic was his devotion to physics. Unlike others in his field, paraphrasing JFK, he "asked not what physics could do for him, but rather what he could do for physics." His showmanship attracted attention, and made his lack of self-aggrandizement all the more striking.

He really did physics for the wonder and joy of it.

4.12 His Influence on Others

Feynman had a profound influence on how physicists calculate, and therefore on what and how much

---

[20] Stanislaw Ulam, private communication and [203]. Von Kármán was the outstanding aerodynamic theoretician of the twentieth century. At the age of 22 (1903) while still in Germany, he started consulting with a locomotive manufacturer (mathshistory. st-andrews.ac.uk/Biographies/Karman/).

[21] The "Feynman Hughes Lectures," are posted online at www .thehugheslectures.info/. Unfortunately this site is unsafe. Volume 4 was titled "Biology, Organic Chemistry, and Microbiology." Evidently Feynman retained his interest in molecular biology. It was clear from this volume, and his time at Los Alamos where he was in charge of the computation group in the theory division, that much as he liked elegance and analytical solutions, only answers mattered, not how they were obtained.

[22] Sigmund Freud attempted an analysis of Leonardo da Vinci based on da Vinci's recurrent "vulture dream." So much more information is available about Feynman, in his autobiographical writings, lectures, and videos on the internet. Perhaps psychoanalysis is possible. Getting insight into how his personality affected his style in physics, which determined what he did — and did not — accomplish, might then be possible.

Figure 4.5. Feynman holding Court at the Shelter Island Conference June 2–4, 1947. Standing (from left to right) are Willis Lamb, Karl Darrow, Victor Weisskopf, George Uhlenbeck, Robert Marshak, Julian Schwinger, and David Bohm. Seated are Robert Oppenheimer (on the arm of the seat), Abraham Pais, Feynman, and Herman Feshbach. (Courtesy of National Academy of Sciences.)

they could accomplish. Physicists listened to him, as Figure 4.5 illustrates.

His influence at a personal level is much more difficult to ascertain because he was so singular. His students didn't grow up to be "little Feynmans." Paraphrasing Galton's quote from Thomas Carlyle's Sartor Resartus: "Great oaks don't grow from cabbage seeds" [81]. Feynman's students had little in common, although they were a bit more eccentric than most — more outsiders than insiders — compared to Gell-Mann's students.

Feynman taught by example. He showed that it was possible to do good work, even great work. His "sum over all paths" formulation of quantum mechanics extended Dirac's action principle for quantum

mechanics in an unexpected direction [69]. Here was real "quantum democracy" at work, each path equally weighted in amplitude, together governing matter as it moved through space and time, making it obvious how quantum and classical mechanics were connected. The future was determined by collective action at the quantum level. What originality! How inspirational! And while I was a student at Caltech, there he was down the hall struggling, figuring things out, understanding. Maybe I could do something too. Hurry up, it's possible!

Feynman taught us how to think for ourselves and judge work, that of others as well as our own. He did this with respect and kindness, never devastating us, like he often ravaged speakers. He helped form our professional superegos.[23] He reinforced our commitment to the values that should be practiced in science. Although Feynman was not religious, he taught us that in science, truth is sacred. He had a commitment to understanding what is true, and expressing it so others could understand it. He showed us the pleasure and satisfaction of understanding.

Feynman's Question

Feynman always went to tea on Thursday afternoon where graduate students gathered before the physics colloquium. Harvey Shepard, a student at Caltech in the early 1960s, relates [194]:

> One day during afternoon tea in Bridge Laboratory at Caltech, Feynman was talking to a group of graduate students. He asked us what research we were working on, and after several responses he said: "Doesn't it bother you knowing that others are working on the same problems using the same approaches? I can't work on something unless I believe that I am doing it in a different way from everybody else."

Pretty intimidating, but not as hard to live by as Edwin Land's maxims, "Don't do anything that someone else can do," and, "Don't undertake a project unless it is manifestly important and nearly impossible." Feynman made the cut.

Advice for the Future

In his Closing Lecture at the 1964 Erice Summer School Feynman advises his students [68]:

> First, learn to calculate ... You will learn by experience and will see how simple it really is ... If you are theorists I would urge you to connect yourselves to nature by calculating ... You should develop a feeling for the subtle interplay between the general and the specific. You should cultivate both. By general and specific I mean this: If you work out a number of examples you may be able to generalize to a general principle. The general and the specific work together. This holds in both directions ... take

> some specific example of [a] theorem and try to figure it out numerically to see whether it works.

I was there in Erice listening to him, and still calculate now (when not writing this book), almost on a daily basis.

A Yardstick

Talking to Feynman could be an incredible experience, like looking at original da Vinci drawings, but better because of the interaction. The back-and-forth felt physical, like stretching your muscles, or playing tennis. It brought out the best in you. Sometimes you would find the unexpected. In the give and take you could gauge how smart you were, and what kind of smarts you had. You stood in the fire; you were annealed.

In Retrospect

It seems absurd that I, inexperienced, knowing so little, should be spending several hours a week as a student with Richard Feynman, who had created so much, and seen even more. For this I am forever grateful.

4.13   Support

Feynman was good to me. He took time to talk, and seemed interested in what I had to say (I kept up with the literature, while he didn't). When I applied for a job at the University of Wisconsin in 1967 to be close to the soon-to-be-completed accelerator at Fermilab, hoping to become an experimentalist again, Lee Pondrom, a member of the facility committee considering my application, requested a letter of recommendation from Feynman. The members were taken aback by Feynman's four-word reply, "You can't go wrong." Feynman usually refused to write recommendations, so I was fortunate, and got an offer.

Problems Working in Biology

After I switched to biology, while working with fruit flies in Seymour Benzer's lab trying to understand their genetics of learning, but still teaching physics, Bob

---

[23] The superego is the inner voice that tells you right from wrong. It is the ethical component of your personality that provides the moral standards by which to live. Like God for believers, it is an uncompromising voice.

Christy, Caltech's Provost, called me into his office and asked, "What are your intentions?" After I told him I intended to switch to neurobiology he said, "Let me tell you my intentions, but first, have you had a sabbatical?" "No," I replied. Christy then told me that when Feynman spent his year in Biology he took a sabbatical. Christy continued:

> I will charge the time you have spent in biology to a sabbatical. At the end of the year I will set up a committee to demote you to a position commensurate with your publications in biology, with a concomitant reduction in salary.

He explained:

> Caltech is a small place that works in a limited number of research areas. Those areas are chosen by the administration. With that focus Caltech's excellence is maintained. We cannot have a professor working on anything that interests him. It sets an unacceptable precedent.

So much for tenure — I had been a tenured Full Professor of Theoretical Physics since 1967. Never mind that the Biology Department had voted to give me 3,000 square feet of space in the Beckman Behavioral Biology Building, still under construction, and money to support two postdoctoral fellows from Harvard's Neurophysiology Department to work with me on problems in hearing, and the development of a cochlear implant.

Notwithstanding Christy's objections, I continued with my plan. Beckman's building was nearing completion, but my space remained unfinished. I went to see Christy who said:

> It is the prerogative of the Biology Department to give you space, but it is my prerogative to supply it with water and electricity.

And he did not, leaving me with 3,000 square feet of cavernously-empty shell space.

Christy did what he could, with the help of Bob Leighton, chairman of the Physics Department, to keep me from switching fields. Somehow Feynman found out about my problems with the administration and yelled at Leighton, who called me in to apologize. A year after I left Caltech, in 1983, Caltech awarded me its "Distinguished Alumnus Award." I suspect Leighton was my advocate.

My instinct, like Feynman's, was to work on problems that were not fashionable, and where novel approaches were possible. Working with the surgeon Bill House, John Pierce and I designed the first multi-electrode cochlear implant in 1973 [175, 176].[24]

When I showed Feynman my work in hearing, his demeanor changed, the pitch of his voice dropped, and he became utterly professorial, like the afternoon I told him about my thesis. There were questions, and answers. I was embarrassed that I hadn't done more.[25] After finishing, he summarized, formally announcing, "Very good." Since he didn't seem to be his usual self, I didn't know if I should believe him.

More than 40 years later I finally found the form of the nonlinear equation governing the traveling waves that spiral within the cochlea [229, 230]. My immediate regret was that I couldn't show Feynman the equation, letting him know that I had finally found it after all.[26]

4.14 Diversions

Jirayr Zorthian

After his struggle with Gravity, Feynman took up drawing seriously with the Armenian-American artist Jirayr Zorthian. Zorthian, once married to the shaving cream heiress Betty Williams, led a free-form lifestyle on his ranch in Altadena, acquired in his divorce settlement. Feynman lived nearby. Feynman and Zorthian met at a party where they made a deal. Zorthian would teach Feynman drawing, and Feynman would teach him physics. Feynman told me Zorthian gave

---

[24] The device used to drive the implant electrodes was similar to the channel vocoder originally used for transmitting speech over telephone lines with much less bandwidth than that required for transmitting unprocessed speech. Channel vocoders are still the heart of current implant stimulation algorithms [123].

[25] My work showed that the WKB (Wentzel-Kramers-Brillouin) approximation, and an approximate "scaling symmetry," held for waves traveling in the cochlea. I also defined a "cochlear transform" (the first example of what later would be called a "continuous wavelet transform") that I used to create pictures of speech that the central nervous system "sees." The Navy eventually used the cochlear transform to listen for submarines.

[26] And I wanted to tell him that the equation was discovered algorithmically, directly from the data. Only the algorithm converting data to equation was invented.

up physics after the first lesson, but Feynman kept on learning how to draw, lesson after lesson. Figure 4.6 is Feynman's sketch of Dirac. Feynman said he drew faces because they were the hardest things to get right. People really know what faces look like.

Figure 4.6. Dirac by Feynman, April 1965. (Caltech Archives.)

Bongo drums

There are many pictures of Feynman playing the bongo drums. Why drums? Feynman was quite upset one morning when he told me about a dinner at Gell-Mann's house the night before. Murray kept hitting keys on his grand piano, one after another, asking Feynman what note he struck. Time after time Feynman couldn't answer. Even though he took piano lessons as a child, he was tone deaf. Feynman's pique surprised me. Murray often needled him, but whenever I had witnessed it, Feynman always rose above it, much like an older brother ignoring the needling of a younger sibling, acting as if they didn't exist.

Feynman also told me of his synesthesia, associating colors with letters and numbers, something that didn't seem to bother him.

4.15   A Dark Side

Nietzsche remarked that if we voyage far from familiar shores, returning years later, we may feel estranged, and view what had been familiar more critically. This is how I now feel as I look back at some of Feynman's behavior. At rare times he had a surprising disregard for the impact of what he said on the feelings of others. I never saw malice, just the unwitting infliction of pain. I recall two unfortunate incidents, and one that was finessed:

Marshall Baker

While I was a graduate student Marshall Baker, a respectable theoretical physicist, a former student of Schwinger's, was invited by Fred Zachariasen to give the Tuesday afternoon particle physics seminar. It was held in a small classroom with tiered rows of seats sloping steeply upward like bleachers in a ballpark. As usual, Feynman and Gell-Mann sat together front row center, Feynman closest to the door, Gell-Mann closer to the center. Lesser luminaries, postdocs, and graduate students sat in rows behind them. Gell-Mann was wearing his customary tweed sports coat with tie, while Feynman, dressed more like a graduate student, impatiently tapped the floor with his omnipresent Hush Puppy shoes. Both of them looked oddly out of place, squeezed into drop-leaf chairs, their paddles out, meant for students taking notes. As Baker started speaking, Gell-Mann surreptitiously picked up a folded *New York Times* lying on the floor beside him, unfolded it, and after snapping it open at eye level began reading right in front of Baker who was standing on the other side of the paper only a yard away. After about a minute, Feynman, who didn't pay much attention to other people or their work, leaned over to Gell-Mann and whispered in his best Far Rockaway accent, "Is this guy smart?" Feynman's voice was hushed, but loud enough so that everyone in the room, including the speaker, could hear it. This was not the first time the seminar attendees had witnessed these two in action. They knew that if Gell-Mann's head nodded up and down behind the paper, Feynman would ask questions. If his head rocked side to side, Feynman wouldn't waste his time. Now Gell-Mann's head nodded up and down, answering the question for everyone except the speaker. What the seminar attendees didn't know was that Baker stuttered when stressed. Feynman started questioning, Baker started stuttering — the more questions the longer the stutter. With Feynman's final question, Baker's stutter wouldn't stop, one ah-bah ah-bah after another. In utter frustration Feynman slammed the palm of his right hand down on the paddle of his drop-leaf chair, shouted "Goddamn it! I can't get a straight answer out of this guy," and stormed out of the classroom, leaving Baker finally speechless.

The next day I happened to walk by Gell-Mann's office. The door was open and I overheard Zachariasen animatedly asking Gell-Mann to give Baker a $100 honorarium as partial compensation for Feynman's atrocious behavior. Gell-Mann seemed sympathetic, but noncommittal.[27]

One of Gell-Mann's Graduate Students

In the second incident, Gell-Mann asked me to chair the thesis committee of one of his graduate students, whom I just will call Bob. Gell-Mann was going to be out of town the day of Bob's oral exam. Feynman was a member of the committee. Bob's thesis was unusual. It considered a simple *interactive field theory* with a *reduced number of spatial dimensions* where certain problems, unapproachable in a more realistic theory with three dimensions, could be solved exactly. I thought Bob's thesis was technically sound, and more interesting than the conventional theses we usually saw. Gell-Mann often worked with toy field theories containing all three spatial dimensions, but *without strongly interacting fields*. Gell-Mann had supervised Bob's research, read his thesis, and was satisfied, as was everyone on the committee, except for Feynman. As Bob presented his results, Feynman's questioning became increasingly hostile. He had a bee in his bonnet. There were no mistakes, but Feynman didn't like the subject matter, and refused to sign off on the thesis. He thought Bob, who had already been at Caltech for four or five years, should pick another thesis problem. I asked Bob to step out of the room, and after much negotiation, Feynman agreed to sign if and when Bob revised his thesis, responding to a list of comments and

questions that Feynman would supply. Bob was devastated. I reported to Gell-Mann upon his return, but he didn't get Feynman to change his mind. Feynman's outburst was not momentary. It tapped deeply into his hatred of abstract theoretical physics, and God knows what else. Bob eventually graduated, but ended up going to medical school in Florida. He now practices geriatric medicine in Los Angeles.

Jenijoy La Belle

Jenijoy La Belle, an attractive young faculty member in the English Department in the 1970s, told me, and later an interviewer, the following story:[28]

> I was going to a meeting in Bridge — the building that he [Feynman] taught in. I was walking up the stairs, and I heard this voice say, "Come back down the stairs." And I went back down and said, "Why did you want me to walk down?" and he said, "So I can watch you walk up the stairs again." I suppose I was wearing a miniskirt, which is what I tended to wear in those days. Then he said, "I'm Richard Feynman." I certainly knew the name, but I didn't connect him with this man in the white shirt and the grey slacks. But I laughed, and he laughed. That's how I met Feynman. ...
> I found this great quotation. Do you know Ilka Chase? She was a novelist and actress in the 1920s and 1930s, and she was in the Bette Davis movie, "Now, Voyager." This wonderful quotation, where at one point she said, "George Moore unexpectedly pinched my behind. I felt rather honored that my behind should have drawn the attention of the great master of English prose." So I thought I would say that I felt honored that mine had attracted the great quantum electrodynamics master.

And then:

> Perhaps it was Oscar Wilde — who defined "flirtation as attention without intention."

Not all women could protect themselves so well.

[27] Speakers at Caltech theory seminars never received an honorarium. I told this story at my talk in honor of Gell-Mann's 80th birthday. Gell-Mann, who was in the audience, said he no longer remembered if this tradition was broken in Baker's case.

[28] Interview by Heidi Aspaturian, Pasadena, California, February-May 2008, April 2009. Oral History Project, California Institute of Technology Archives (resolver.caltech.edu/CaltechOH:OH_LaBelle_J).

4.16 Towards the End

When the muon was discovered by Neddermeyer and Anderson in 1937 it appeared to be exactly like the electron, only heavier, provoking Isidor Rabi to ask, while eating with friends at a New York Chinese restaurant, "Who ordered that?" Whenever I had entered Feynman's office I saw, carefully scrawled in small crabbed letters in the upper left-hand corner of his slate blackboard, his version of Rabi's question — "Why does the muon weigh?" Shortly before I left Caltech in the summer of 1981, while in his office, I realized that his question had been erased. I felt Feynman had surrendered. It was a moment of great sadness.

Earlier in February of that year Feynman had submitted his first paper on QCD to *Nuclear Physics* [74]. As he characterizes it, "QCD field theory with six flavors of quarks with three colors, each represented by a Dirac spinor of four components, and with eight four-vector gluons, is a quantum theory of amplitudes for configurations each of which is 104 numbers at each point in space and time." After many simplifications, using only two spatial dimensions (remember Bob, p. 63), he "describes some small steps" in obtaining a "qualitative picture of the salient features of the QCD behavior" at low energies. Feynman had met his match. This would also be his last paper on QCD, marking the end of an era, and a unique style of theoretical physics.

Feynman turned his attention to computation. In 1980 John Hopfield came to Caltech and revived interest in neural networks and analog computation. In 1981 Feynman, Carver Mead, and Hopfield taught a one year course, "The Physics of Computation." From the course catalogue:

> Computation is a physical process or "behavior" carried out through the operation of the laws of physics on a very complex system. Common physical principles of computation emerge in the microphysics of computational devices — whether gate, Josephson junction, neuron, or enzyme. When large numbers of simple devices are made into a computer or nervous system capable of large-scale parallel processing, new physical problems arise ... .

Viewing computation in this more general sense was revolutionary, perhaps even now. Here Feynman finally thought about biology without wasting his talents.

In the Shadow of Giants

Feynman was a giant, and among giants, one of a kind. Unlikely as it might seem, as a child I grew up in a veritable forest of giants, albeit of a different kind: giants from Vienna, Galicia, and other parts of the Austro-Hungarian empire. There were intellectuals, engineers, mathematicians, doctors, and survivors of two World Wars, first fleeing West, then East, both times on foot, losing everything, sometimes their lives. Finally I met Feynman from Far Rockaway. He seemed like just another giant in my pantheon of giants, but with different interests and sensibilities. Giants, like Greek Gods, are not perfect, and you accept them for what they are.

4.17 Final Thoughts

It is important not to let the way in which Feynman amused himself detract from his remarkable intellect. He never confused the two, and we shouldn't either. There is a notion that Feynman's life rises above and beyond his work. The public sees him through books like *Surely You're Joking, Mr. Feynman!* [75], *Tuva or Bust,* and many interviews. But what is revealed, entertaining as it may be, does him a great disservice, and misses the point. He was not remarkable because he was a good comedian or great performer. There are plenty of those. Oscar Wilde famously said to André Gide, "I have put all my genius into my life; I have put only my talent into my works." For Feynman, it was quite the opposite.

Theoretical physicists search for universality, for the same laws acting over all space and time. We have our Cosmological Principle.[29] A different universality exists in the creative process, governing work in all fields where creativity exists. In Agnes de Mille's biography of the choreographer Martha Graham, she writes [153]:

---

[29] The Cosmological Principle, as I learned it from H. P. Robertson, states the no matter where you stand in the Universe, the statistical distribution of objects in the sky, and their properties, will be the same.

The greatest thing [Graham] ever said to me was in 1943 after the opening of "Oklahoma!," when I suddenly had unexpected, flamboyant success for a work I thought was only fairly good, after years of neglect for work I thought was fine. I was bewildered and worried that my entire scale of values was untrustworthy. I talked to Martha. I remember the conversation well. It was in a Schrafft's restaurant over a soda. I confessed that I had a burning desire to be excellent, but no faith that I could be. Martha said to me, very quietly: "There is a vitality, a life force, an energy, a quickening that is translated through you into action, and because there is only one of you in all of time, this expression is unique. And if you block it, it will never exist through any other medium and it will be lost. The world will not have it. It is not your business to determine how good it is nor how valuable nor how it compares with other expressions. It is your business to keep it yours clearly and directly, to keep the channel open. You do not even have to believe in yourself or your work. You have to keep yourself open and aware to the urges that motivate you. Keep the channel open."

Feynman could have said as much to us.

Physics progresses through cooperation. Most physicists are cogs in a wheel, dutifully learning what others have done, modifying their work accordingly, and publishing their results. Feynman was less coupled, a cowboy from Queens working things out for himself, then sharing with others.

Feynman had an unsurpassed excitement and curiosity about life that spilled over to his work. As a physicist his style was direct and concrete, shorn of philosophical thought. Feynman saw the physical world for what it was, not for what he wanted it to be.

There are physicists that discover, like Rutherford, and those that invent, like Dirac. Feynman will be remembered first and foremost as an inventor. We honor Feynman for the tools he gave us, his honesty, curiosity and passion in work. We cherish the time he spent with us. Even after all these years we miss him dearly.

4.18 Epilogue

Events described here occurred 60 years ago. To put that time interval in perspective, going back another 60 years from the time I entered college, Röntgen discovered X-Rays, Becquerel discovered radioactivity the following year, Thomson discovered the electron the year thereafter, and three years later Planck quantized radiation. Just six years before X-Rays were discovered, Heinrich Hertz concluded his experiments demonstrating the existence of electromagnetic waves.

Because measurement has become increasingly expensive, and theory remarkably complex, discovery in *fundamental* physics has progressively become more difficult, and is now almost nonexistent.[30] Feynman got in just under the wire. Except for gravity and the strong interactions, the uncharted physics he faced could be understood in relatively simple terms, and he figured out how to do it.

---

[30] Research in all the sciences has become progressively less "disruptive" [167].

Chapter 5
Murray Gell-Mann

Yuval Ne'eman (left) looks on in March 1964 as Murray Gell-Mann holds up a picture of an Ω^- decay, confirming the existence of SU(3) as a symmetry for hadrons. SU(3) was independently proposed by Ne'eman and Gell-Mann. (Caltech Archives.)

Initial sections of this chapter originated with a talk I gave at the "Conference in Honour of Gell-Mann's 80th Birthday" [227], Gell-Mann in attendance, at Nanyang Technology University on February 24, 2010.

5.1   Prologue

In the fall of 1964 Dan Kevles moved to Caltech from Princeton as a young Assistant Professor of History, specializing in the history of science. As an undergraduate he had majored in physics. Shortly after his arrival, and my return from CERN, I barged into his office, told him that elementary particle physics was in great flux and tremendously exciting; history was in the making, just waiting for him to record. And much of it involved Richard Feynman and Murray Gell-Mann, whose offices were just 300 feet away. And quarks had just been discovered!

My excitement was not contagious. Dan lectured me, saying that no one can recognize what is historically important while it was still happening. One must wait many years to understand the historical significance of events. What he didn't add is that it is also convenient for historical figures to be unavailable to contradict historians who document their actions, and sometimes their motives.

Well, today I'll risk it. With Murray Gell-Mann in the audience to keep me honest, I'm going to tell you about the Murray I saw in action, and a little bit about the history of quarks and how he viewed them.

5.2   An Early Beacon

Murray, a belated "Happy Birthday!" I learned a lot from you, and for that I am very grateful. We go way back, even further than you realize. In the summer of 1957, after a hard day's work as a counselor at a day camp in Detroit, I came across an article you coauthored in *Scientific American* which summarized the state of elementary particle physics [87]:

> At present our level of understanding is about that of Mendeleev, who discovered only that certain regularities in the properties of the elements existed. What we aim for is the kind of understanding achieved by Pauli, whose

exclusion principle showed why these regularities were there, and by the inventors of quantum mechanics, who made possible exact and detailed predictions about atomic systems.

This article appeared just three months before Sputnik, when physics wasn't yet fashionable. At the time I was about to start my junior year at the University of Michigan as a math major, but was thinking of switching to physics in graduate school. Here was a big green light saying: "Go!"[1]

In my senior year I went to see my quantum mechanics professor Paul Hough for advice on which graduate school to attend. This was the Hough who would become *the* Hough of the Hough-Powell bubble-chamber digitizer, and the Hough transform in image processing. His comment: "Bethe is at Cornell, where I come from, but he's getting old. There're a couple of young guys at Caltech — Feynman and Gell-Mann — why don't you go there." So I did.

Caltech

It was wonderful to be at Caltech in the early 60's. Carl Anderson, who had used cloud chambers to discover the positron and then the muon, was the avuncular chairman of the Physics Department. His small department included six soon-to-be Nobel Prize winners: Richard Feynman, William Fowler, Murray Gell-Mann, Sheldon Glashow, Rudy Mössbauer, and Ken Wilson; Yuval Ne'eman, Jun J. Sakurai, and Fred Hoyle were visitors; and Sidney Coleman, Hung Cheng, and Roger Dashen fellow graduate students. If that wasn't enough, I could always go across campus and talk with Max Delbrück or Linus Pauling, phenomenologist par excellence. I got the impression that it was possible to do things. Discoveries could be made.

Money was pouring into particle physics, helped now by fear of the U.S.S.R. and Sputnik's success. Pictures from bubble chambers were just beginning to provide an enormous wealth of information. I still remember driving an hour and a half across Los

---

[1] In a manner of speaking we go back even further. Murray's father Arthur Isidore Gellman was born in Czernowitz Galicia, part of the Austro-Hungarian Empire, only 55 miles east of Pechenizhyn where both my mother Rachela Zweig (née Freilich) and father Alfred Zweig were born. It's probably no accident that neither Murray nor I have a middle name.

Angeles to my first APS meeting at UCLA in 1961. In a dark, cavernous and half-empty auditorium three speakers, Bogdan Maglić, Bill Walker, and Harold Ticho showed slides demonstrating the existence of the first three vector-meson resonances, the ω, ρ, and K^*. APS meetings seemed pretty interesting!

At nearly the same time Murray gave a seminar on the "Eightfold Way" at Caltech. It was fantastic. Murray was feeling his oats. As far as I was concerned, his classification of particles into irreducible representations of SU(3) was obviously correct, although the scientific community did not fully accept his assignments until the Ω^- was discovered three years later. His output in 1961 and 1962 was phenomenal, with one new paper appearing every two months. Nine of those papers were written with seven different collaborators (Appendix 5.A, p. 83).

5.3   Murray's Toy Field Theories

Field theory had been successfully used to create a relativistic quantum theory of electricity and magnetism, but Murray wasn't certain that field theory would ultimately prove to be the proper framework for a theory of the strong interactions. However, he thought much could be learned from "toy field theories" (Murray's term), initially their symmetries, later algebraic properties of the electromagnetic and weak interaction currents, even though the field theories from which they were abstracted were flawed.

Murray had a long history with toy field theories dating back to 1957 when he abstracted symmetry relations from a toy strongly-interacting field theory with fields corresponding to real particles. The abstract of his paper titled "Model of the Strong Couplings" begins with [88]:

> An attempt is made to construct a crude field theory of hyperons and K particles, which are assumed to have spin 1/2 and spin 0, respectively.

The hadrons in that model were governed by "Global Symmetry." He continues:

> Supposing that the model we have presented has elements of truth, we may add the following remarks:

> (1) The symmetry properties of the model may be correct even though the use of field theory is unjustified. For this reason an analysis purely in terms of the symmetry group of the theory is in order.

Here Murray constructs an interactive field theory, that he knows might not be correct, and explores the consequence of a symmetry operating within that theory. Liking what he sees, he assumes the symmetry is correct, and throws away the rest. Careful as he was, the symmetry in this case was not the right one, but the idea of working with a toy field theory from which to abstract concepts that might be true was novel and ultimately fruitful. Practicing theoretical physics this way was his most original idea.

5.4   The Eightfold Way

Four years later in the Eightfold Way Murray proposed, independently of Yuval Ne'eman,[2] that SU(3) be used to classify strongly interacting particles [90, 159], rather than Murray's earlier "Global Symmetry" [88], that was adopted by Lee and Yang [143].[3] The name, adopted from the Eightfold Path of Buddhism, refers to the eight-member families to which many hadrons belonged.

Murray used *mathematical* particles l and $\bar{L}$ as fundamental fields:

> For the sake of a simple exposition, we begin our discussion of unitary symmetry [SU(3)] with "leptons" [l and $\bar{L}$], although our theory really concerns the baryons and mesons and the strong interactions. The particles we consider here for mathematical purposes do not necessarily have anything to do with real leptons, but there are some suggestive parallels.

---

[2] While an intelligence colonel in the Israel Defense Forces, and the Israeli military attaché in London, Ne'eman studied physics under Abdus Salam at Imperial College, completing his Ph. D. thesis in only two years wherein he discovered SU(3) symmetry at the age of 36 (like Murray, he graduated from high school at 15). Ne'eman's paper was received for publication by Nuclear Physics on February 13, 1961. The Eightfold Way Synchrotron Report was dated a month later, March 15, 1961.

[3] Global Symmetry implied that the Gell-Mann–Okubo mass formula (Eq. 5.1) be replaced by $(M_\mathcal{N} + M_\Xi)/2 = (M_\Lambda + 3M_\Sigma)/4$, where the number "3" multiplies M_Σ rather than M_Λ.

After using l and $\bar{L}$ to construct states that transformed like real particles, Murray reassures the reader that:

> We shall attach no physical significance to the l and $\bar{L}$ "particles" out of which we have constructed the baryons. The discussion up to this point is really just a mathematical introduction to the properties of unitary spin.

The Eightfold Way was widely circulated as Synchrotron Report CTSL-20, and later published in a collection of papers about SU(3) and quarks [100].

Ideas from the 1961 Eightfold Way paper were crafted into a formal field theory paper loosely related to the Sakata model, and not the mathematical l and $\bar{L}$. From Section IV of the formal paper [91]:

> We generalize the Fermi-Yang description to obtain the symmetrical Sakata model and abstract from it as many physically meaningful relations as possible.

Commutation relations for the electromagnetic and weak currents, evaluated at the same time, were given, relations that could be checked experimentally. Unlike his earlier toy field theory [88], this one was "free"; it didn't include strong interactions between hadrons.

Largely ignored in the West, the idea of using U(3), a slightly larger group than the SU(3) it contained, was actively pursued in Japan starting in early 1959 [128, 216]. The Japanese abstracted this symmetry from the Sakata model, as Murray would do two years later. Physics in Japan after Admiral Perry, as art in Russia after Ivan the Great, progressed in universes parallel to the West's, loosely connected to it, the Japanese more creative than the Russians.

The Gell-Mann–Okubo Mass Formulae

In the Eightfold Way Murray used a particular type of broken SU(3) symmetry to derive a mass relation for baryons in the baryon octet [90]:

$$\frac{M_{\mathcal{N}} + M_{\Xi}}{2} = \frac{3M_{\Lambda} + M_{\Sigma}}{4}, \qquad (5.1)$$

but states that "there is no particular reason to believe, however, that the analogous sum rules for mesons are obeyed."

Okubo extended Murray's work and derived an equation relating mass M to isospin I, and strangeness $\mathbb{S}$ for particles belonging to the same SU(3) irreducible representation [165],

$$M = a + b \times \mathbb{S} + c \times [I(I+1) - 1/4 \times \mathbb{S}^2], \quad (5.2)$$

where a, b, and c are constants determined from 3 of the 4 isospin multiplets in an octet (the mass of the fourth multiplet is then determined). For the baryon octet Okubo's equation reduces to Eq. 5.1. But Eq. 5.2 also applies to the pseudoscalar meson octet, where it didn't work, confirming Murray's reluctance to view this relation as universal.

However, someone noticed that if the mass M is replaced in Equation 5.2 by its square, M^2, the relation worked very well for the pseudoscalar mesons, so that's what everyone did, including Murray. While he was on sabbatical from Caltech teaching at MIT in the spring of 1963 he argued that:

> A mass rule for bosons will always be in terms of the mass squared. This is because in all field theories the mass of a fermion always enters linearly, while that for a boson always comes in quadratically. [A fermion is a particle with half-integral spin (1/2, 3/2, ...), a boson has integral spin (0, 1, 2, ...).]

This quote is taken from Murray's lecture notes compiled by J.A. Campbell, R. Logan and C.E.W. Ward. I was familiar with the argument, not happy with it, but reluctantly adopted the use of mass squared for both the pseudoscalar and vector mesons.

In a footnote Okubo credits Feynman for Eq. 5.1 [165]:

> A similar formula has already been suggested by R. P. Feynman at Gatlinburg Conference held in 1958.

Gell-Mann was with Feynman at the Gatlinburg Conference where parity violation was being discussed, but didn't reference Feynman.[4]

---

[4] Interview of Robert Marshak by Charles Weiner on October 4, 1970, Niels Bohr Library & Archives, AIP, College Park, MD USA.

Years later after Feynman had accepted the existence of constituent quarks he told me that mass formulae for both mesons and baryons should be linear in mass, not mass squared, but didn't explain why, or why a linear mass relation for the pseudoscalar mesons worked so poorly. At the present time vector-meson mass relations work better when they are linear in mass ("A Better Vector Meson Mass Relation," p. 100); quadratic mass relations must be used for the pseudoscalar mesons.

5.5 Mathematical Quarks for a Toy Field Theory

Nature Reads the Book of Free Field Theory

In February 1964 Murray introduced a toy *free* field theory (no strong interactions) with fields corresponding to *mathematical particles* he called quarks. In this final reincarnation of his toy field theories Murray replaced the p, n, and Λ of the Sakata model with three quarks, attaching "no physical significance" to them [93]:

> We assign to the triplet t [of quarks] the following properties: spin $\frac{1}{2}$, [charge] $z = -\frac{1}{3}$, and baryon number $\frac{1}{3}$. We then refer to the members $u^{\frac{2}{3}}$, $d^{-\frac{1}{3}}$, and $s^{-\frac{1}{3}}$ of the triplet as "quarks" ... . A formal mathematical model based on field theory can be built up for the quarks exactly as for p, n, Λ in the old Sakata model ... . All these [current commutation] relations can now be abstracted from the field theory model and used in a dispersion theory treatment.

The commutation relations of the electromagnetic and weak interaction currents were what mattered, not how the currents were constructed, be it with all the know hadrons in 1957 [88], the p, n, and Λ of the Sakata model in 1962 [91], or quarks.

Specifying the strong interactions was unnecessary since the currents are only related to the electromagnetic and weak interactions. The electromagnetic interactions, for example, are solely dependent on electrical charges, and their motion, not on where the charges might come from, or what other interactions the charge carriers might have.

Since everything but the current commutation relations was discarded, and those relations were the same as those given in 1962 [91], *using quarks led to no new physics*. However, the toy field theory was a *free* field theory; no more was present than what was needed for abstraction — it was "elegant."

Whereas previously Murray had been justified in throwing everything but the commutation relations away, this time — by not realizing that quarks were the constituents of hadrons — he was "throwing out the baby with the bath water."

Were Murray's Quarks Real?

For Murray, quarks — like the free field theory in which they existed — remained a convenient temporary fiction. As he eloquently explains in May 1964, four months after his quark paper was published [94]:

> We use the method of *abstraction* from a Lagrangian field theory model. In other words, we construct a mathematical theory of the strongly interacting particles, which may or may not have anything to do with reality, find suitable algebraic relations that hold in the model, postulate their validity, and then throw away the model. We compare this process to a method sometimes employed in French cuisine: A piece of pheasant meat is cooked between two slices of veal, which are then discarded. ...
>
> Such particles [quarks] presumably are not real but we may use them in our field theory model anyway. Since the quark model is mathematically the simplest, we shall in fact employ it in the next section, as in ref. 11 [Murray's first quark paper], for our process of abstraction.

Quarks allow Murray to formulate his concept of abstraction in its purest form. Why specify strong interactions, as he did in his first 1957 toy field theory paper ([88]), if those specifications are unnecessary, and might be incorrect? Who needs the fields in a field theory to correspond to real particles, especially if you worry that field theory might be flawed, and think that the bootstrap is probably correct? Who needs all those l and $\bar{L}$ particles of the Eightfold Way if only three mathematical quarks will do.

Although Murray persistently argued that quarks could not be real, he once considered the possibility that they might be real strongly interacting particles (unlike his *Physics Letters'* quarks), rejected the idea, and *told us why*. In his 1966, opening talk at the *XIII*[th] *International Conference on High Energy Physics* he asks [95]:[5]

> Now what is going on? What are these quarks? ... It is hard to see how deeply bound states of such heavy real quarks could look like $\bar{q}q$, say, rather than a terrible mixture of $\bar{q}q, \bar{q}q\bar{q}q$, and so on. ... The idea that mesons and baryons are made primarily of quarks is difficult to believe, since we know that, in the sense of dispersion theory, they are mostly, if not entirely, made up out of one another. The probability that a meson consists of a real quark pair rather than two mesons or a baryon and antibaryon must be quite small. Thus it seems that whether or not real quarks exist, the q and $\bar{q}$ we have been talking about are mathematical entities. ...
>
> If the mesons and baryons are made of mathematical quarks, then the quark model may perfectly well be compatible with the bootstrap hypothesis, that hadrons are made up out of one another.

Here Murray treats quarks as if they were ordinary hadrons, and uses proof by contradiction to argue that it makes no sense to think that hadrons "are made primarily of quarks." But this specious argument misses the point. All would be well if quarks are *real elementary particles with their own laws of interaction* that form the basis of a quantum field theory in which hadrons with their own interactions are created as emergent phenomena. Then no contradiction would arise.[6] Several years later Heisenberg used a similar argument to conclude that quarks could not be real particles (Section 2.12, p. 27).[7]

---

[5] I remember thinking at the time that Murray was summarizing the conference before it began! That he was able to do so, and have people love him for it, indicated his charm, up-to-date knowledge, and mastery of the field.

[6] In fact, it turned out that "$\bar{q}q$" is "a terrible mixture of $\bar{q}q, \bar{q}q\bar{q}q$, and so on," but not as Murray meant it.

[7] Valentine Telegdi provides an independent description of our differing views on the reality of quarks: Telegdi, Valentine L. Interview by Sara Lippincott. Pasadena, California, March 4 and 9, 2002. Oral History Project, California Institute of Technology Archives: oralhistories.library.caltech.edu/146/.

Murray's Subsequent View of Quarks

Remarkably, Murray continued to believe that quarks were mathematical even after the SLAC deep-inelastic scattering experiments indicated that electrons were scattering off point particles inside protons and neutrons [21, 25]. At the end of February 1972, Murray delivered a set of lectures in Schladming Austria titled "Quarks" [97]. In these lectures Murray spoke of "constituent quarks," but viewed his quarks as "current quarks."[8] Murray began with:

> In these lectures I want to speak about at least two interpretations of the concept of quarks for hadrons and, the possible relations between them.
>
> First I want to talk about quarks as "constituent quarks". *These were used especially by G. Zweig (1964) who referred to them as aces* [my italics].

It would have been more precise to say: "*These were introduced by G. Zweig (1964) ... .*"; constituent quarks had not existed as a tool in a toolbox for me, or anyone else, to use.

Murray goes on to say:

> The whole idea is that hadrons act as if they are made up of quarks, but the quarks do not have to be real.

No, that was not the whole idea. Murray continues:

> There is a second use of quarks, as so-called "current quarks" which is quite different from their use as constituent quarks ... .
>
> We have to be very careful then to abstract as much as we can so as to learn as much as we can from the current quark picture, but not to abstract too much, otherwise first of all experiments may prove us wrong, and secondly that it may involve us with the existence of actual quarks, maybe even free quarks — and that, of course, would be a disaster.

Murray's current quarks were not constituent quarks, but constituent quarks were also current quarks

---

[8] Aces were both constituent and current quarks (Section 6.8: "Weak-Interaction Predictions," p. 102).

used to describe the weak and electromagnetic interactions (Section 6.8, p. 102). Murray didn't mention the current quark aspect of constituent quarks.

And finally:

> If quarks are only fictitious there are certain defects and virtues. The main defect would be that we never experimentally discover real ones and thus will never have a quarkonics industry. The virtue is that then there are no basic constituents for hadrons — hadrons act as if they were made up of quarks but no quarks exist — and, therefore, *there is no reason for a distinction between the quark and bootstrap picture: They can be just two different descriptions of the same system, like wave mechanics and matrix mechanics* [my italics].[9] In one case you talk about the bootstrap and when you solve the equations you get something that looks like a quark picture; in the other case you start out with quarks and discover that the dynamics is given by bootstrap dynamics. ...
>
> If we go too far ... and try to construct a complete Fock space for quarks and antiquarks on a light-like plane, abstracting the algebraic properties from free quark-theory, we are in danger of ending up with real quarks, and perhaps even with free real quarks as I mentioned before. In our work, we are always between Scylla and Charybdis; we may fail to abstract enough, and miss important physics, or we may abstract too much and end up with fictitious objects in our models turning into real monsters that devour us.

This was Murray's vision. He understood that he always must find Scylla and Charybdis, and then, like Jason, sail between them. Except when it came to quarks, this is exactly what he did.

5.6 Later Years

Murray's work eventually became less concerned with experiment, and more with theory. I walked into his office one day and asked, "Murray, you're so good at phenomenology, why aren't you doing it?" He replied, "I'm not interested in it any more." I was shocked. It was like Picasso in his prime giving up painting. In fact the world of high energy physics had changed. Phenomenology was no longer central. It was time to ask questions like "What is the force between two quarks?"[10] The challenge was to create a theory for constituent quarks. It was the end of an era.

Over the years Murray and I drifted apart. Murray worked on foundations of quantum mechanics, then complexity, and linguistics. I switched to neurobiology, or as Murray put it with a smile, "cutting up cats."

Caltech held a birthday party for Murray in January, 1989 when he turned 60, and I flew in from Los Alamos to attend. Murray was surprised to see me, and asked me why I came. I was surprised by his question, saying that he loomed large in my graduate student days, and for that time I honor him. When he turned 80 I flew to Singapore to help celebrate his birthday. This time he just seemed happy to see me.

The historian Dan Kevles — whose office I barged into 60 years earlier, asking him to record history in the making — true to his word, went on to research the past and write about George Ellery Hale in the Gilded Age, and Robert Millikan, a founder and first president of Caltech. Eventually, Dan did broaden his vision of what historians do. In 1998 he wrote a book about Caltech's then sitting president, David Baltimore [134].

---

[9] However, Gell-Mann's view of Geoffrey Chew and his vision for the bootstrap were not uniformly positive, as revealed a year earlier in a discussion session at the 1971 International Conference on Duality and Symmetry in Hadron Physics [96]:

> GELL-MANN: ... For example, Geoffrey Chew firmly believes that the bootstrap method will derive everything about the hadrons, including the underlying symmetries. Perhaps even the three plus one character of space.
>
> AUDIENCE: general chuckling.

In reality, bootstrapping hadronic symmetries was a distinct possibility, as believable as any other bootstrap calculation, as demonstrated in 1963 by Abers, Zachariasen, and Zemach (see quote on p. 89 in Section 6.2).

---

[10] Ken Young, one of Murray's brightest students, lamented in his dinner speech at Murray's 80th birthday party that Murray hadn't told him to compute the force between two quarks when he asked Murray for a thesis problem in 1970. Murray didn't because, for him, quarks were not real particles. Two years later that problem was assigned to David Politzer by Sidney Coleman, and to Frank Wilczek by David Gross. Politzer, Wilczek and Gross received the 2004 Nobel Prize "for the discovery of asymptotic freedom in the theory of the strong interaction," i.e., that the force between two quarks becomes weaker the closer they get.

5.7 Mr. Physics

Murray was an exceptionally clear thinker, better at understanding the current state of theoretical particle physics than anyone else, always distilling the essence of what was known and writing it down in context so others could understand and use it. A very good example is provided by his 1956 paper that introduced the quantum number strangeness and the Gell-Mann–Nishijima charge formula to Western audiences (Eq. 5.5, p. 74) [85]. Previously these idea were primarily known in Japan where their discovery was initiated by Tadao Nakano and Kazuhiko Nishijima in 1953 [155], and completed by Nishijima in 1954 and 1955 [161, 162]. Murray's paper exhibits a clarity of thought that he continued five years later with the Eightfold Way [90]. These papers remain a pleasure to read and exemplify what I believe was Murray's greatest contribution: his unsurpassed ability to sniff out ideas, understand the direction in which the field should move, and simply but elegantly communicate that to others. Murray became *the* particle theorist others turned to for understanding. Murray wanted to be, and became, "Mr. Physics" of particle physics, for almost a generation.

Fear of Mistakes

Despite being Mr. Physics, Murray was not as relaxed and self-confident as might be expected. He told me the following story about an event that had made a great impression on him: As an MIT graduate student he attended a physics colloquium given by a distinguished theoretical physicist. All the well-dressed MIT professorial heavy-weights were sitting in the front row listening attentively. The speaker proposed a new theory, and made a prediction. It was a beautiful talk, but at the very end during the question period a disheveled experimentalist at the back of the auditorium stood up waving a paper and announced that he had just measured the quantity the speaker had predicted, and the prediction was wrong.

Murray was always careful, afraid of mistakes. He wanted to understand what everyone was doing; no grubby experimentalist was going to unexpectedly announce the result of an experiment that Murray didn't know about. How else could he summarize the state of particle physics in the *opening* talk at the 1966 *XIIIth International Conference on High Energy Physics*? Deep down, as opposed to Feynman, Murray was very insecure.

Contributions

Murray's papers, most with others, helped drive the field forward, witness his 1961–62 publications listed in Appendix 5.A, p. 83. His writings kept at the Caltech Archives are in 123 boxes, 54 feet in linear extent.[11]

The number of contributions Murray made was remarkable, as recognized in his 1969 Nobel Prize that was not awarded for a singular piece of work, but rather "for his contributions and discoveries concerning the classification of elementary particles and their interactions." The prize was not awarded for quarks, whose significance was not widely recognized at the time.

How can Murray's unique contributions, as I witnessed them and described here, be put in some perspective? Science is a social enterprise, and society recognizes individuals who influence the work of others. Murray was concerned with describing reality, making predictions that could be tested, and providing a theoretical framework that enabled others to expand on his vision. Murray thereby set an agenda for almost an entire generation of physicists, dominating our field like no other.

By abstracting from toy field theories, Murray identified algebraic relations among operators, the equal-time current commutation relations — holding even when the strong interactions were present — that he postulated as being true at a time when particle physics was in a state of great confusion, with no one really understanding where the truth lay. Since the matrix elements of these operators were measurable, his assumption was testable, and some relations held with surprising accuracy [5, 211].

If Ernst Mach at the end of the 19th century could insist that atoms weren't real, perhaps it's not surprising that Murray continually denied the reality of quarks, but of course for different reasons.

5.8 A Dark Side

For all his eminence, Murray had his issues with credit and the appropriation of other people's ideas

---

[11] archives.caltech.edu/

without sufficient attribution. It was a recurrent problem. Three examples follow, starting with strangeness $\mathbb{S}$ and the Gell-Mann–Nishijima charge formula, continuing with the V–A theory of the weak interactions, and ending with quarks.

Strangeness and the Gell-Mann–Nishijima Charge Formula

The quantum number strangeness $\mathbb{S}$, its selection rules in weak interactions, and the Gell-Mann–Nishijima charge formula first appeared in a Western journal in April 1956, with Murray as author [85]. All three had a long history in Japan's premier physics journal, *Progress of Theoretical Physics*, published in English and available everywhere particle physics was practiced. This history was not described by Murray, but is outlined here to better understand the issue of attribution for the discovery of strangeness and its application to the Gell-Mann–Nishijima charge formula.

Nakano-Nishijima's November 1953 Paper — An Equation for Charge

For any strongly interacting particle X, and its antiparticle $\bar{X}$, define a function $n(X)$ as the number of Xs minus the number of $\bar{X}$s for a system of particles. Nakano and Nishijima noticed that [155]:

$$n(V_1) - n(\Pi) = b - 2(q - I_3), \qquad (5.3)$$

where V_1, Π, q, I_3, and b are the Σ baryon, K meson, charge Q, isospin projection I_Z, and baryon number B. Solving for Q in this equation gives

$$Q = I_Z + \frac{B + n(K) - n(\Sigma)}{2}.$$

Strangeness had not yet been introduced, but this equation sets Nishijima well on the way to its discovery. $n(K) - n(\Sigma)$ will become strangeness $\mathbb{S}$, yielding the Gell-Mann–Nishijima charge formula (Eq. 5.5, [161]).

Nishijima's July 1954 Paper — The Concept of Strangeness

Building on the 1953 paper of Nakano and Nishijima, Nishijima defined a new quantum number "η," conserved by the strong interactions, which he used to explain the perplexing pattern of allowed and

forbidden reactions of strongly interacting particles, i.e., their associated production (Eq. 2.3, p. 15).

He also showed that the charge of a hadron could be written as

$$q = I_3 + \frac{b + \eta}{2}. \qquad (5.4)$$

Because η was identical to Murray's later-defined strangeness $\mathbb{S}$, Eq. 5.4 became the famous Gell-Mann–Nishijima charge formula,

$$Q = I_Z + \frac{B + \mathbb{S}}{2}. \qquad (5.5)$$

Murray's name "strangeness" was adopted by the high energy physics community, and η was forgotten.

Nishijima's March 1955 paper — Strangeness-changing Selection Rules

Based on the observed pattern of allowed and forbidden hadronic weak decays, Nishijima proposed a selection rule limiting the changes in η in such reactions [162],

$$\Delta\eta = 0, \pm 1, \qquad (5.6)$$

and observed that it was "an analogue of the isospin selection rule $\Delta I = 0, \pm 1$ for the electromagnetic interaction."

Murray's April 1956 Paper — Appropriating Nishijima's Work

The Introduction to Murray's paper begins with [85]:

> The purpose of this communication is to present a coherent summary of the *author's theoretical proposals* [1] concerning the new unstable particles [my italics].

His footnote [1] explains:

> Much of the work to be presented is contained in the following publications: M. Gell-Mann: *Phys. Rev.*, **92**, 833 (1953); M. Gell-Mann and A. Pais: *Proceedings of the Glasgow Conference*, 1954, and in an unpublished note: M. Gell-Mann: On the Classification of Particles (circulated August, 1953). In these references, however, the proposals under discussion are mentioned only briefly and are sometimes buried in a mass of other material. Here they

are treated alone and in some detail. Practically the same proposals have been put forward in Japan. See Tadao Nakano and Kazuhiko Nishijima: *Prog. Theor. Phys.* **10**, 581 (1953) and K. Nishijima (to be published).

The "proposals under discussion," purportedly in Murray's three references, are strangeness (Murray's "Section 3·1"), the Gell-Mann–Nishijima charge formula (his "Equation (3.1)"), and selection rules for strangeness in weak hadronic decays (his "Equation (4.1)"). The odd thing is that these proposals don't appear anywhere in the three references. It's not that they are "mentioned only briefly and are sometimes buried in a mass of other material:" the truth is that they don't appear at all. A new quantum number that would have corresponded to strangeness is never mentioned.

This inconsistency is easily verified because Murray's 1953 "unpublished note" is now available in *Murray Gell-Mann, Selected Papers*, pp. 35–37 [101]. His Glasgow Conference paper with Pais may be downloaded from inspirehep.net/literature/1304427. His 1953 publication, a half-page Letter to the Editors at the Physical Review, only concerns "Isotopic Spin and New Unstable Particles" [83].

Actually, Murray had taken a different tack; in his unpublished note he writes [101]:

Let us now represent each particle with which we are concerned as equivalent to a number of nucleons or anti-nucleons plus a number of τs [Ks] or $\bar{\tau}$s [$\bar{K}$s].

He then ends up with a rectangular 3×5 table containing as many equations as hadron multiplets that might exist (15 in all, counting antiparticles), rather than one unifying equation (Eq. 5.5).

Without a quantum number, this is as far as Murray gets. In the Glasgow Conference paper he and Abraham Pais drop Murray's "representation of particles" and simply list six equations for the hadrons that were well established at that time. For example, the sixth equation was:

$$Q = I_Z - \frac{1}{2} \text{ for the } \Xi^- \text{ and a hypothetical}$$

$$\Xi^0 \ (I = \frac{1}{2}, \text{ doublet}).$$

The d'Espagnat-Prentki Equation: $Q = I_Z + U/2$

Jacques Prentki, future editor of *Physics Letters*, and Bernard d'Espagnat wrote four papers ([46, 47, 48, 49]) in 1955–56 trying to understand the consequences of what Murray and Pais said in their Glasgow Conference paper. The last of these four papers is in the same issue of *Suppl. del Nuovo Cimento* in which Murray's 1956 paper appears. Their paper starts with [49]:

The model of Gell-Mann [1] has a very great virtue in that it agrees with experiment. But, as it stands, it contains quite obviously a lot of arbitrariness in that the relation $Q(I_Z)$ varies in a strange way from particle to particle: A general formula is:

$$Q = I_Z + U/2, \tag{5.7}$$

where U is in some cases 0, in some cases 1 or -1. These values are precisely chosen in each case so as to account for the metastability of each particle.

Their quote "[1]" is to Murray's 1953 publication [83] and his Glasgow Conference paper with Pais [84]. Eq. 5.7 is not given in Murray's papers; rather it is d'Espagnat and Prentki's summary of Murray's many equations for the hadron multiplets. Evidently they were not aware of Nishijima's papers.

Recall Feynman's quote (p. 52): "In general we look for a new law by the following process. First we guess it." Now we have a better idea of what he meant. Neither d'Espagnat or Prentki, despite four tries, or Gell-Mann, even trying for a year or more, were able to understand the behavior of the "New Unstable Particles." For that they would have had to guess that a conservation law existed for a new quantum number of the strong interactions. Their inability to do so attests to the difficulty of "guessing." The hard part was to look at the problem in the right way so they would be led to a place where guessing would be easier. The 1953 paper of Nakano and Nishijima [155] put Nishijima on the right path with their Eq. 5.3. Murray's idea of "represent[ing] each particle ... as equivalent to a number of nucleons or anti-nucleons plus a number of τs or $\bar{\tau}$s" led him to a dead end.

Far from presenting "a coherent summary of [Murray's own] theoretical proposals," the 1956 paper merely restates Nishijima's results in Murray's notation, and follows Nishijima in exploring their consequences, with no indication of Murray himself having ever contributed to the introduction of strangeness or the Gell-Mann–Nishijima charge formula.

Improper Attribution

When Murray references Nishijima's papers by writing "Practically the same proposals have been put forward in Japan $\cdots$ K. Nishijima (to be published)," he gives the impression that Nishijima's work would be published contemporaneously with his own, or later. However, the primary results in Murray's paper, the introduction of the quantum number strangeness and the Gell-Mann–Nishijima charge formula, had been published by Nishijima almost two years earlier. Murray's other relation ($\Delta \mathbb{S} = \pm 1$)[12] had been published by Nishijima more than a year earlier. Nowhere does Murray acknowledge that his thinking about strangeness was influenced by Nishijima's publications.

Nishijima's work was brilliant. It was not easy to figure out how to think about the strong interactions as a host of new particles with seemingly bizarre properties were being discovered. He never got the recognition he deserved.

Murray and Parity Violation

The Feynman chapter described the difficulties involved in creating a theory for parity violation. This section continues the story when Murray becomes involved. Experiment showed that, given a picture of a nucleus undergoing β-decay, it was possible to determine whether or not that picture was taken through a mirror [215]. The resulting attempts to formulate a theory for such disintegrations underscores the importance of understanding when experiments are wrong.

Lunch with Sudarshan and Marshak

The V–A universal theory of weak interactions was first discovered by E. C. George Sudarshan and Robert Marshak, who detailed their discovery to Murray at a "summit" meeting Marshak had arranged at a lunch in Los Angeles in the first week of June, 1957 [150]. The lunch group included Sudarshan, Marshak, and

Caltech's Gell-Mann, Felix Boehm, Aaldert Wapstra and Berthold Stech. In an extensive presentation:[13]

> He [Sudarshan] made the observation that the data [concerning parity violation] was internally inconsistent. He also singled out the experiments which were most likely to be mistaken. He suggested that the weak decay interaction was of the universal form V–A ... . Gell-Mann was enthusiastic about Sudarshan and Marshak's discovery.

Murray describes the same meeting in an interview with Sara Lippincott,:[14]

> At that time, Robert Marshak and his student, George Sudarshan, came to visit. Art [Rosenfeld] and I went to meet with them, and I think with Felix Boehm, at RAND, in Santa Monica. We had lunch together. And they told us about their work, in which they had figured out that these experiments could really be wrong — that they could be criticized and perhaps were actually wrong — lending a lot of strength to this very beautiful, fantastically simple theory of the weak interactions. I liked that, and I liked the strengthening of confidence that this might work.

Based on an interview with Murray on February 19, 1997, George Johnson describes Murray's account of the summit meeting (p. 154, [132]):

> Sudarshan was even happier to hear that Boehm's experiments could indeed be taken to support V–A. He [Sudarshan] and Marshak anxiously asked Murray if he was planning to write a paper. Probably not, he said. He and Art Rosenfeld, an old colleague from Chicago, were in the midst of producing a long, laborious review of weak interaction physics. Murray thought he would content himself with a paragraph suggesting that V–A would be the last stand for the Universal Fermi Interaction ... .

---

[12] Murray did not include $\Delta \mathbb{S} = 0$ because he did not consider leptonic decays.

[13] E. C. George Sudarshan in Section I of *Seven Science Quests Symposium*, November 6–7, 2006, quest.ph.utexas.edu/sudarshan_vminusa.html. Also, see p. 153 of [132].

[14] Sara Lippincott, Santa Fe Institute, New Mexico, July 1997. Oral History Project, California Institute of Technology Archives, resolver.caltech.edu/CaltechOH:OH_Gell-Mann_M.

However, in a talk honoring Sudarshan on the occasion of his 60th birthday, Marshak writes [150]:

> Murray claims in his 1983 Catalunya (Spain) talk on "Particle theory from *S*-matrix to quarks" [and to Johnson, directly above] that, at the "summit" luncheon meeting, he mentioned the V–A theory as a possible "last stand"; however, that claim does not jibe with the recollections of the other four participants in the "summit" meeting [lunch] — Boehm, Stech, Sudarshan and myself. Gell-Mann's Catalunya remarks also do not jibe with Feynman's account of the genesis of the Feynman–Gell-Mann version of V–A theory. Incidentally, Stech was very quiet at the "summit" meeting and when I inquired, years later, why he had not extended the chirality invariance principle — about which he had shown so much familiarity in his 1955 paper with Jensen — to quickly opt for V–A after parity violation was announced, he informed me that, as a Cal Tech post-doc in 1957, he had mentioned the idea to Gell-Mann who discouraged him from pursuing the chirality invariance approach.

In the Gell-Mann–Rosenfeld article, written after the "summit" meeting, they conclude [86][15]:

> Despite all of the objections to it at the present time, the UFI [Universal Fermi Interaction] hypothesis with V, A coupling seems so attractive that it should perhaps be borne in mind until definitely disproved by experiment.

Note that Gell-Mann and Rosenfeld refer to a V,A coupling, not Sudarshan's more specific, and correct, V–A coupling, as the attractive choice. At this point Murray still had not adopted V,A, let alone V–A.

Meanwhile, after talking to "the boys at Caltech" (quote starting page 43), Feynman in an American Institute of Physics taped oral history interview said [70]:

> I wrote this thing up [Feynman's V–A theory of the weak interactions]. Then Gell-Mann had

come back from somewhere, and we talked it over. It was his original idea that V may be wrong, and he was uncomfortable. And this is, again, something not for publication. May I tell you something privately?

As Johnson relates [132], when Murray heard that Feynman was planning to write a paper on V–A to submit to the *Physical Review*,

> Well then, Murray thought, I'm going to write it up too. He felt bad since he had told Marshak and Sudarshan that he wasn't going to compete with them and write a full-scale paper. But it wasn't exactly what he considered a solemn oath. If Dick was going to rush into print, he had no choice. It was simple self-preservation.

Not wanting to be left out, Murray put his reservations aside, and jumped right in. He had not done the laborious work of analyzing the negative experiments, and so couldn't be confident that these experiments were probably wrong. As Murray related to Lippincott (Footnote 14 of this chapter):

> They [Sudarshan and Marshak] started writing on their ideas, which went further, in the sense that they criticized some of these experiments and showed how they might be wrong.

A Shotgun Wedding

The story of how Feynman and Murray came to write their famous V–A paper together is a sad one, related to me by Caltech's Provost, Bob Bacher. Bob had been head of the bomb physics division at Los Alamos. He was responsible for the assembly of the plutonium core of the "Fat Man" for the Trinity test, an assembly that took place in the front master bedroom of the abandoned McDonald Ranch House in the New Mexico desert. He had seen Feynman in action at Los Alamos, and later, as chairman of Caltech's Physics Department, was responsible for bringing him to Caltech. With Feynman's encouragement, he then hired Murray three years later.

Bob heard that Feynman and Murray were writing two separate papers on parity violation. Murray was building on the V–A theory of Sudarshan and Marshak. Feynman had independently settled on V–A for the weak interactions, just inventing it on his

---

[15] Footnote 1 of [86] says "The survey of literature pertaining to this review was completed in July, 1957." However, the paper refers to a September, 1957 private communication, so changes to the paper were possible up to that time.

own (Section 4.2: "Before Meeting Feynman: Parity Violation," p. 41). Bob thought that the publication of two separate papers on the same topic by two faculty members who occupied adjacent offices would be "unseemly," and asked them to combine their papers. Otherwise it might look as if they were fighting with each other, and that would look bad for Caltech. Feynman was unhappy with Bob's request, but complied.

Later Murray became very upset by Feynman's recounting of Feynman's invention of V–A in *Surely You're Joking, Mr. Feynman!* [75]. As Murray was reading the book he came across a section that described Feynman's feelings when he first realized that his V–A theory was correct:

> I was very excited. It was the first time, and the only time, in my career that I knew a law of nature that nobody else knew.

As James Gleick recounts (pp. 411, 490 [104]):

> Gell-Mann's rage could be heard through the halls of Lauritsen Laboratory, and he told other physicists that he was going to sue. For later editions of the paperback Feynman added a parenthetical disclaimer: "Of course it wasn't true, but finding out later that at least Murray Gell-Mann — and also Sudarshan and Marshak — had worked out the same theory didn't spoil my fun.

Feynman had not been claiming credit. He was just excited about his having invented the V–A theory on his own. Earlier in 1963 Feynman described the history of V–A ([152], p. 477):

> The V–A theory that was discovered by Sudarshan and Marshak, publicized by Feynman and Gell-Mann ... .

Here Feynman was being overly generous. Unlike Murray, he actually did invent the V–A theory on his own, independent of Sudarshan and Marshak.

The Sudarshan–Marshak formulation of the V–A theory [198] differed from that invented by Feynman. Marshak writes [150]:

> Oppie [Oppenheimer] had received our V–A preprint in September [1957] and had told me

in October — during a Washington encounter — that he preferred our chirality-invariant argument for V–A [199] to the two-component Klein-Gordon approach of the Feynman–Gell-Mann preprint. Nine years later — shortly before his death — Oppie rediscovered our Padua-Venice paper and wrote me: "It is a beautiful paper and, for whatever good it is, even at this late date I read it with excitement and great pleasure."

Marshak continues [150]:

> [Feynman] was aware that Sudarshan and I had first proposed the V–A theory but did not know what Gell-Mann would say. Again, as in the case of Oppenheimer, Feynman tried to make amends in later years; Dick wrote me a letter in 1985, in which he said: "It was great seeing you and talking. I hope some day we can get this straightened out and give Sudarshan the credit for priority that he justly deserves. ... these matters all vex me — and I wish I had not caused you and Sudarshan such discomfort. At any opportunity I shall try to set the record straight — as I have always done — but nobody believes me when I am serious."

When did Sudarshan and Marshak work out their version of the V–A theory? According to Marshak [150] and the reference given in Footnote 4 of this chapter:

> By the time of the Seventh Rochester Conference in April 1957, it was clear to both Sudarshan and myself that the only possible Universal Fermi Interaction for weak processes was V–A (with a left-handed neutrino) and not a combination of S and T (with a right-handed neutrino), as was widely believed [86].

This was one or two months before the "summit" meeting with Murray in the first week of June, 1957. Because of missteps by Marshak, the Sudarshan-Marshak V–A publication was delayed until April 1958 while the Feynman–Gell-Mann paper appeared three months earlier.

There was an important difference between the two. The Feynman–Gell-Mann paper simply *asserted* V–A, providing no evidence, while the Sudarshan-Marshak paper *inferred* V–A from the existing data and pointed out the experiments that were most likely in error. Those experiments were eventually repeated and gave the results Sudarshan and Marshak had predicted. When quoted, the Feynman–Gell-Mann paper is almost always given precedence over the one written by Sudarshan and Marshak.

Attribution

Feynman and Murray were aware of the Sudarshan-Marshak V–A paper when they wrote their own paper. Footnote 5 of the Feynman–Gell-Mann V–A paper states [62]:

> A universal V, A interaction has also been proposed by E. C. G. Sudarshan and R. E. Marshak (to be published).

It's curious that Feynman and Murray did not credit them with using V–A, only V,A, since the theory Sudarshan presented to Murray at the "summit" meeting used the specific combination V–A for the weak interaction coupling, as did all the Sudarshan-Marshak publications.

The Feynman–Gell-Mann paper concludes with the Acknowledgements:

> The authors have profited by conversations with F. Boehm, A. H. Wapstra, and B. Stech. One of us (M. G. M.) would like to thank R. E. Marshak and E. C. G. Sudarshan for valuable discussions.

"Valuable discussions" do not accurately describe the content of Murray's conversation with Sudarshan and Marshak, where Sudarshan told him that a Universal Fermi Interaction with a V–A coupling is the right theory of β-decay, despite its disagreement with several experiments.

Feynman, Glashow [103], Marshak, and I give priority of discovery to Sudarshan. Murray is silent on the subject.

How Did Murray Come Up With Quarks?

A Fortuitous Lunch with Robert Serber

I was asked to give a talk on the origins of the constituent quark model at the "Baryon 1980" conference in Toronto, Canada, July 14–16, 1980 [226]. My original intent was to describe both my and Murray's road to quarks. I wanted to consult Murray before my talk, but his wife Margaret had recently been diagnosed with cancer. Bob Serber was thanked in Murray's 1964 quark paper for "stimulating" Murray's thinking during a visit to Columbia University, so I wrote to Bob instead.

He replied:

> Box 260, Cruz Bay
> St. John, V.I. 00830
>
> July 8, 1980
>
> Dear George:
> I'll be glad to give you my recollections of my connection with the origination of the quark model.
> A few months before the '64 paper was published Murray came to Columbia to give a Colloquium on particle physics, and the two of us had lunch together at the Faculty Club. I had been thinking about SU(3) as a result of Gian Carlo Wick's having given a series of lectures on group theory and the SU(3) group, and during lunch I described the "quark" model to Murray: three quarks for the baryons, which gave octet and decuplet, and quark-antiquark for the mesons, which gave the octet. Murph Goldberger has told me that in a conversation not too much later Murray told him that the idea was new to him and that he hadn't thought of anything of the kind before. This is something you can check with Murph. Murray immediately asked about the charges on the quarks, and after a couple of minutes of thought and scribbling, came up with the fractional charges. It was a surprise to me; I hadn't made the observation before, and the credit is certainly due to Murray.
> During his colloquium talk, later in the afternoon, Murray mentioned our luncheon talk and made some remarks about the new particles. I don't remember just what he said. Perhaps T. D. Lee could help you there; his recollection might be better than mine.
> Baqi Bég told me recently that he remembers a discussion that afternoon in which we were trying to come up with a name for the particles.

He says that something close to "quark" was suggested; I think he said it was "quirk." I recall only that there was some talk of a name, but Bég says that he remembers the conversation very clearly, so he would be another good source of information.

Then there is Murray himself – but I sympathize with your reluctance to bother him in the present unhappy circumstances. But you are closer than I, and better able to judge.

I have tried to answer in detail, as you requested. If you have further questions I will do my best to answer them. I will be at this address until July 27th, when I leave for a week's meeting at Los Alamos (you could reach me there c/o Louis Rosen), and after Aug 2nd back at Columbia.

With best regards,
Bob Serber

The original hand-written letter is reproduced as Figure 5.2 on pp. 83-84 in Appendix 5.B.

After Serber told Murray about quarks, Murray realized that he could use Serber's quarks as the mathematical particles in a toy free field theory from which to construct currents and their equal-time commutation relations. He published this observation almost a year later [93].

I included Serber's letter in the first draft of my Baryon 1980 paper. When I showed the draft to Murray he was furious and said: "Why would you ever want to do anything as stupid as that!" He didn't dispute Serber's account, but added "It is none of your business and you don't know anything about it." I wrote to Murph and his reply, also reproduced in Appendix 5.B, confirms Serber's account. Although I was sure Serber's recollections were accurate, Murray's remark that "It is none of your business" led me to remove all mention of Serber. The final draft of the Baryon 1980 paper said:

> The quark model had two independent births. I will describe the one I witnessed. Perhaps Murray Gell-Mann will one day illuminate the other.

Murray chose not to. Since this is no longer possible, Serber's letter is included here.

After a June 4th 1983 telephone interview with Bob Serber [43], Robert Crease and Charles Mann told Murray Serber's account of Murray's conversation with Serber at the Columbia Faculty Club, quoting Serber as asking Murray: "Why don't you consider that?" Crease and Mann, now quoting Murray, tell us the response Murray gave them to Serber's question: [43]:

> "So I *showed* him [Serber] why I hadn't considered it," Gell-Mann told us later. "It was a crazy idea. I grabbed the back of a napkin and did the necessary calculations to show that to do this would mean that the particles would have to have fractional electric charges — $-1/3, +2/3$, like so — in order to add up to a proton and neutron with a charge of plus [one] or zero."

If we are to believe Serber, and I do, that wasn't Murray's reason. Murray didn't consider quarks before his lunch with Serber because he hadn't thought of them. It appears that Murray told Crease and Mann that his calculation was performed to show Serber that quarks had fractional charge, something Murray already knew, but didn't explore because it was "a crazy idea." That's not what Serber's letter says: "Murray immediately asked about the charges on the quarks, and after a couple of minutes of thought and scribbling, came up with the fractional charges." According to Serber, Murray didn't know the charges and did the calculation to find out. Unless misquoted, Murray was skirting the truth.

Improper Attribution

Murray thanked Serber in his 1964 quark paper [93]:

> These ideas were developed during a visit to Columbia University in March 1963; the author would like to thank Professor Robert Serber for stimulating them.[16]

This clearly doesn't do justice to Serber's contribution as he remembers it:

---

[16] This note of thanks is reminiscent of Murray's acknowledgement in his, and Feynman's, V–A paper: "One of us (M. G. M.) would like to thank R. E. Marshak and E. C. G. Sudarshan for valuable discussions."

During lunch I described the "quark" model to Murray: three quarks for the baryons, which gave octet and decuplet, and quark-antiquark for the mesons, which gave the octet.

As Jeffrey Mandula puts it:[17]

The Serber–Gell-Mann path to quarks was almost Biblical. As in Genesis, Serber said "Let there be triplets," and there were triplets, and Murray called the triplets "quarks."

Since Serber invented quarks, and at the time it clearly was a very clever idea (although it turned out to be much more than that), he either should have been a coauthor on a paper with Murray, or Murray should have referenced him for representing hadrons as combinations of quarks. Murray appropriated Serber's idea of quarks without proper attribution, and became famous for it.

Amnesia

In Murray's book *The Quark and the Jaguar* [98], neither Serber nor I are mentioned. In 2015, Murray and Harald Fritzsch edited a collection of papers on quarks for a book titled *50 Years of Quarks* [228]. It starts off with Murray's 1964 *Physics Letters* article [93], moves to Murray's 1972 Schladming lecture [97], continues with a talk that I gave on quarks at CERN in 2013, followed by many other papers written by various authors. Neither of the 1964 CERN Reports, nor the Erice Summer School Lectures are reproduced. The first CERN Report, appended here as Appendix B, certainly should have been included.

5.9   Comparing Feynman with Gell-Mann

The contrast between Feynman and Gell-Mann — both geniuses in their own way — couldn't be greater. How they practiced physics was profoundly different, as was the nature of their contributions, the gratification they derived from work, and their integrity. Each had his own style appropriate for a certain type of problem that arose at a particular time in history.

---

[17] Private communication.

Murray had a remarkable ability to identify promising ideas embedded in a sea of confusion. He clarified, expanded, and communicated them. His knack was taking a problem too hard to solve exactly and saying something about it that, although limited in scope, was true (the equal-time current commutation relations). Like Niels Bohr, he was more of a phenomenologist than a theorist, organizing observations and incomplete theoretical ideas. The Eightfold Way is a good example, correctly clustering hadrons into representations of SU(3), without understanding why that should be possible.

Feynman, like Dirac, was a theorist. They wrote down equations that described nature, paying little attention to experiment until they were finished. As was true for Leibniz and the notation for calculus he created, they also found ways to take existing theory and reframe it to made computations simpler, Dirac with his slick "bra-ket" notation for quantum mechanics, Feynman with "Feynman diagrams" for quantum field theory and condensed matter physics.

The reactions Feynman and Gell-Mann had to the discovery of parity violation were tellingly different (for Feynman, see the long quote starting on page 42). When prodded by his sister, Feynman simply wrote down the V–A theory of β-decay using left-handed neutrinos (referred to as "$(1 + \gamma_5)v$" below), while Murray wrote a lengthy review paper with Arthur Rosenfeld outlining all possibilities [86]. Although Sudarshan had explained to Murray which experiments were wrong, and that V–A was the correct theory ("Lunch with Sudershan and Marshak," p. 76), Murray and Rosenfeld ignored Sudarshan's assessment and adopted a conservative position [86]:

We have looked so far at possibility (a) above, $(1 − \gamma_5)v$, which couples right-handed v and left-handed $\bar{v}$ [*not the V–A theory*]. Possibility (b), which is $(1 + \gamma_5)v$, couples the other two states instead [*this is V–A!*]. We shall see below that (b) is excluded by experiment, while (a) is strongly supported. We may therefore discuss the longitudinal neutrino in terms of (a) only.

So despite Sudarshan's warning, they picked the wrong theory to discuss. Although experiments were central to their paper, rather than carefully examining them, *they hedged their bets* (see the quote on p. 77 that is footnoted).

Although Feynman was not a phenomenologist, he did look very carefully at the Caltech experiment, and concluded that it strongly supported V–A ("The Theory of Parity Violation," p. 43), convincing him to accept the theory of β-decay that he had previously invented while just "playing around." Murray, by contrast, had access to the same experimentalists, who worked on the same floor where he had his office at Caltech, but apparently didn't consult them.

One month after the Gell-Mann–Rosenfeld review paper was published, the Feynman – Gell-Mann V–A theory of the weak interactions, with left-handed neutrinos, was published.

Gell-Mann, like Feynman, missed the importance of spontaneous symmetry breaking for particle physics. This idea had its origins in solid state physics, a field to which Feynman contributed, but which Murray disparaged, calling it "squalid state physics."[18]

The Bottom Line

Both Feynman and Gell-Mann were great communicators, but in entirely different ways. In rereading Murray's papers today, I still marvel at his writing. It is polished, lucid, and engaging, saying things clearly and carefully. In conversation both you and he would be conscious of his intellectual prowess, and his desire to dominate.

By contrast, Feynman was always down-to-earth, explaining things in the simplest possible terms, delighting in his performance, making you *think* you understood.

Another striking difference was their attitude toward observation. As a phenomenologist, Murray paid close attention to the latest experimental results, always trying to understand the changes in theory that might be required, although he was not always successful (witness V–A). Feynman on the other hand didn't start with observation: it wasn't his style, and he didn't have to. His self-confidence and predilection to strike out on his own were great assets.

Murray drove his field continuously, sculpting and creating in response to the work of others. His method of abstraction, while interesting and important at the time, did not address the central problem of the strong interactions. Feynman's contributions were more singular and sporadic. Each looks best when judged by a different metric. For Feynman use the "l_∞ norm" which finds the maximum, although it isn't clear which of his many contributions comes out on top. For Murray the summation of many smaller contributions provided by the equally weighted "l_1 norm" is most appropriate.[19]

Feynman's legacy is greater, longer lasting, and more profound than Murray's. The problems Feynman solved required unsurpassed physical intuition, creativity, and mathematical virtuosity. His contributions were *sui generis*, and unique in their expression. History will always remember Feynman for the utility of Feynman diagrams and the beauty of his path-integral formulation of quantum mechanics that connects the quantum and classical worlds. Murray, ironically, will be remembered for inventing quarks.

---

[18] In the mid 1970s, when Caltech's particle physics group was in decline, I tried to hire Kenneth Wilson, one of Murray's brilliant students who I knew from my graduate student days. Murray was not supportive, saying "he works on such creepy things." Ken was awarded the 1982 Nobel Prize in Physics "for his theory for critical phenomena in connection with phase transitions." In 1974 Ken founded Lattice Gauge Theory with his paper "Confinement of quarks" [213].

---

[19] The l_∞ and l_1 norms of a vector are two different ways of assigning it a length. The l_1 norm is the sum of the absolute values of its components; the l_∞ norm is the maximum of those values.

Appendices for Chapter 5

5.A   Gell-Mann's 1961–62 Publications

1. "The Reaction $\gamma + \gamma \rightarrow \nu + \bar{\nu}$," *Phys. Rev. Lett.* **6**, 70 (1961).
2. "The Eightfold Way: A Theory of Strong Interaction Symmetry," *Caltech Synchrotron Report CTSL-20* (1961).
3. "Broken Symmetries and Bare Coupling Constants" (with F. Zachariasen), *Phys. Rev.* **123**, 1065 (1961).
4. "Form Factors and Vector Mesons" (with F. Zachariasen), *Phys. Rev.* **124**, 953 (1961).
5. "Gauge Theories of Vector Particles" (with S. L. Glashow), *Ann. Phys.* **15**, 437 (1961).
6. "Symmetry Properties of Fields," Proceedings of Solvay Congress (1961).
7. "Symmetries of Baryons and Mesons," *Phys. Rev.* **125**, 1067 (1962).
8. "Experimental Consequences of the Hypothesis of Regge Poles" (with S. C. Frautschi and F. Zachariasen), *Phys. Rev.* **126**, 2204 (1962).
9. "Decay Rates of Neutral Mesons" (with D. Sharp and W. G. Wagner), *Phys. Rev. Lett.* **8**, 261 (1962).
10. "Factorization of Coupling to Regge Poles," *Phys. Rev. Lett.* **8** 263 (1962).
11. "High Energy Nuclear Scattering and Regge Poles" (with B. M. Udgaonkar), *Phys. Rev. Lett.* **8**, 346 (1962).
12. "Elementary Particles of Conventional Field Theory as Regge Poles" (with M. L. Goldberger), *Phys. Rev. Lett.* **9**, 275 (1962).

5.B   Serber's Letter

The following provides supporting material for the description of Murray's lunch with Serber given in "How Did Murray Come Up With Quarks?," p. 79. Serber's hand-written letter to me comes first, a typed transcription was previously given on page 79:

Box 260, Cruz Bay
St. John, V.I. 00830

July 8, 1980

Dear George:

I'll be glad to give you my recollections of my connection with the origination of the quark model.

A few months before the '64 paper was published Murray came to Columbia to give a colloquium on particle physics, and the two of us had lunch together at the Faculty Club. I had been thinking about SU_3 as a result of Gian Carlo Wick's having given a series of lectures on group theory & the SU_3 group, and during the lunch I described the "quark" model to Murray: 3 quarks for the baryons, which gave octet & decuplet, and quark-antiquark for the mesons, which gave the octet. Murph Goldberger has told me that in a conversation not too much later Murray told him that the idea was new to him and that he hadn't thought of anything of the kind before. This is something you can check with Murph. Murray immediately asked about the charge on the quarks, and after a couple of minutes of thought and scribbling, came up with the fractional charges. It was a surprise to me; I hadn't made the observation before, and the credit is certainly due to Murray.

During his Colloquium talk, later in the afternoon, Murray mentioned our luncheon talk and made some remarks about the new particles. I don't remember just what he said. Perhaps T.D. Lee could help you there; his recollection might be better than mine.

Bob: Beg told me recently that he remembers a discussion that afternoon in which we were trying to come up with a name for the particles. He says that something close to "quark" was suggested; I think he said it was "quirk". I recall only that there was some talk of a name, but Beg says that he remembers the conversation very clearly, so he would be another good source of information.

Then there is Murray himself - but I sympathize with your reluctance to bother him in the present unhappy circumstances. But you are closer than I, and better able to judge.

I have tried to answer in detail, as you requested. If you have further questions I would do my best to answer them. I will be at this address until July 27th, when I leave for a week's meeting at Los Alamos (you could reach me there c/o Louis Rosen), and after Aug 2nd back at Columbia.

With best regards,
Bob Serber

Figure 5.2. Letter I received from Bob Serber in 1980 describing his lunch with Gell-Mann when he told him about quarks.

To confirm Bob's recollections I contacted Goldberger and Baqi Bég. Both replied. First, Goldberger's response:

> George,
>
> My recollection is the following: Murray spent the spring term of 1963 at MIT and gave a series of lectures about SU(3) and the Eightfold Way among other things. He went to Columbia in Feb 63 and came back with the report that he had been asked by Bob Serber following a seminar there why he never attached any significance to the fundamental triplet realization of the group, and my recollection is that he said he had never previously given it very serious consideration. I'll see if I can find my notes from his lectures to see if there was any further elaboration at that time.
>
> Murph

Murph's original hand-written note is reproduced as Figure 5.3.

Goldberger found notes compiled by J. A. Campbell, R. Logan and C. E. W. Ward of Murray's lectures while he was a Visiting Professor at MIT during the spring term of 1963. Murray is quoted as saying:

> Finally, it is of interest to look at the number of elements that have to go into the construction of a theory of elementary particles. In conventional field theory, we have to introduce a field for every particle and one for its antiparticle, a total of 60 or 70 in all. In building our higher symmetry theory we introduced three fermions [particles with half-integral spin] S^+, D^+, B^0, three mesons [particles with integral spin] s^+ d^+, d^0, and their antiparticles — a total of twelve elements. This is still a lot, and we may

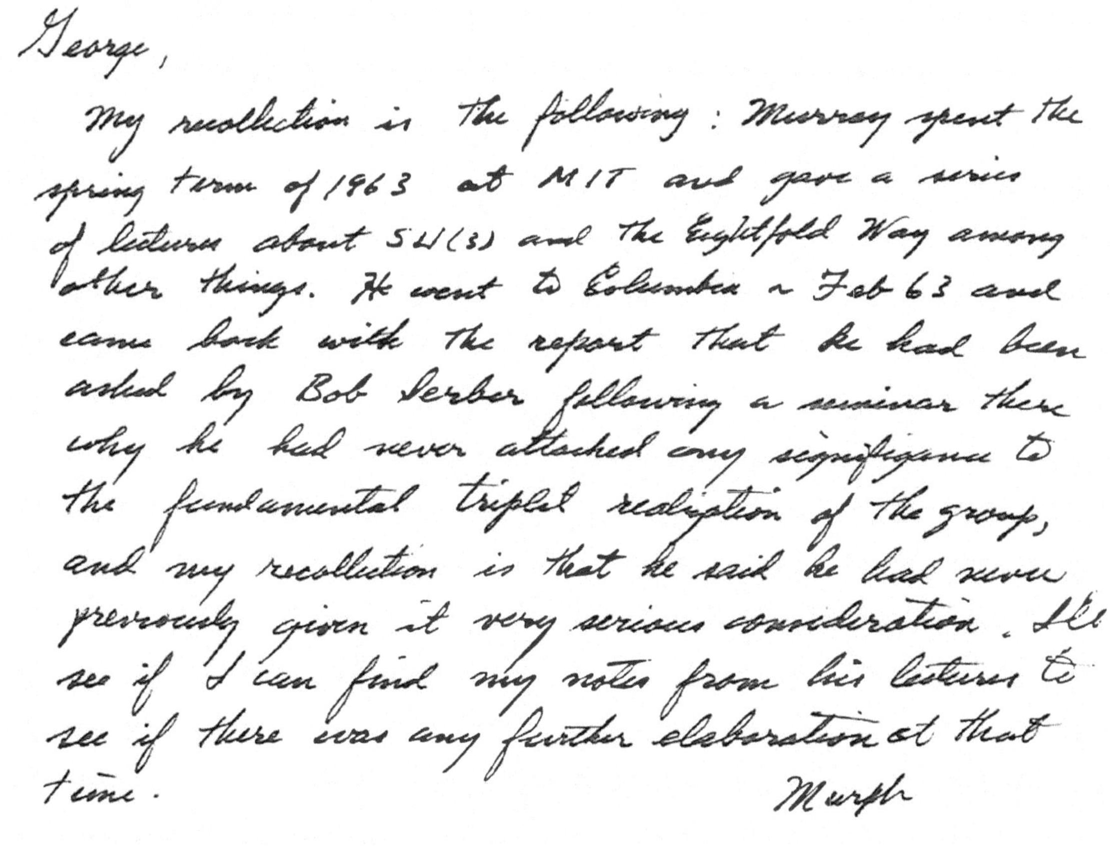

George,

My recollection is the following: Murray spent the spring term of 1963 at MIT and gave a series of lectures about SU(3) and the Eightfold Way among other things. He went to Columbia in Feb 63 and came back with the report that he had been asked by Bob Serber following a seminar there why he had never attached any significance to the fundamental triplet realization of the group, and my recollection is that he said he had never previously given it very serious consideration. I'll see if I can find my notes from his lectures to see if there was any further elaboration at that time.

Murph

Figure 5.3. Letter I received from Goldberger in 1980 recalling Murray's lunch with Serber.

look around to find out what the simplest possible system may be. If you try it, you find that only one, or two, or three, particles can't produce the observed combinations of quantum numbers. However, four will do the trick. A triplet of fermions f and a single baryon β can be combined in a way similar to what we have done earlier;

$$\text{Mesons: } \bar{\beta}\beta \qquad \bar{f}f$$
$$(\underset{\sim}{1} \otimes \underset{\sim}{1} = \underset{\sim}{1}) \qquad (\underset{\sim}{3} \otimes \underset{\sim}{3})$$
$$\text{Baryons: } \beta \qquad \bar{f}f\,\beta$$
$$(\underset{\sim}{1}) \qquad (\underset{\sim}{3} \otimes \underset{\sim}{3})$$

Of course, there is still what I may call "Chew's hope," where we may start with possibly only the mass of the pion, and derive all the rest.

Quarks, or the use of SU(3) triplets to construct the baryons were not mentioned. In fact Murray rules quarks out when he says "you find that only one, or two, or three, particles can't produce the observed combinations of quantum numbers." As Serber told Murray, three quarks do the job.

Concerning the name "quark," Bég writes:

I was <u>not</u> actually present at the meeting at which the quark was named; however, I did hear a very lucid account of the discussion — the day after — from Murray Gell-Mann. It appears that Bob Serber asked about whether the fundamental (defining) representation of flavor SU(3) could be physically realized. Murray, who had been giving talks about the Eightfold Way (representations of the group: SU(3)/Z_3) said that that would be a funny <u>quirk</u>. At this point Serber suggested calling the triplets <u>quorks</u>; this name was actually used by Murray in a lecture at Columbia University. As we all know, the next change of vowels was facilitated by the work of James Joyce.

Chapter 6
A Deeper Layer of Reality

Myself as a college student. (Joachim Nachbar)

Man asked God for a riddle,
and God obliged:
"What is green, hangs from a tree, and sings?"
Stumped,
man asked God for the answer,
and God replied:
"A herring!"
"A herring? But why is it green?"
"Because I painted it green."
"But why does it hang from a tree?"
"Because I put it there."
"And why does it sing?"
"If it didn't sing,
you would have guessed it was a herring."[1]

6.1 Prologue

We think of nature as continuous. Euclid's geometry uses lines, not dotted lines, and calculus would not exist if the world we sensed was at microscopic scale where particles wildly collide, relics from the Big Bang.

A hint that our sense of continuity is superficial came from the observation of erratic motions of tiny particles. As a child you might have observed what the Roman poet Titus Lucretius Carus described more than 2,000 years ago in his poem "On the Nature of Things."[2]

> Observe what happens when sunbeams are admitted into a building and shed light on its shadowy places. You will see a multitude of tiny particles mingling in a multitude of ways … .

But you might not have had his interpretation:

> Their dancing is an actual indication of underlying movements of matter that are hidden from our sight … . It originates with the atoms which move of themselves. Then those small compound bodies, that are least removed from

the impetus of the atoms, are set in motion by the impact of their invisible blows, and in turn cannon against slightly larger bodies. So the movement mounts up from the atoms, and gradually emerges to the level of our senses, so that those bodies are in motion that we see in sunbeams, moved by blows that remain invisible.

What were these invisible particles whose existence could only be inferred by their action on larger objects? It was just matter of drilling down.

6.2 Physics Before Quarks: a Précis

Classical physics reached its zenith at the turn of the 20th century. To Lord Kelvin everything was in accord with observation except for "two clouds," as he noted on Friday April 27, 1900 at his Royal Institution of Great Britain lecture titled "Nineteenth Century Clouds over the Dynamical Theory of Heat and Light" [135]. The first cloud concerned the "luminiferous aether" and the Michelson-Morley experiment. It was soon lifted by Henri Poincaré, Hendrik Lorentz, and Albert Einstein, culminating in Einstein's 1905 Special Theory of Relativity.

Kelvin's second cloud was theory's inability to correctly compute the specific heat of gases (item 4, p. 30). This cloud expanded over a period of 25 years as the puzzling wave nature of particles became apparent, and atoms could be observed only indirectly through their spectral lines. Confusion reigned until Heisenberg formulated a quantum theory of the hydrogen atom with a remarkable insight: Think only in terms of quantities that can be observed [115]. The discovery of quantum mechanics had been delayed by physicists making classical models of quantum mechanical atoms they had never seen, rather than finding the equations that governed the light they emitted. The lesson for future physicists: *Only work with observables.*

The behavior of atoms, ions, electrons, and light was first captured in quantum mechanics, and more fully in a quantum field theory of electromagnetic interactions. But when it came to a field theory for strong interactions, the cardinal question — *what were its particles?* — was never asked. Hadrons were chosen by default; they were conveniently available.

---

[1] A parable from Eastern Galicia, part of the Austro-Hungarian Empire until 1918, related to me as a teenager by my aunt Anna Goldschlag, a survivor of World War I, and then the horrors of World War II.

[2] T. L. Carus, *De Rerum Natura* Book II (∼ 60 BCE).

In the 1950s the primary observables of the strong interactions were scattering amplitudes describing the possible outcomes of scattering one hadron off another. As an alternative to field theory, the bootstrap, based entirely on scattering amplitudes, was developed in response to the proliferation of hadrons. The force between two hadrons as they scattered was created by the exchange of hadrons. That force could be strong enough to bind the initial hadrons into other hadrons which, in turn, could be exchanged or scattered. In this mushrooming manner hadrons created one another.

By the early 1960s it appeared that theoretical particle physics had drifted far from field theory, focusing instead on either hadron symmetries, or the bootstrap. *The bootstrap was even being used to derive the existence of these symmetries.* In the abstract to their 1963 paper titled the "Origin of Internal Symmetries," Ernest Abers, Fredrik Zachariasen, and Charles Zemach write [2]::

> Internal symmetries such as isotopic spin [isospin] are not necessarily arbitrary constraints to be imposed at the beginning of a calculation. The bootstrap requirement that all particles be determined as composite states of one another leads naturally to symmetric solutions for masses and coupling constants.

The next step was to bootstrap "symmetry breaking." The bootstrap appeared to be the future of theoretical physics [35].

Meanwhile, progress in quantum field theory never materialized, not because that type of theory was too difficult to handle, as was commonly believed, but because its fields were incorrectly chosen. Heisenberg's requirement that one only work with observables, appropriate in 1925, misled many theorists some 30 years later because what they observed — hadrons — were not the particles that would form the basis of a quantum field theory that could describe them. The particles for that theory — constituent quarks — had not been discovered.

When constituent quarks were first discovered in January 1964, only the consequences of their existence — hadrons with their quantum numbers, masses, and interactions — could be observed. How then were these constituents discovered, their quantum numbers

determined, and their reality established? What predictions followed? These questions are answered in the following sections, fleshing out the overview presented in Chapter 2.

Symmetry

Feynman defines what is meant by a symmetry in Chapter 4 of *The Character of Physical Laws* [66]:

> Professor Weyl, the mathematician, gave an excellent definition of symmetry, which is that a thing is symmetrical if there is something that you can do to it so that after you have finished doing it, it looks the same as it did before. That is the sense in which we say that the laws of physics are symmetrical; that there are things we can do to the physical laws, or to our way of representing the physical laws, which make no difference, and leave everything unchanged in its effects.

Rotate a sphere about its center and it remains a sphere, even though the points on the sphere are mapped to other points. The sphere is symmetrical. The same idea generalizes to the laws of physics which are expressed in terms of equations. Transform the equation $x^2 = 1$ by flipping the sign of x and it remains unchanged. The equation is symmetrical under a parity transformation.

Symmetries of Particles

Particle theory consists of dynamics, as represented by equations controlling the evolution of a system in time, and the symmetries those equations possess. Symmetries initially played a central role in hadron physics because understanding their dynamics was too difficult. The dynamical equations were not directly related to observation. Not so for symmetries. The symmetry between protons and neutrons in the nucleus, and then in scattering reactions, referred to as charge independence (left column, p. 13), was discovered experimentally. Charge independence told you that interchanging protons and neutrons *in the strong interaction equations* left those equations unchanged, but was silent on the structure of those equations.

Symmetries were used to organize the ever increasing number of hadrons into clusters with identical quantum numbers, except for charge, and similar mass. Since different symmetries grouped hadrons differently, the symmetry of the strong interactions could be identified by observing how hadrons clustered, without knowing the equations that determined how hadrons interacted and evolved in time.

The Mathematics of Particle Symmetries

Where there's symmetry, there's a "symmetry group." Eugene Wigner, a Hungarian physicist who had studied mathematics with David Hilbert, and was one of Max Born's assistants at the University of Göttingen, realized that Heisenberg's nucleon $\mathcal{N} = (p, n)$ was a "representation" of a symmetry group, where the group was SU(2), the "special unitary group of degree 2." The pion $\pi = (\pi^+, \pi^0, \pi^-)$ was another SU(2) representation, but with three dimensions. Operations in SU(2), analogues of rotations in the symmetry group SO(3) of the sphere, transform linear combinations of members of a representation into one another.[3]

Representations that can not be broken down into two or more smaller representations, whose members transform only among themselves, are called "irreducible representations," or "multiplets." Hadrons in two or more irreducible representations, if considered together, form a "reducible representation." The $\mathcal{N}$, π, and K are irreducible representations of SU(2). SU(2) symmetry is also called "isospin symmetry," and irreducible representations of SU(2) are "isospin multiplets."

SU(2) — Isospin Symmetry

Associated with each irreducible representation of SU(2), is a vector $\vec{I}$ in a three-dimensional "isospin space," and a nonnegative integer, or half integer, I. The projection of $\vec{I}$ onto the Z axis of this space is designated by I_Z, and these projections are quantized. There are $2I+1$ projections corresponding to the $2I+1$ hadrons in the multiplet. The projections have a maximum value of I corresponding to $\vec{I}$ pointing "up" and a minimum value $-I$, corresponding to $\vec{I}$ pointing "down," and any value $I-1, I-2, \ldots$ in between these extremes. As I_Z decreases in steps of 1 from its maximum value, so do the charges of the corresponding

hadrons in the multiplet. The quantum number I also determines the length $\sqrt{I(I + 1)}$ of the vector $\vec{I}$.

Both $\vec{I}$ and I_Z are of interest because they are quantum numbers for the strong interactions. Their values for a collection of hadrons are the "sums" of their individual values, and those values are conserved. Summing isospin projections is simply adding their values arithmetically. The rule for summing values of $\vec{I}$ is more complex because they are vectors that can point in different directions (Appendix A.5: "Addition of Isospin, Spin, or Angular Momentum," p. 149).

The nucleon $\mathcal{N}$ is assigned an isospin vector $\vec{I}$ with $I = 1/2$ which can point either in the proton ($I_Z = 1/2$) or neutron ($I_Z = -1/2$) direction. Similarly, the π^+, π^0, and π^- are different manifestations of the pion π whose isospin quantum numbers are $I = 1$ and $I_Z = (1, 0, -1)$, corresponding to $\vec{I}$ pointing up, sideways, or down. Like the proton and neutron, all three pions enjoy charge independence; the nuclear force treats them equally. The photon of the electromagnetic interactions distinguishes them.

As previously indicated, hadrons in the same isospin multiplet are close in mass. The proton and neutron in the nucleon isospin doublet have masses 938.2 and 939.5 MeV. The small mass difference indicates that SU(2) symmetry is not perfect.

When additional strongly interacting particles were discovered starting in the late 1940s, it was also possible to place them into isospin multiplets of SU(2). For example, the eight baryons listed in Figure 2.2 on page 16 form four irreducible representation of SU(2) with $I = (1/2, 0, 1, 1/2)$ and strangeness $\mathbb{S} = (0, -1, -1, -2)$, corresponding to the $\mathcal{N}$, Λ, Σ, and Ξ.

Looking back again to Figure 2.2, note that four kaons are listed with identical, or nearly identical, masses. While three pions form an irreducible representation of SU(2) with $I = 1$, the kaons do not form an irreducible representation with $I = 3/2$ because their strangeness differs. They form a reducible representation of SU(2), which can be broken into two irreducible representations, the $K = (K^+, K^0)$ and $\bar{K} = (\bar{K}^0, K^-)$ with strangeness $\mathbb{S} = 1$ and -1, respectively, both with $I = 1/2$.

In the language of group theory, SU(2) symmetry is conserved by the strong interactions, but broken by both the electromagnetic and weak interactions. However I_Z is conserved by the electromagnetic

---

[3] The definition of a group, and a simple example of how operations in a group transform members of one of its representations, is given in Appendix A.5, left column, p. 150.

interactions, while I is not. Two examples of symmetry breaking are given by Σ^0 decay:

$$\Sigma^0_{I=1,J_Z=0} \quad \rightarrow \quad \Lambda_{I=0,J_Z=0} + \gamma, \text{ and}$$

$$\rightarrow \quad p_{I=1/2,J_Z=1/2} + e^- + \bar{\nu}_e,$$

where the isospin I of the baryon is changed in both reactions, but I_Z is only changed in the weak decay.

SU(3) Symmetry

When strangeness was added to the other quantum numbers of the strong interactions, the ideas of Heisenberg and Wigner were extended. Charge independence became "charge and strangeness independence," which made SU(3) the symmetry of the strong interactions. Unlike SU(2), this symmetry was badly broken, because members with different strangeness in the same irreducible SU(3) representation were no longer close in mass, and strong interactions between hadrons depended on their strangeness [90, 159].

The four spin-1/2 baryon representations of SU(2), the $\mathcal{N}, \Lambda, \Sigma$, and Ξ, formed one irreducible representation of SU(3), the baryon octet. When an addition neutral pseudoscalar meson, the η, was discovered in 1961, the $\pi, K, \bar{K}$, and η, with $I = (1, 1/2, 1/2, 0)$ and strangeness $\mathbb{S} = (0, 1, -1, 0)$, became a second eight-dimensional irreducible representation of SU(3), the pseudoscalar meson octet. Representations of SU(3) were decoupled from the quantum numbers of their members.

SU(6) Symmetry – Combining SU(3) with Spin

The clustering of hadrons with similar mass and quantum numbers was hierarchical. Within each "top" cluster, a further clustering was possible. The further down the hierarchy, the closer the clustering in mass, but the less diverse the quantum numbers. The existence of these clusters and sub-clusters indicated that a hierarchy of symmetries, with their associated symmetry groups, was at play.

The nucleon $\mathcal{N}$ with its proton and neutron were one such sub-cluster. The $\mathcal{N}$ with the three other spin-1/2 baryon isospin multiplets, the Λ, Σ, and Ξ, formed a higher cluster, an irreducible representation of the group SU(3) — the baryon octet. Four spin-3/2 baryon isospin multiplets with $I = (3/2, 1, 1/2, 0)$ and $(\mathbb{S} = 0, -1, -2, -3)$,

the Δ, Σ^*, Ξ^*, and Ω, formed a second irreducible representation of SU(3) — the baryon decuplet. These two representations where clustered into an irreducible representation of SU(6), where members of that representation had different spins.

A different hierarchy existed for mesons. At its highest level the octet of pseudoscalar mesons with spin 0 was combined with a nonet of vector mesons with spin 1 to form another irreducible representation of SU(6).

The existence of these hierarchies, and the quantum numbers the particles in their representations contain, is derived in Section 6.5: "Hadrons from Aces with Spin," p. 96. Those hierarchies are made explicit in Appendix 6.A, p. 116.

Symmetries of Wave Functions

The Pauli Exclusion Principle

When an electron is attracted to a helium nucleus, it doesn't continuously spiral inwards releasing a stream of radiation, as it would classically. It jumps toward the nucleus in discrete steps from one orbit or "state" to the next, emitting a single photon with each transition, until it reaches the "ground state."

A second electron would try to repeat the process, *only entering the ground state if the two electron spins are in opposition*. This difference in spin orientation is necessary to make the two electrons distinguishable.

This restriction on the electron's spin follows from the "Pauli exclusion principle" which says that two or more identical electrons cannot occupy the same quantum state, i.e., have the same set of quantum numbers. Since the electron spin can only have one of two possible spin projections, only two electrons with different spin projections can occupy the ground state.

Technically, the Pauli exclusion principle is enforced by requiring that the wave function of a collection of electrons be antisymmetric under the interchange of any two of them. For example, if the wave function of a two-electron system is $\psi(e_1, e_2)$, where e_i represents the spin and spatial coordinates of the ith electron ("Schrödinger's Wave Function," p. 9), then if ψ is antisymmetric, $\psi(e_1, e_2) = -\psi(e_2, e_1)$. But if the two electrons are identical, $\psi(e_1, e_1) = -\psi(e_1, e_1)$, implying $\psi(e_1, e_1) = 0$, i.e., the wave function must vanish: A physical system with two identical electrons can not exist.

If the Pauli exclusion principle did not hold, electrons would all collect in an atoms's ground state, and chemistry as we know it would not exist. Pauli's principle is responsible for the diversity of the elements in the periodic table, and prevents gravity from pulling you, and your surroundings, to the center of the Earth.

The Pauli exclusion principle was generalized to hold for all particles with half-integral spin, and became a special case of the "spin-statistics theorem."

The Spin-Statistics Theorem

The spin-statistics theorem relates the spin of particles to the symmetries of their wave functions.

It says:

- The wave function of a system of identical fermions (particles with spin $S = 1/2, 3/2, ...$) changes sign when the positions of any two fermions are interchanged. One says fermions obey "Fermi-Dirac statistics."
- The wave function of a system of identical bosons (particles with spin $S = 0, 1, ...$) remains the same when the positions of any two bosons are interchanged. Bosons obey "Bose-Einstein statistics."

The number of identical bosons in a quantum state is not restricted by the spin-statistics theorem; witness the many identical spin-1 photons in a laser beam in the same quantum state, all in phase, combining their effects coherently.

Just as the Pauli exclusion principle constrains the electronic structure of the elements, so the spin-statistics theorem constrains the kind of baryons created from quarks.

6.3   Down the Rabbit Hole

The ϕ-decay Anomaly

The ϕ meson was produced in the reaction

$$K^- + p \rightarrow \Lambda + \phi,$$

with the ϕ then decaying into two kaons,

$$\begin{aligned} \phi &\rightarrow K^+ + K^-, \text{ or} \\ &\rightarrow K^0 + \bar{K}^0. \end{aligned}$$

Since the observables were Λ, K, and $\bar{K}$, how did one know that a ϕ was produced rather than, for example, a new heavy proton resonance "p^*," created in the reaction

$$K^- + p \rightarrow K^- + p^*,$$

that decayed into Λ and a K^+,

$$p^* \rightarrow \Lambda + K^+?$$

To see what resonances were created, a "Dalitz plot" was constructed.[4]

Dalitz Plot

In the two-body elastic scattering reaction $K^- + p \rightarrow K^- + p$ the total energy of the final state $K^- p$ must equal that of the initial state. In the three-body inelastic scattering reaction, $K^- + p \rightarrow \Lambda + K^+ + K^-$, the energies of the two pairs, $K^+ K^-$ and ΛK^+, can take on a continuum of values. Each inelastic scattering event that produces a K^+ and K^- has particular $M^2(K^+ K^-)$ and $M^2(\Lambda K^+)$ values, where $M^2(K^+ K^-)$ is the mass squared of the $K^+ K^-$ system. A scatter plot of these two mass-squared values, called a Dalitz plot, is given in Figure 6.2. Each interaction contributes one point to the plot. The conservation of energy and momentum restricts points to lie within the ellipsoidal region shown. Measurement errors are responsible for points whose location lies outside this region.

The points in the Dalitz plot are projected onto the abscissa and ordinate to create histograms of the mass-squared distributions of the $K^+ K^-$ and ΛK^+ systems. The tall thin vertical pencil of $K^+ K^-$ events standing on the abscissa clearly demonstrates the existence of the ϕ that decayed into $K^+ + K^-$. The narrow width of the peak indicates that the ϕ was long-lived, which is surprising, as we shall see. While the mass-squared distribution of the K^+ and K^- was concentrated around a single value, the ΛK^+ mass-squared distribution was spread out in value. The absence of a peak in the $M^2(\Lambda K^+)$ histogram on the ordinate indicates that a p^* resonance was not formed.

---

[4] For an explanation of Dalitz plots see www.slac.stanford.edu/slac/sass/talks/BrianL.pdf or warwick.ac.uk/fac/sci/physics/staff/academic/gershon/talks/gershon-BadHonnef.pdf
(alternatively, type "Introduction to Dalitz Plot Analysis – University of Warwick" into Google and click on "Introduction to Dalitz Plot Analysis").

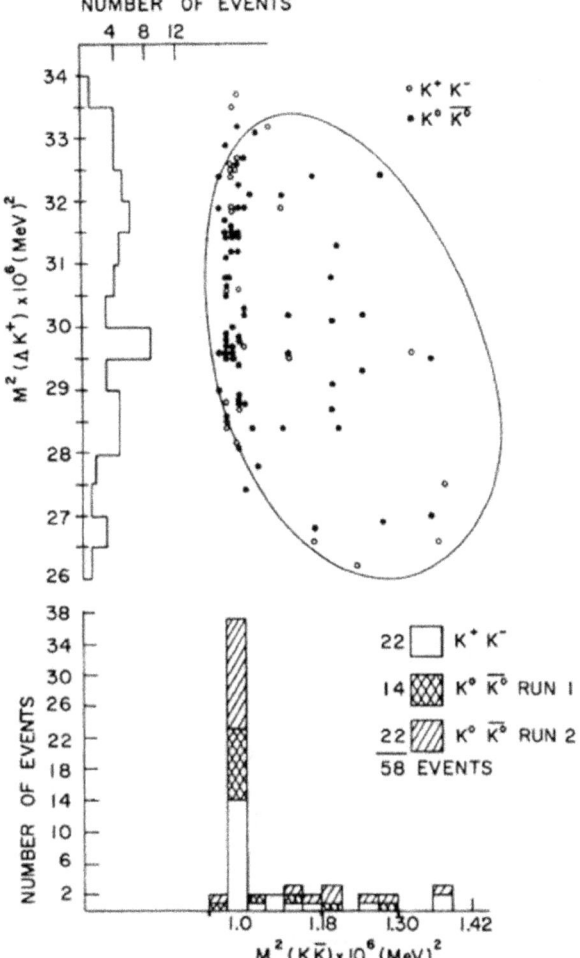

An analogous analysis can be performed when a $K^0 + \bar{K}^0$ are produced, and $K^0\bar{K}^0$ points have been added to the Dalitz plot.

Although clearly seen decaying into $K\bar{K}$, the ϕ was expected to decay primarily into $\rho\,\pi$, with the ρ then decaying into $\pi\,\pi$,

$$\phi \;\to\; \rho + \pi, \text{ leading to}$$
$$\to\; \pi + \pi + \pi.$$

But now look at Figure 6.3 where the three-π mass distribution is displayed. There is no statistically significant evidence for a peak above a smooth "background"; there is no indication that ϕ decayed to $\rho\,\pi$!

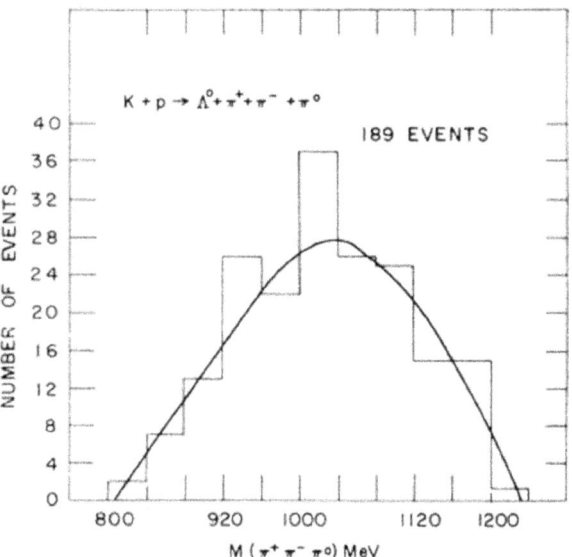

Figure 6.3. The $M(\pi^+\pi^-\pi^0)$ mass distribution from the reaction $K^- + p \;\to\; \Lambda + \pi^+ + \pi^- + \pi^0$ taken from [41]. The absence of a large peak above background (solid line) at the ϕ mass of 1019 MeV indicates the suppression of the decay $\phi \;\to\; \rho + \pi$ (which would have been followed by $\rho \to \pi + \pi$).

Figure 6.2. The Dalitz plot for the reaction $K^- + p \to \Lambda + K + \bar{K}$ taken from [41]. The mass-squared distributions of $K\bar{K}$ and ΛK^+ are projected on the abscissa and ordinate. Peaks in these histograms correspond to resonances. The tall $K\bar{K}$ peak on the abscissa indicates the existence of a meson resonance, called the ϕ, that decays into either $K^+ + K^-$ or $K^0 + \bar{K}^0$. Its extremely narrow width indicates that it was long-lived, i.e., that its decay was very suppressed for a heavy strongly interacting particle. The absence of a statistically significant peak in the $M^2(\Lambda K^+)$ histogram on the ordinate shows that an excited proton, that would have decayed into $\Lambda + K^+$, was not produced.

The authors of the paper commented on this unexpected absence:

The observed rate [for ϕ to $\rho\,\pi$] is lower than ... predicted values by one order of magnitude;

however the above estimates are uncertain by at least this amount so that this discrepancy need not be disconcerting.

But I was very disconcerted. Since the experiment was exceptionally clean, I had to accept its results, but not the authors' conclusion. Even without any calculations the dearth of $\rho + \pi$ events, compared to the number of $K + \bar{K}$ events at the ϕ mass, was remarkable because the decay $\phi \to K + \bar{K}$ was *a priori* very unlikely. The K and $\bar{K}$ momenta were almost zero in the rest frame of the ϕ; that's why ϕ's points in the Dalitz plot were so close to the ellipsoidal boundary encircling the kinematically allowed events. That meant very little "phase space" was available for this mode of decay.[5] Limited phase-space suppressed ϕ into $K + \bar{K}$. The phase space available for ϕ decay into $\rho + \pi$ was much greater because the ρ and π momenta were much larger, implying that $\rho + \pi$ should be much more prevalent than $K + \bar{K}$, but just the opposite was true.

My calculations indicated that the $\rho \pi$ mode was suppressed by at least *two* orders of magnitude,[6] an unprecedented suppression for strong interactions.

---

[5] The phase space available when parking a car is greatest when backing it into a parking space, rather than driving it in forward. The arc of acceptable angles is larger going in backwards. This asymmetry in parking direction is related to an asymmetry built into cars — rotatable wheels are only in front. Analogously, the more momenta available to a particle in a decay, the larger its "arc" in phase space, and therefore the more likely the decay will take place. Since the momenta of the K and $\bar{K}$ were very small, the decay was highly improbable. Nevertheless, only this decay was observed, not the $\rho + \pi$ decay with its much larger phase space.

[6] A simple estimate I made was:

$$\frac{\text{Decay rate to } K\bar{K}}{\text{Decay rate to } \rho\,\pi} \propto \left(\frac{p_{K\bar{K}}}{p_{\rho\,\pi}} \right)^3$$

$$\approx 1/4 \text{ (expected)}$$
$$\gtrsim 35 \text{ (observed)}.$$

Here p_{ij} is the momentum of either particle i or j in the rest frame of the ϕ. A second estimate giving a similar result was based on a paper by Gell-Mann, David Sharpe and Bill Wagner (Feynman's notetaker in his quantum gravity class) that computed the rate at which the ω decayed into three pions [92].

6.4   The Deeper Layer

In the summer and fall of 1963, because of the unexpected suppression of ϕ decay, I examined essentially all data related to hadron resonances in several hundred papers and preprints — some right, some wrong, many inconclusive. It gradually became apparent that *there were two layers of reality*. The first contained hadrons, with their highly unstable resonances. The deeper layer, which was responsible for the first, was only indirectly observable through the hierarchy of hadron representations and quantum numbers, the many relations between hadron masses, the pattern of hadron weak decays, and the suppression of ϕ decay. As in Plato's cave,[7] where only shadows of the real world appeared, hadrons were "shadows," the "real world" their elementary constituents, governed by laws of their own.

The problem was to ferret out the constituents from their shadows — the composites — so that a theory for hadrons could be formulated in terms of a few elementary particles, as originally proposed in two CERN Reports and a series of lectures at the 1964 Erice Summer School [220, 221, 223]. The first CERN Report (24 pages, dated January 17, 1964) was never published, but was widely distributed at the time. It is reproduced in Appendix B, p. 155.[8] The longer second CERN Report (80 pages, dated February 21, 1964), although eventually published ([221]), is most conveniently obtained as a download from the CERN Document Server.[9] The 1964 Erice Summer School Lectures, "Fractionally Charged Particles and SU_6," were published the following year, but are best downloaded from ResearchGate.[10]

Constituents

Fermi and Yang suggested that the pion was not an elementary particle, but rather a composite, a bound state of a nucleon and antinucleon (Section 2.3: "Why So Many Hadrons?," p. 17). Sakata extended that model in 1956 to include strangeness, using the first three baryons and their antiparticles to form mesons and

---

[7] Book VII of Plato's *Republic*, www.age-of-the-sage.org/greek/philosopher/myth_allegory_cave_plato.html.

[8] cds.cern.ch/record/352337?ln=en.

[9] cds.cern.ch/record/570209?ln=en.

[10] www.researchgate.net/search.Search.html?type=publication&query=Fractionally%20Charged%20Particles%20and%20SU6.

other baryons. By 1961 enough was known about baryons to see that Sakata's model was incorrect because some of the baryons it predicted had not been found.

But the idea that hadrons had constituents fascinated me. As described in Section 2.7: "Constituents," p. 23, I created baryons from three aces

$$\mathcal{A} = (p_0, n_0, \Lambda_0), \qquad \text{(from 2.6)}$$

each with baryon number $B = 1/3$, that consisted of an isospin doublet $\mathcal{N}_0 = (p_0, n_0)$ with $I_Z = (1/2, -1/2)$ and strangeness $\mathbb{S} = 0$, and an isospin singlet Λ_0 with $I = 0$ and strangeness $\mathbb{S} = -1$.[11] Like their baryon namesakes, the parity of the three aces was set to 1.

Alternatively, aces and antiaces were often written with subscripts and superscripts,

$$\mathcal{A} = (A_1, A_2, A_3) \text{ and } \bar{\mathcal{A}} = (A^1, A^2, A^3).$$

This notation was used when creating baryon wave functions (p. 121), and hadronic couplings.

The Gell-Mann–Nishijima charge formula,

$$Q = I_Z + \frac{B + \mathbb{S}}{2}, \qquad \text{(5.5 revisited)}$$

is an empirical linear relation between a hadron's quantum numbers. Since these quantum numbers are additive, the relation also holds for hadron constituents, enabling a calculation of ace charges from their three other quantum numbers, yielding

$$Q = (2/3, -1/3, -1/3). \qquad \text{(2.13 revisited)}$$

These charges are the same as those of Sakata's constituents, but shifted down by 1/3, as they must be according to the Gell-Mann–Nishijima charge formula, because the constituents' baryon numbers B are all shifted down by 2/3.

Antiace quantum numbers were the negative of their ace counterparts giving ace-antiace pairs the quantum numbers of the vacuum, as is apparent in the meson decay graph of Figure 2.6, p. 25, where a virtual antiace-ace pair $\bar{A}'A'$ emerges from the vacuum.

Ace Interactions

The interactions between aces were unknown. Although the strong interactions were the obvious and most conservative choice, another possibility was considered in the second CERN Report [221]:

> There may be an interaction, stronger than the strong interactions, which governs the behavior of aces causing them to bind to form mesons and baryons. In this model the strong interactions would be viewed as "some kind of van der Waals force." Just as two isolated electrons do not interact with a van der Waals force, so two aces do not interact strongly.

It took almost a decade to realize that massless neutral spin-1 "gluons," coupling to a new quantum number called "color," created the force between quarks that resulted in the formation of hadrons (p. 135).

Naming the Constituents

The name "ace" had two origins, one from the Latin "as" meaning "unit," which I used to represent the unit from which hadrons were created. Aces were also picked because there are four aces in a deck of cards. Four different units were chosen because I thought a correspondence existed between the elementary particles of the strong interactions, and those of the weak interactions — the four leptons. However, in 1963 only three aces were needed to create the hadrons.[12] Today there are six leptons and six quarks.

Sometimes I called aces "concrete quarks," Murray's name (Section 2.12, p. 27), to emphasize their reality and to distinguish them from Gell-Mann's "mathematical" or "current quarks" [93], although *aces were chimeric,* also functioning as current quarks when they were used to compute hadronic electromagnetic and weak interactions (Section 6.8: "Weak-Interaction Predictions," p. 102). After a year or two aces were referred to as "constituent quarks" by the physics community.

---

[11] The notation for the first three aces, (p_0, n_0, Λ_0) or (p_0, n_0, λ_0), was adopted by the Russians and the Japanese, the latter dropping the subscript 0. However, this notation was eventually replaced by Gell-Mann's (u, d, s) — up, down, and strange.

[12] I would never have called the constituents aces if I thought there were only three. The letters "a,b,c," an alphabet for three aces, would have been the natural choice. I thought more than four aces and four leptons were possible. In that case, additional aces were to be represented graphically by regular polygons with an increasing number of sides, and increasing areas.

6.5   Hadrons from Aces with Spin

Counting spin projections, there were six constituents: three aces whose spins could point either up or down,

$$\mathbb{A} = (p_0\uparrow,\ p_0\downarrow,\ n_0\uparrow,\ n_0\downarrow,\ \Lambda_0\uparrow,\ \Lambda_0\downarrow\,).$$

When spin was added to aces in the 1964 Erice Summer School Lectures [223], SU(3) was extended to SU(6). $\mathbb{A}$ formed a six-dimensional irreducible representation of SU(6).

Low mass mesons were created from linear combinations of all possible $\mathbb{A}\bar{\mathbb{A}}$ pairs called "deuces," where each $\mathbb{A}$ designated any one of its six members. The low mass baryons came from linear combinations of $\mathbb{A}\mathbb{A}\mathbb{A}$ triplets or "treys," where those combinations were totally symmetric under the interchange of any two $\mathbb{A}$s. More massive mesons, some with higher spin, were created from $\mathbb{A}\mathbb{A}\bar{\mathbb{A}}\bar{\mathbb{A}}$ ([223]). The quantum numbers of a hadron were simply the sums or products of its constituents' quantum numbers, taking angular momentum into account for parity.

In the lowest-mass mesons and baryons aces didn't orbit around one another, so a hadron's spin was simply the sum of its constituent's spins. If ace and antiace spins pointed in the same direction the corresponding meson had spin 1, and was called a "vector meson." If the spins pointed in opposite directions, a spin-0 "pseudoscalar meson" was formed. If the three ace spins in a baryon pointed in the same direction, a member of the spin-3/2 baryon "decuplet" was created. When two of the three spins canceled, a spin-1/2 baryon in the "octet" was formed.

The strong-interaction symmetry that resulted from creating hadrons out of three spin-1/2 aces with two possible spin projections was called "ace-spin symmetry" [223], a more descriptive term than SU(6).

Mesons From Deuces with Spin

The total number of mesons created from three aces spinning either up or down is:

$$(3\,\text{aces}\ \times\ 2\,\text{spins}) \times (3\,\text{antiaces}\ \times\ 2\,\text{spins})$$
$$= 6 \times 6\ \text{mesons with spin}$$
$$= 36\ \text{mesons, counting spin projections.} \qquad (6.1)$$

These 36 mesons with spin form a reducible representation of SU(6) which is the sum of two irreducible representations, one with 35 mesons, the other with only 1,

$$\text{SU(6)} : \ 6 \otimes \bar{6} = 35 \oplus 1.$$

As we will now confirm, the 35 mesons, counting spin projections, form two clusters of eight spin-0 and nine spin-1 mesons. Just like an isospin multiplet with isospin I has $2I + 1$ members (p. 90), so a meson with spin S has $2S + 1$ possible spin projections. Therefore a spin-0 or spin-1 meson has either one or three possible spin projections, yielding the following decomposition of the mesons in the 35 dimensional SU(6) irreducible representation:

$$(8\ \text{spin-0 mesons}\ \times 1\ \text{spin})$$
$$+\ (9\ \text{spin-1 mesons}\ \times 3\ \text{spins})$$
$$=\ (8 + 27)\ \text{mesons with spin}$$
$$=\ 35\ \text{mesons, counting spin projections.}$$

Note that eight pseudoscalar and nine vector mesons were correctly predicted, not nine pseudoscalar and eight vector mesons. Ace-spin symmetry connects each SU(3) hadron representation with the spin of its members.

The pseudoscalar and vector mesons, not counting spin projections, form a 17 dimensional reducible representation of SU(3), the sum of three irreducible representations,

$$\text{SU(3)} : \ 17 = 8 \oplus 8 \oplus 1,$$

where each of the SU(3) eight-dimensional irreducible representations is the sums of four SU(2) irreducible representations,

$$\text{SU(2)} : \ 8 = 3 \oplus 2 \oplus \bar{2} \oplus 1$$

The four spin-0 isospin multiplets comprising the octet are:

$$8\ \text{pseudoscalar mesons}$$
$$=\ (3\ \text{isospin-1}\ \pi s) + (2\ \text{isospin-1/2}\ Ks)$$
$$+\ (2\ \text{isospin-1/2}\ \bar{K}s)$$
$$+\ (1\ \text{isospin-0}\ \eta_8),$$

where, for example, (3 isospin-1 πs) represents the three pion states (π^+, π^0, π^-). The η_8 is also called the η.

Correspondingly, the five spin-1 isospin multiplets in the nonet are:

8 ⊕ 1 vector mesons

= (3 isospin-1 ρs) + (2 isospin-1/2 K^*s)

   + (2 isospin-1/2 $\bar{K}^*$s)

   + (1 isospin-0 ω_8) + (1 isospin-0 ω_1), or

   + (1 isospin-0 ω) + (1 isospin-0 ϕ).

The two isospin-0 vector mesons that are observed (ω and ϕ) are a mixture of the two isospin-0 vector mesons found in the SU(3) octet and singlet (ω_8 and ω_1). Mixing occurs because SU(3) symmetry is broken by making the Λ_0 heavier than the p_0 and n_0, whose masses remain the same. Increasing the Λ_0 mass separates the $\Lambda_0 \bar{\Lambda}_0$ from the $p_0 \bar{p}_0$ and $n_0 \bar{n}_0$ in the mesons' wave functions, thereby creating the ω and ϕ.[13]

All the mesons in the 35-dimensional irreducible representation of SU(6) had been observed by 1963.

Meson Spin, Parity, and Charge Conjugation

If an ace-antiace pair has orbital angular momentum $\vec{L}$, total spin $\vec{S}$, and total angular momentum $\vec{J} = \vec{L} + \vec{S}$, then:

- $|L - S| \leq J \leq L + S$,
- Parity $P = (-1)^{L+1}$,
- Charge conjugation $C = (-1)^{L+S}$.

These relations are discussed in "A Primer on Quarks," A.5 and A.6, pp. 149 and 151.

After discounting many erroneous measurements, the 35 mesons with $L = 0$ precisely corresponded to all the observed low-mass mesons — *correctly matching spin, isospin, charge, strangeness, parity, and charge conjugation.* The existence of the other heavier SU(6) isosinglet representation containing the pseudoscalar meson η_1, also called η', had not been established.

Baryons from Treys with Spin

The construction of baryons from three aces (treys) was more complicated than that of mesons from

deuces because of the spin-statistics theorem (p. 92). Baryon wave functions had to be antisymmetric under the interchange of any two identical aces. However enforcing this constraint led to the creation of the wrong set of baryons. The right set was obtained by insisting the wave functions be symmetric! The choice was to abandon aces, or assume that there was something missing in the theory, a degree of freedom that, when added to the wave function and included in the exchange process, would make the wave function totally antisymmetric. I chose the latter possibility, constructing baryon wave functions from totally symmetric combinations of three aces with two possible spin projections. The missing degree of freedom was the new quantum number color (p. 135).

The total number of baryons created from three aces spinning either up or down is:

(3 aces × 2 spins) × (3 aces × 2 spins)

   × (3 aces × 2 spins)

= 6 × 6 × 6 baryons with spin

= 216 baryons, counting spin projections.

These 216 baryons form a reducible representation of SU(6), which consists of four irreducible representations, with 56, 70, 70, and 20 members,

SU(6) : $6 \otimes 6 \otimes 6 = 56 \oplus 70 \oplus 70 \oplus 20$.

The 56-dimensional representation was chosen because its baryons were in agreement with those found experimentally, although the baryon wave functions were totally symmetric in their aces, just the opposite of what is required by the spin-statistics theorem. The 56 consists of eight spin-1/2 and 10 spin-3/2 baryons that have two and four spin projections, respectively, yielding a decomposition of the 56 dimensional SU(6) irreducible representation into two SU(3) irreducible representations,

(8 spin-1/2 baryons × 2 spins)

   + (10 spin-3/2 baryons × 4 spins)

= (16 + 40) baryons with spin

= 56 baryons, counting spin projections.

Note that an octet of spin-1/2 and decuplet of spin-3/2 baryons were correctly predicted, not a decuplet of spin-1/2 and octet of spin-3/2 baryons.

The octet and decuplet of baryons, not counting spin projections, form an 18 dimensional reducible

---

[13] The wave functions are $\omega_8 = (p_0\bar{p}_0 + n_0\bar{n}_0 - 2\Lambda_0\bar{\Lambda}_0)/\sqrt{6}$ and $\omega_1 = (p_0\bar{p}_0 + n_0\bar{n}_0 + \Lambda_0\bar{\Lambda}_0)/\sqrt{3}$, as represented graphically in Figure 6.13, p. 120. Since SU(3) symmetry is broken by making the Λ_0 heavier than the p_0 and n_0, while leaving the latter two indistinguishable, ω_8 and ω_1 mix to separate $\Lambda_0\bar{\Lambda}_0$ from $p_0\bar{p}_0 + n_0\bar{n}_0$, creating $\omega = (p_0\bar{p}_0 + n_0\bar{n}_0)/\sqrt{2}$ and $\phi = \Lambda_0\bar{\Lambda}_0$.

representation of SU(3), a sum of two irreducible representations,

$$\text{SU(3)}: \ 18 = 8 \oplus 10,$$

where each of the two SU(3) irreducible representations are sums of four irreducible representations of SU(2),

$$\text{SU(2)}: \ 8 \ = \ 2 \oplus 1 \oplus 3 \oplus 2,$$
$$10 \ = \ 4 \oplus 3 \oplus 2 \oplus 1.$$

The four isospin multiplets comprising the octet are:

8 spin-1/2 baryons

$= (2 \text{ isospin-1/2 } N\text{s}) + (1 \text{ isospin-0 } \Lambda)$

$+ (3 \text{ isospin-1 } \Sigma\text{s}) + (2 \text{ isospin-1/2 } \Xi\text{s}).$

Correspondingly, the decuplet's four isospin multiplets are:

10 spin-3/2 baryons

$= (4 \text{ isospin-3/2 } \Delta\text{s}) + (3 \text{ isospin-1 } \Sigma^*\text{s})$

$+ (2 \text{ isospin-1/2 } \Xi^*\text{s}) + (1 \text{ isospin-0 } \Omega).$

With the exception of the Ω, all the baryons in the 56-dimensional irreducible representation of SU(6) had been observed by 1963.

The eight spin-1/2 baryon wave functions are represented graphically in Figure 6.15, p. 122.

Note that if baryons were created from three aces without spin, $3 \times 3 \times 3 = 27$ states would have been formed, consisting of four irreducible representations of SU(3),

$$\text{SU(3)}: \ 3 \otimes 3 \otimes 3 = 10 \oplus 8 \oplus 8 \oplus 1,$$

and we would not have had any principle to select only one 8 and the 10. Ace-spin symmetry was necessary.

The hierarchies of meson and baryon symmetries and their representations outlined in this section are detailed in Appendix 6.A, p. 116. Every detail mattered when deciding whether to believe that aces are real particles inside of hadrons. Guidance for making this decision is given in Appendix 6.B: "Evaluating Theories," p. 118.

The Origin of Symmetry Breaking

Aces were responsible for the SU(2), SU(3), and SU(6) symmetries of hadrons. The SU(2) and SU(3) symmetries for aces were broken by ace mass differences,

which in turn created the symmetry breaking observed for hadrons, as stated in the abstract of the first CERN Report [220]:

> The breaking of this symmetry [SU(3) and SU(2)] is assumed to be universal, being due to mass differences among the aces.

This idea is evidently in concert with current thinking. Steven Weinberg writes [209]: "The only things [in QCD] that can violate the proton-neutron symmetry are the masses of the quarks."

SU(6) symmetry breaking, the largest of the three, had a dynamical origin: The interaction energy between aces depended on their relative spin orientation.

Exotic Hadrons

So far only hadrons created from deuces and treys have been considered. More "exotic" hadrons were predicted in the second CERN Report:

> We would expect that baryons are built not only from the product of three aces, AAA, but also from $\bar{A}AAAA$, $\bar{A}\bar{A}AAAAA$, etc., where $\bar{A}$ denotes an anti-ace. Similarly, mesons could be formed from $\bar{A}A$, $\bar{A}\bar{A}AA$, etc. For the low mass mesons and baryons we will assume the simplest possibilities, $\bar{A}A$ and AAA, that is, "deuces and treys."

Note that particles with fractional charge were not predicted to exist, and haven't been found.

Hadrons with high spin could be created by giving aces angular momentum, or by introducing additional deuces or treys. The former possibility was discussed in the second CERN Report, the latter in the 1964 Erice Summer School Lectures where exotic mesons were created from aces with spin, i.e., $\mathbb{A}\mathbb{A}\bar{\mathbb{A}}\bar{\mathbb{A}}$, now called "tetraquarks." The $6 \times 6 \times \bar{6} \times \bar{6} = 1296$ dimensional reducible representation of SU(6) was decomposed into its irreducible representations of SU(6), and then SU(3) ([223], p. 227).

6.6   Zweig's Rule for Masses

Zweig's rule for couplings did not predict the suppression of ϕ decay (p. 25); it was an after-the-fact explanation that required postulating the existence of hadron constituents with particular quantum numbers, together with the assignment of those constituents to the vector and pseudoscalar mesons. With those inputs, could the observed relationships between meson masses be predicted? Would there be more outputs than inputs? What about baryon masses, and the relationship between meson and baryon masses?

The mass of a deuce or trey was assumed to be the sum of their ace masses plus their *pairwise* ace interaction energies [220]:

Mass(deuce or trey)

$= \Sigma$ ace masses $+ \Sigma$ pairwise ace interaction energies.

Specifically, the mass of a deuce $A_i A^j$, where A^j is the antiace of A_j, i.e. $A^j = \bar{A}_j$, was given by

$$M_{A_i A^j} = M_{A_i} + M_{\bar{A}_j} + E_{A_i \bar{A}_j}, \qquad (6.2)$$

where M_{A_i} is the mass of A_i and $E_{A_i \bar{A}_j}$ is the interaction energy between A_i and $\bar{A}_j$.[14]

The mass of a trey $A_i A_j A_k$ was given by

$$M_{A_i A_j A_k} = M_{A_i} + M_{A_j} + M_{A_k} + E_{A_i A_j *} + E_{A_i * A_k} + E_{* A_j A_k}, \qquad (6.3)$$

where $E_{A_i A_j *}$ is the interaction energy between aces A_i and A_j with ace $* = A_k$ a spectator.

The SU(2) and SU(3) symmetries were broken by making n_0 slightly heavier than p_0, and Λ_0 significantly heavier than n_0,

$$M_{p_0} \lesssim M_{n_0} \ll M_{\Lambda_0}.$$

Empirically, interaction-energy differences were less than their corresponding mass differences, and the interaction energy between an ace and antiace or two aces differed if the two spins pointed in the same or opposite direction.

---

[14] In 1963 the sign of the interaction energy was not known. We now know $E_{A_i \bar{A}_j}$ is positive, representing the energy of gluons and quark-antiquark pairs that surround A_i and $\bar{A}_j$, and that the first two ace masses are very light. Since only hadron mass differences were predicted, the sign of the interaction energy, and the magnitude of ace masses, did not affect mass relations, but might have affected the sign of their errors.

Note that hadron masses were never predicted, only mass inequalities or linear mass relations that could be written in terms of ace mass differences.[15]

Meson Mass Relations

In analogy to the nucleon $\mathcal{N} = (p, n)$, define $\mathcal{N}_0 = (p_0, n_0)$ when the mass difference between p_0 and n_0 can be ignored.

Mass Inequalities Following From $M_{\mathcal{N}_0} \ll M_{\Lambda_0}$

The assignment of deuces to mesons and the mass inequalities of aces, implied meson mass inequalities:

For the meson octet:

$$M_\pi \ll M_K \ll M_\eta$$
$$137 \qquad 496 \qquad 550.$$

For the meson nonet:

$$M_\rho \approx M_\omega$$
$$750 \qquad 784,$$
$$M_\rho \ll M_{K*} \ll M_\phi$$
$$750 \qquad 888 \qquad 1018,$$

where masses, unless explicitly dated, were those known in 1963. These inequalities simply reflected the mass differences between strange and non-strange aces.

Since ace-spin symmetry predicted that there would be nine vector mesons in two irreducible SU(3) representation, the single Gell-Mann–Okubo mass formula for the octet of pseudoscalar mesons in one irreducible representation (Eq. 5.2. p. 69) was replaced by two vector meson mass relations:

$$M_\omega \approx M_{\rho^0}$$
$$784 \qquad 750$$
$$782.65 \pm 0.12 \quad 775.26 \pm 0.25 \text{ in } 2023,$$

$$M_{K*0} \approx \frac{M_{\rho^0} + M_\phi}{2}$$
$$888 \qquad 884$$
$$895.55 \pm 0.2 \quad 897.36 \pm 0.13 \text{ in } 2023, \qquad (6.4)$$

---

[15] For pseudoscalar mesons linear relations in the squares of masses were used ("The Gell-Mann–Okubo Mass Formulae," p. 69).

where the 2023 values are from ref. [168]. Thus the constituent assignment for vector mesons based on the suppression of ϕ decay, and the equation for a deuce mass given in Eq. 6.2, led to the prediction of two successful, if imperfect, mass relations.

Averaging Interaction Energies for Meson Mass Relations

When physicists worked with symmetry groups, assumptions were made about the nature of symmetry breaking. That's how the Gell-Mann–Okubo mass formula was obtained for mesons and baryons in irreducible representations (Eq. 5.2, p. 69). Those relations were recovered with aces by making assumptions about their interaction energies. For example, the Gell-Mann–Okubo mass formula for pseudoscalar mesons (Eq. 5.2) followed from Eq. 6.2 if an averaging relation held for interaction energies,

$$E_{\Lambda_0 \mathcal{N}_0} = \frac{E_{\mathcal{N}_0 \bar{\mathcal{N}}_0} + E_{\Lambda_0 \bar{\Lambda}_0}}{2}, \tag{6.5}$$

where $\mathcal{N}_0$ is either p_0 or n_0. The origin of this empirical relation is still puzzling. It must follow directly from QCD.

A Better Vector Meson Mass Relation

The differences between the left and right-hand side of the two equations given above in Eqs. 6.4 were related in the first CERN Report:

$$\frac{M_\omega - M_{\rho^0}}{4} = \frac{M_\phi + M_{\rho^0}}{2} - M_{K^{*0}}$$

$$8.5 \pm ? \qquad - \quad 4 \pm ? \text{ in 1963}$$

$$1.85 \pm 0.07 \qquad 1.81 \pm 0.24 \text{ in 2023.}$$

$$\text{(2.8 revisited)}$$

While this relation was not encouraging in 1963, it constituted a prediction that is now borne out, and provides evidence that for vector mesons, mass relations are linear in mass. This is evident when Eq. 2.8 is rewritten as an equation for the ω mass in terms of the three other masses. It yields $M_\omega = 782.50$ compared to its 2023 measured mass $M_\omega = 782.66 \pm 0.13$ [168]. Using squares of masses gives $M_\omega = 820.76$, so a quadratic mass formula doesn't work nearly as well.

Baryon Mass Relations

Mass Inequalities Following From $M_{\mathcal{N}_0} \ll M_{\Lambda_0}$

The less strangeness a baryon had in an irreducible SU(3) representation, the heavier it would be, because Λ_0 had less strangeness than p_0 or n_0, while being heavier. Therefore swapping a Λ_0 for a p_0 or n_0 in a baryon was predicted to decrease the baryon's strangeness while increasing its mass, as observed in both the baryon octet and decuplet:

For the baryon octet:

$$M_{\mathcal{N}} \ll M_\Lambda \approx M_\Sigma \ll M_\Xi$$
$$939 \qquad 1115 \qquad 1193 \qquad 1315.$$

The large Σ-Λ mass splitting in the octet, where both baryons had the same strangeness, indicated that interaction energy in the baryon octet, while relatively small, could not be ignored.

For the baryon decuplet:

$$M_\Delta \ll M_{\Sigma^*} \ll M_{\Xi^*} \ll M_\Omega$$
$$1237 \qquad 1382 \qquad 1533 \qquad ? \; ,$$

where the Ω mass was not known in 1963.

Mass Inequalities Related to $M_{p_0} \lesssim M_{n_0}$

The less charge a baryon had in an irreducible SU(2) representation (isospin multiplet), the heavier it would be, because n_0 had less charge than p_0 while being heavier. Therefore swapping an n_0 for a p_0 in a baryon was predicted to decrease the baryon's charge while increasing its mass, as observed in the octet:

$$M_p \lesssim M_n$$
$$938.2 \qquad 939.5,$$
$$M_{\Sigma^+} \lesssim M_{\Sigma^0} \lesssim M_{\Sigma^-}$$
$$1189 \qquad 1193 \qquad 1197,$$
$$M_{\Xi^0} \lesssim M_{\Xi^-}$$
$$1311 \qquad 1318.$$

Similar predictions were made for the baryon decuplet electromagnetic mass splittings, but they were unknown in 1963. They are still poorly determined, but where available, behave as predicted.

Averaging Interaction Energies for Baryon-Octet Mass Relations

The Gell-Mann–Okubo mass formula for the baryon octet (Eq. 5.1, p. 69) followed from Eq. 6.3, p. 99 with averaging relations similar to Eq. 6.5, p. 100 for mesons. For example, the average of the interaction energy between two Λ_0s in a trey and two non-strange aces was assumed to be approximately equal to the interaction energy between a strange and non-strange ace,

$$\frac{E_{\Lambda_0 \Lambda_0 *} + E_{\mathcal{N}_0 \mathcal{N}_0 *}}{2} \approx E_{\Lambda_0 \mathcal{N}_0 *} \approx E_{\mathcal{N}_0 \Lambda_0 *}.$$

Similar relations were assumed for treys with spectators in the other two positions. Then the Gell-Mann–Okubo mass formula for the baryon octet followed.

A novel relation between the masses of the Σ charge states was found from the averaging relation restricted to p_0 and n_0,

$$\frac{E_{p_0 p_0 *} + E_{n_0 n_0 *}}{2} \approx E_{p_0 n_0 *} \approx E_{n_0 p_0 *}.$$

Then,

$$\frac{M_{\Sigma^+} + M_{\Sigma^-}}{2} \approx M_{\Sigma^0}$$

$$1193.4 \pm 0.3 \qquad 1193.2 \pm 0.7 \text{ in 1963}$$
$$1193.41 \pm 0.04 \qquad 1192.642 \pm 0.038 \text{ in 2023.}$$

A Better Baryon Mass Relation

Other interaction-energy dependent mass relations were predicted in the second CERN Report. However, the most compelling baryon mass formula, given in the first CERN Report, only assumed two-body interactions between aces in baryons. It related mass differences in three of the four isospin multiplets of the baryon octet (the fourth multiplet has only one member). The starting point was the mass formula for treys (Eq. 6.3, p. 99), and the representation of the baryon octet in terms of treys (Eq. 6.15, p. 121). No assumptions about interaction energies were made; they all canceled.

As is easily seen in the graphical representation, of the eight baryons [Figure 6.15, p. 122], six have similar triangular structure, and the mass differences of their corresponding baryons were the ones related:

$$(M_{\Xi^-} - M_{\Xi^0}) = (M_{\Sigma^-} - M_{\Sigma^+}) - (M_n - M_p)$$

$$5.6 \pm 1.4 \qquad 7.0 \pm 0.5 \text{ in 1963}$$
$$6.85 \pm 0.21 \qquad 6.79 \pm 0.08 \text{ in 2023.}$$

$$\text{(2.9 revisited)}$$

Agreement has significantly improved with time, and the equation is still satisfied within errors 60 years later.

Sidney Coleman and Sheldon Glashow had also obtained this relation, *but assumed SU(3) symmetry was exact* in order to get it [39]. As expected, the magnetic moment relations they obtained with the same assumptions were qualitatively, but not quantitatively, correct. There is no reason to believe that the mass relation should be any more accurate. Coleman and Glashow got the right answer for the wrong reason!

Relating Meson and Baryon Masses

Since aces were constituents of both mesons and baryons, the properties of these hadrons could be related. For example, ignoring interaction energy differences compared to ace mass differences, as previously assumed, $M_\Lambda \gg M_n$ implied $M_{\Lambda_0} \gg M_{n_0}$, which correctly predicted

$$M_K \gg M_\pi,$$
$$M_{K^*} \gg M_\rho.$$

This followed from Eq. 6.2, p. 99 for deuce masses, and the assignment of deuces to mesons given in Figures 6.12 and 6.13, pp. 119 and 120.

Similarly $M_n \gtrsim M_p$ implied $M_{n_0} \gtrsim M_{p_0}$, which correctly predicted

$$M_{K^0} \gtrsim M_{K^+},$$
$$M_{K^{*0}} \gtrsim M_{K^{*+}}.$$

6.7 Zweig's Rule for Couplings

Just as Faraday saw his idiosyncratic lines of force propelling particles of charge along their length (Maxwell's quote in Section 6.14, p. 107), I pictured hadron couplings and their decays with Zweig diagrams and, alternatively, as "Tinkertoy" constructions. Both types of visualization are discussed in "Couplings Computed Graphically," p. 121.

Zweig Diagrams

As shown in Figure 2.6, p. 25, and exemplified for ϕ decay in Figure 6.4, I conjectured that when hadrons decayed through the strong interactions their constituents flowed into their decay products. The ϕ was expected to decay primarily into $\rho + \pi$, but this decay mode was not observed because ϕ's constituents were not present in ρ or π. Had the K meson been slightly heavier, the decay $\phi \rightarrow K + \bar{K}$ would have been forbidden by the conservation of energy, and ϕ decay would have been even more strongly suppressed.

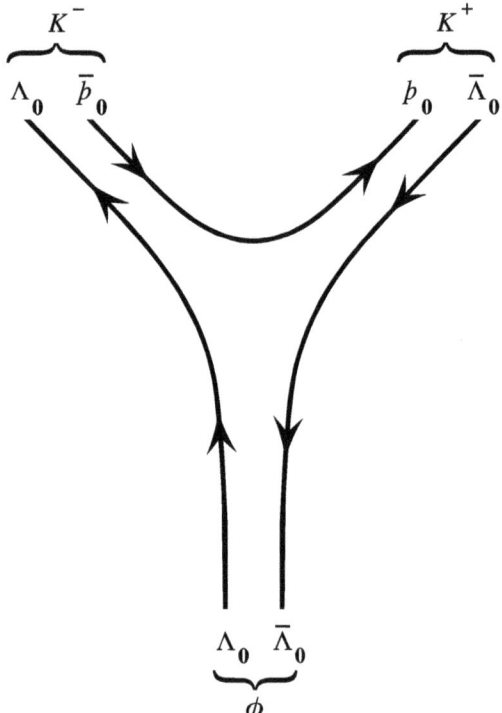

Figure 6.4. The Zweig diagram for ϕ decaying into $K^- + K^+$. As in Feynman diagrams, time runs upward in the graph. Antiparticles are particles running backward in time. Since aces are represented by lines pointing up, antiaces point down. As the Λ_0 and $\bar{\Lambda}_0$ in the ϕ separate a virtual $\bar{p}_0 p_0$ materializes from the vacuum, separates, and binds with the Λ_0 and $\bar{\Lambda}_0$ to create the final K^- and K^+. Although allowed, this decay has little phase space, giving the ϕ a narrow width, as previously seen in Figure 6.2, p. 93. (Sampson Wilcox, Research Laboratory of Electronics at MIT.)

Since particles and antiparticles annihilate, you might have expected the Λ_0 and $\bar{\Lambda}_0$ in the ϕ to do so. Instead they separate and combine with a virtual $\bar{p}_0 p_0$ or $\bar{n}_0 n_0$ pair, creating the final decay products. Aces in decaying hadrons were conserved — they could not "eat each other."[16]

Zweig's rule specified not only what was forbidden by the strong interactions, but also how strong those interactions would be when they were allowed. In Zweig diagrams mesons were represented by linear combinations of pairs of lines pointing in opposite directions (deuces), baryons by linear combinations of triplets of lines pointing in the same direction (treys). Each line had to begin and end in an external particle. The coupling constant for a set of mesons was the weighted sum of all possible diagrams where all lines were connected, the weight for each diagram equal to the product of the normalization constants of the deuces in the mesons being coupled [Figure 6.17, p. 123]. The square of the coupling constant was proportional to the probability of interaction or decay.

A different, but mathematically equivalent, graphical computation of hadron couplings was given using a Tinkertoy construction as exemplified in Figures 6.17 and 6.18, pp. 123 and 124 taken from the second CERN Report. The childhood origin of this calculus is related in Section 6.14: "Thinking Graphically," p. 107.

6.8   Weak-Interaction Predictions

In addition to being the constituents of hadrons, aces were also quantum fields used to construct electromagnetic and weak-interaction currents (second CERN Report). In this role they were identical to what Gell-Mann would later call "current quarks" (p. 71) from which he constructed current commutation relations. Weak-interaction currents formed from aces were used to predict hadron decay rates and selection rules. Like the mass and coupling relations, Gell-Mann did not predict these because he didn't construct hadrons from quarks.

---

[16] The suppression of ϕ decay, Zweig's rule, was later called the OZI rule [127, 166]. Unlike Zweig's rule, the OZI rule didn't specify the relative strengths of reactions that are allowed ("Couplings Computed Graphically," p. 121).

The $\Delta\mathbb{S} = \Delta Q$ Rule

One of the most striking predictions of the CERN Reports was the absence of a particular class of β-decays. If hadrons are created from elementary constituents, then all hadronic electromagnetic and weak interactions must result from the electromagnetic and weak interactions of their constituents. The first CERN Report states that when a neutron ($n \sim n_0 n_0 p_0$) β-decayed,

$$n \to p + e^- + \bar{\nu}_e,$$

it really was one of its two n_0 aces that decayed,

$$n_0 \to p_0 + e^- + \bar{\nu}_e.$$

Similarly, the strangeness-changing β-decay of the lambda ($\Lambda \sim p_0 n_0 \Lambda_0$) resulted from the decay of its Λ_0,

$$\begin{aligned} \Lambda_0 &\to p_0 + e^- + \bar{\nu}_e, \text{ or} \\ &\to p_0 + \mu^- + \bar{\nu}_\mu. \end{aligned} \quad (6.6)$$

The n_0 and Λ_0, not the n and Λ, were governed by the V–A theory of the weak interactions. More generally, all hadron weak interactions were expected to result from the weak interaction of their ace constituents.

Naively one expected that both Σ^+ and Σ^- would undergo strangeness changing β-decays at similar rates,

$$\frac{\text{Rate of}\{\Sigma^+ \to n + e^+ + \nu_e\}}{\text{Rate of}\{\Sigma^- \to n + e^- + \bar{\nu}_e\}} = \mathrm{O}(1),$$

where $\mathrm{O}(1)$ stands for a number "of order 1." All quantum numbers, like charge and baryon number, that needed to be conserved, are conserved in both reactions, and the phase space available to the decaying particles is identical..

However, expressing the baryons in terms of aces, it becomes obvious that the $\Sigma^+ \sim p_0 p_0 \Lambda_0$ cannot β-decay into $n + e^+ + \nu_e$ because the Λ_0 must decay, and an e^+ cannot be a β-decay product of a Λ_0. Therefore,

$$\text{Rate of}\{p_0 p_0 \Lambda_0 \to n_0 n_0 p_0 + e^+ + \nu_e\} = 0.$$

More generally, in the β-decay of Λ_0, the strangeness $\mathbb{S}$ of the ace increases by 1, as does its charge Q (Eq. 6.6). Therefore, if the strangeness of a hadron changes in β-decay, $\Delta\mathbb{S} = \Delta Q$ for that reaction. Since

$\Delta\mathbb{S} = -\Delta Q$ in Σ^+ β-decay, this reaction is forbidden, as expected from the first CERN Report.[17]

Beta-decays of the Σ^+ to neutrons had not been seen in 1963. No Σ^+ β-decays have been reported since that time, while Σ^- β-decay proceeds at the expected rate (Table 3 of the second CERN Report). The measured limit on the branching ratio is

$$\frac{\text{Rate of}\{\Sigma^+ \to n + e^+ + \nu_e\}}{\text{Rate of}\{\Sigma^- \to n + e^- + \bar{\nu}_e\}} < \mathrm{O}(0.005) \text{ in 2023,}$$

so the Σ^+ β-decay is suppressed by more than a factor of 200. The assumption that hadrons should be thought of in terms of aces had predictive power.

The $\Delta\mathbb{S} = \pm 1, 0$ Rule

If hadrons decay weakly to other hadrons through the decay of their constituents, then strangeness cannot change by more that one unit in any decay. So both

$$\begin{aligned} \Xi^- &\to \Lambda + \pi^-, \text{and} \\ \Xi^- &\to \Lambda + e^- + \bar{\nu}_e, \end{aligned}$$

are allowed, but both

$$\begin{aligned} \Xi^- &\nrightarrow n + \pi^-, \text{and} \\ \Xi^- &\nrightarrow n + e^- + \bar{\nu}_e, \end{aligned}$$

are forbidden, as noted by Nishijima (Eq. 5.6, p. 74), but not fundamentally understood until the introduction of aces. $\Delta\mathbb{S} = \pm 2$ decays remain undetected.

6.9 How to Look for Aces

A method for "seeing" aces was suggested in the second CERN Report

Since aces lived inside of protons and neutrons, that was the place to look (Section XI in [221]):

> high-momentum-transfer experiments may be necessary to detect aces.

The electron would be the best projectile. Being a point particle without strong interactions, it had no structure that could be confused with that of the nucleon, and its lifetime, unlike the muon's, was infinite. Light in

---

[17] The branching ratio of Σ^- β-decay is give in Table 3 of the second CERN Report, where the absence of Σ^+ β-decay from the Table indicates that it was predicted not to occur.

the form of photons emitted from the electron's charge would interact with charges in the nucleon, "illuminating" it, allowing one to see the fractional charges within, if the momentum p transferred by the photon was large enough. The corresponding wavelength λ of light would then be small enough ($\lambda = h/p$) to create a high-resolution "image" of the nucleon's internal structure. Such experiments were eventually performed using electrons and neutrinos as point projectiles. ("Seeing Is Believing," p. 128).

These high-momentum-transfer reactions would be violent, knocking constituents out of the nucleon, together with their "hangers on," in a process that is now called "deep inelastic scattering."

6.10   A Fortuitous Accident

Where did I get the idea of dividing a baryon into three pieces?

In the Sakata model baryons were predicted to come in SU(3) representations containing 15 and 3 members. Those representations were not observed. Instead, representations with 8 and probably 10 members were found. In October 1963, after attending the "Sienna International Conference on Elementary Particles," and settling down at CERN, I returned to this vexing problem. As described in 1980 [226]: "One day while I was looking at a review paper by E. Behrends,

J. Dreitlein, C. Fronsdal and B. Lee [13] I came across their multiplication table for SU(3) representations [reproduced here as Figure 6.5, with antiparticle representations denoted by an asterisk " * " rather than a bar " ⁻ " above them]:

> Almost reflexively I started to work out the irreducible representations used to classify the baryons in the Sakata model (with its sakatons $\mathcal{S}$). To decompose $3 \otimes 3 \otimes 3^*$ [$\mathcal{S} \otimes \mathcal{S} \otimes \bar{\mathcal{S}}$], I took $D^3(1,0)$ [$\mathcal{S}$] in the second row and multiplied it by $D^3(1,0)$ [$\mathcal{S}$] in the first column to get $6+3^*$. Then taking $D^6(2,0)$ [the 6 of $6+3^*$] in the fourth row and multiplying it by $D^3(1,0)$ [$\mathcal{S}$] in the first column I got $10+8$, which I immediately recognized as the wrong answer. The product $6 \otimes 3$ had been formed instead of $6 \otimes 3^*$. There was no $3^* = D^3(0,1)$ [$\bar{\mathcal{S}} = D^3(0,1)$] column. ... The lack of symmetry and confusing notation of the Table had misled me into multiplying $((3 \otimes 3)^6 \otimes 3)$ [$(\mathcal{S} \otimes \mathcal{S})^6 \otimes \mathcal{S}$] instead of $(3 \otimes 3)^6 \otimes \bar{3}$ [$(\mathcal{S} \otimes \mathcal{S})^6 \otimes \bar{\mathcal{S}}$] of the Sakata model.

Although the 10 and 8 representations were not the correct ones in the Sakata model, they were almost certainly empirically correct. Therefore baryons had to be created from three aces using $(\mathcal{A} \otimes \mathcal{A})^6 \otimes \mathcal{A}$. This was a Eureka moment. The elementary particles from which hadrons are created had been discovered.[18] It

Complete designation	Abbr. design	Highest weight	Fig. no.	Isotopic content	Basic	$\otimes D^3(1,0)$	$\otimes D^6(2,0)$	$\otimes D^8(1,1)$	$\otimes D^{10}(3,0)$
$D^1(0,0)$	1	$(0,0)$		0	ψ	3	6	8	10
$D^3(1,0)$	3	$\frac{1}{6}(\sqrt{3},1)$	2(a)	$0,\frac{1}{2}$	ψ_a	$6+3^*$	$10+8$	$15+6^*+3$	$15'+15$
$D^3(0,1)$	3*	$\frac{1}{6}(\sqrt{3},-1)$	2(b)	$0,\frac{1}{2}$	ψ^a	$8+1$	$15+3$	15^*+6+3^*	$24+6$
$D^6(2,0)$	6	$\frac{1}{3}(\sqrt{3},1)$	2(c)	$0,\frac{1}{2},1$	ψ_{ab}	$10+8$	$15'+15+6^*$	$24+15^*+6+3^*$	$24+21+15^*$
$D^6(0,2)$	6*	$\frac{1}{3}(\sqrt{3},-1)$	2(d)	$0,\frac{1}{2},1$	ψ^{ab}	15^*+3^*	$27+8+1$	24^*+15+6^*+3	$42+15+3$
$D^8(1,1)$	8	$\frac{1}{3}(\sqrt{3},0)$	2(e)	$0,\frac{1}{2},\frac{1}{2},1$	$\psi_a{}^b, \chi_A$	$15+6^*+3$	$24+15^*+6+3^*$	$27+10+10^*+8+8+1$	$35+27+10+8$
$D^{10}(3,0)$	10	$\frac{1}{2}(\sqrt{3},1)$	22	$0,\frac{1}{2},1,\frac{3}{2}$	ψ_{abc}	$15'+15$	$24+21+15^*$	$35+27+10+8$	$35+28+27+10$
$D^{10}(0,3)$	10*	$\frac{1}{2}(\sqrt{3},-1)$		$0,\frac{1}{2},1,\frac{3}{2}$	ψ^{abc}	24^*+6^*	$42^*+15^*+3^*$	$35^*+27+10^*+8$	$64+27+8+1$

Figure 6.5. Using this table from [13] to find the baryon representations in the Sakata Model, I noticed that the table contained the entry 10+8, precisely the numbers I wanted. That entry corresponded to $D^6(2,0) \otimes D^3(1,0) = (3 \otimes 3)^6 \otimes 3$ [$(\mathcal{S} \otimes \mathcal{S})^6 \otimes \mathcal{S}$]. Here the 10+8 in the table has been italicized and boxed.

---

[18] Work on aces for the CERN Reports [220, 221] was almost finished by Thanksgiving 1963 when Ricardo Gomez, a Caltech Research Fellow I had worked with on the K-decay experiment at Berkeley (p. 22), came to visit me at CERN. When I told him what I was doing, he quietly smiled and said I was "crazy." He wanted to go to the Bataclan, the one and only strip club in town (Calvin's Geneva). Much to my surprise, we skipped a very long line and were seated at a small circular table up front with a free bottle of champagne. The manager thought Ricardo was "the" *Ricardo Gomez*, the famous bicycle racer, and Ricardo did nothing to disabuse him of this idea.

would take the physics community almost a decade to recognize this, and find the corresponding equations. Learning how to solve them is an ongoing process.

6.11   Difficulties

Ferreting out aces from data was not as straightforward as it might seem. Although aces resolved the ϕ-decay anomaly, and made many successful predictions, some predictions contradicted experiment, and even violated basic theoretical principles ("Were aces real?," p. 106).

Problems With Experiment

Accurately measuring hadron properties was difficult. Not knowing the errors on measurements was an even more serious problem. Without errors measurements were more than useless — they could be misleading. The Particle Data Group, which has been publishing measurements on an annual basis since 1964, has examined how its values have changed with time.[19] Their results for the neutron lifetime, and a ratio of two constants appearing in β-decay, are shown in Figure 6.6. The unreliability of measurements, as exemplified in this figure, was not fully appreciated in 1963. Using measurements as the starting point for theory was a dangerous business.

Problems With the Hadron Spectrum

Matt Roos's 1963 "definitive" compilation of particles, published in the prestigious *Reviews of Modern Physics,* listed 26 meson resonances [179]. *We now know that 19 of these resonances do not exist, and of those 19, seven are "exotic," i.e., cannot be created from $A\bar{A}$ pairs.* My training in both experimental and theoretical physics was helpful in separating "wheat from chaff." Others did not have the training, interest, or patience.

Why Were So Many Experiments Wrong?

When a new technology is introduced it takes time before it is understood and procedures for its use are standardized. The bubble chamber catalyzed an explosion in the number of hadrons discovered in the early 1960s, and created many false alarms. It was invented

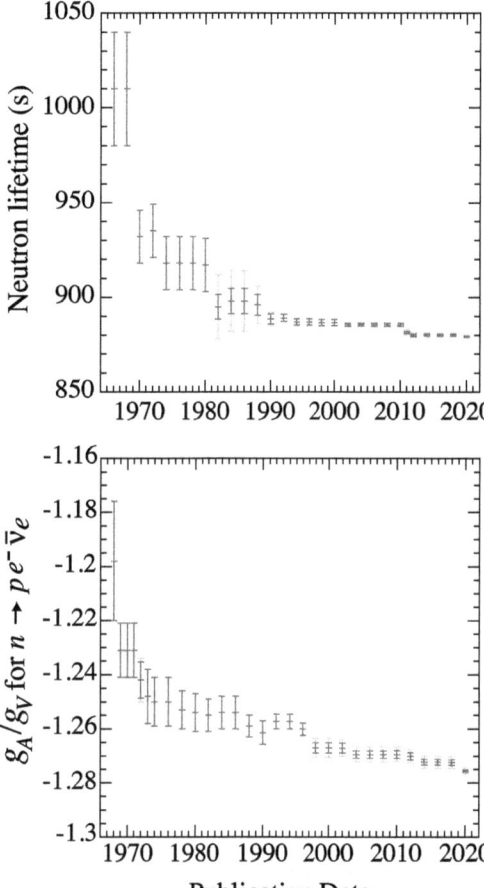

Figure 6.6. Upper graph: neutron lifetime as a function of the year it was published. When there are two pairs of error bars for a point, the larger one includes an estimate of nonstatistical errors. Lower graph: the time variation of g_A/g_V, the ratio of the two constants in the V–A theory of β-decay (p. 43). In the β-decay of *elementary* particles like the muon, or the ace n_0, this ratio is -1. Values tend to move monotonically in time with the errors on adjacent points overlapping, or nearly overlapping, as if the values reported were biased by their previously measured values. (Two "History plots," courtesy of the Particle Data Group.)

by Donald Glaser who had been Carl Anderson's graduate student at Caltech. Anderson had used cloud chambers in the 1930s to discover the positron and muon. Charged particles traveling through supersaturated vapor in a cloud chamber created condensation nuclei that formed a thin line of fine droplets which

---

[19] See History Plots Fig.1 at pdg.lbl.gov/2024/figures/figures.html.

traced the path of every charged particle. In a bubble chamber [Figure 6.7] the tracks were made by a trail of bubbles in a superheated liquid whose pressure was abruptly dropped. From these tracks events could be reconstructed, sometimes signaling discovery.

Figure 6.7. A retired bubble chamber resurrected as a sculpture at Fermilab. During operation it was filled with superheated liquid hydrogen. As charged particles traveled through it they created bubbles when the pressure in the chamber was suddenly dropped by increasing its volume with a huge piston (centered below the chamber in the picture). Those bubbles traced the paths of particles. The chamber shown is 15' in diameter and was pulsed about three times a second. (Public domain via Wikimedia Commons.)

Problems arose because statistical fluctuations, or rare "background events," could mimic the signatures of new particles. In a rush towards the limelight some physicists published bumps in mass distributions whenever they saw one, hoping that a new particle had been discovered. Physicists who had made the same mass plots in other experiments and who hadn't seen bumps didn't publish, leaving the published literature with biased information.

6.12 Were Aces Real?

Aces behaved like real particles. They had mass, spin, and *dynamics*. When hadrons interacted their aces flowed from hadron to hadron, while ace-antiace pairs were created from the vacuum. Aces within hadrons were bound by springs, sometimes vibrating or rotating together. In mesons, ace-antiace pairs rotated with orbital angular momentum $\vec{L}$ that added to their combined spins $\vec{S}$ to give the resulting meson a spin of $\vec{J} = \vec{L} + \vec{S}$. The mass of mesons created this way depended on the product $\vec{L} \cdot \vec{S}$ [221]. How "mathematical" were these interactions?

Finally, the electromagnetic and weak interactions of hadrons were successfully attributed to the electromagnetic and weak interactions of their ace constituents. It was hard to imagine that aces weren't real. I applied the "duck" test: If it walks like a duck, swims like a duck, and quacks like a duck, it's probably a duck.

Arguing against their reality was the existence of the first resonance discovered in 1953, the spin-3/2 Δ^{++} (Figure 2.7 and Eq. 2.11, pp. 32 and 33):

$$\Delta^{++} = p_0 p_0 p_0.$$

Here *three identical aces occupied the same quantum state, in violation of the spin-statistics theorem* (p. 92). The Δ^{++} wave function was symmetric under the exchange of any two of its aces, not antisymmetric!

How can the apparent violation of the spin-statistics theorem for aces be explained? If aces were simply mathematical constructs used to create toy free field theories — Gell-Mann's quarks — there was no problem. If aces were real particles – as Zweig's aces were — the problem remained.

Although a serious problem, it was not a show stopper. Rutherford's nucleus, with all its positive charge concentrated at its center, violated the laws of classical physics, as did Bohr's theory of the atom with its quantized electron orbits. These problems were eventually solved with the introduction of the nuclear force and the creation of quantum mechanics. I thought a solution to the spin-statistics problem also would be found.

6.13 Conclusion of the CERN Reports

In winding up, the second CERN Report asks:

> Are aces particles? If so, what are their interactions? Do aces bind to form only deuces and treys? What is the particle (or particles) that is responsible for binding the aces? Why must one work with masses for the baryons and mass squares for the mesons? And more generally, *why does so simple a model yield such a good approximation to nature* [my italics]?

Now we know that aces exist as real particles, and understand the laws of their interaction, but how aces combine to form the *low-mass* hadrons with their remarkable regularities remains somewhat of a mystery. The mass of individual hadrons can be computed, but why the observed *relationships* between these masses exist is still not clear (Eqs. 2.8 and 2.9).

Empirically, mass squares for the pseudoscalar mesons must still be used.[20] While almost all vector meson mass relations still work with mass squares, linear relations work as well, and the best relation must now be linear (p. 100). The spin-1/2 and spin-3/2 baryon mass relations still stand as linear.

The last most interesting question, "why does so simple a model yield such a good approximation to nature," remains largely unanswered.

In Summary What was the reader of the CERN Reports to think in 1964? After dutifully listing various possibilities, the last line of the second CERN Report lists the one the author wanted to be true:

> There is also the outside chance that the model is a closer approximation to nature than we may think, and that fractionally charged aces abound within us.

Indeed they do.

6.14 Thinking Graphically

Quarks' humble beginnings were rooted in a tradition of thinking graphically. Graphical thinking has a long history in physics. Most familiar is Faraday's use of graphical representations of electric and magnetic forces. He thought of electromagnetic fields as permeating all of space with "lines of force" that accelerated charge and rotated magnets. Lines of force replaced "action at a distance," a concept abhorrent to Newton, even though his theory of gravity was based on it ("The Problem with Gravity," p. 11).

Faraday's views were described and interpreted mathematically by James Clerk Maxwell in his Preface to Volume 1 of "A Treatise on Electricity and Magnetism," published in 1873. Maxwell compared Faraday's idiosyncratic graphical representation of the electromagnetic field with the mainstream mathematical works of Laplace, Lagrange, Poisson, and Gauss. Maxwell writes:

> Before I began the study of electricity I resolved to read no mathematics on the subject till I had first read through Faraday's *Experimental Researches on Electricity.* I was aware that there was supposed to be a difference between Faraday's way of conceiving phenomena and that of the mathematicians ... As I proceeded with the study of Faraday, I perceived that his method of conceiving the phenomena was also a mathematical one, though not exhibited in the conventional form of mathematical symbols. I also found that these methods were capable of being expressed in the ordinary mathematical forms, and these compared with those of the professed mathematicians.
>
> For instance, Faraday, in his mind's eye, saw lines of force traversing all space where the mathematicians saw centres of force attracting at a distance: Faraday saw a medium where they saw nothing but distance: Faraday sought the seat of the phenomena in real actions going on in the medium, they were satisfied that they had found it in a power of action at a distance impressed on the electric fluids. When I had translated what I considered to be Faraday's

---

[20] Theoretical arguments involving a relatively new type of symmetry — "spontaneously broken approximate symmetry" — suggest that mass squares should be used for pseudoscalar mesons [208]. The symmetry would be exact if the pseudoscalar mesons had zero mass (and the quark masses were also zero). The very small mass of the pion is taken as evidence that spontaneous symmetry breaking holds approximately for the pion, and is also assumed for the other pseudoscalar mesons. This approximate symmetry implies that the mass squares of the pseudoscalar mesons are proportional to a weighted *linear* combination of the masses of their quark constituents, leading to the Gell-Mann–Okubo mass formula with mass squares for the pseudoscalar mesons. However, it is not clear why spontaneous symmetry breaking should work equally well for all the pseudoscalar mesons, since their masses vary by almost a factor of four, and the largest of these masses, the η mass, is not small.

ideas into a mathematical form, I found that in general the results of the two methods coincided.

Maxwell showed that mathematicians using action at a distance, and Faraday using fields with lines of force, would get physically identical results.[21] Seventy-five years later Freeman Dyson would show the equivalence of Feynman's graphical formulation of QED using Feynman diagrams of particles traveling forwards and backwards in time, and Schwinger's more mathematical Lagrangian approach that was closer to quantum field theory.

The calculus I developed represented hadrons and their couplings with deuces and treys. It had both a graphical and mathematical form that were computationally equivalent. An example of a coupling computed graphically in the second CERN Report is reproduced in Figure 6.17 on p. 123.

The Origin of Graphical Calculus

In the hard sciences it is possible to separate the results of research from the psychology of the researcher, and it is conventional to maintain this separation in publications. I'm going to break this rule to make the CERN Reports more accessible, and to provide insights into the early childhood education that encouraged me and others to think graphically.

The physics papers I wrote before and after 1964 were conventionally professional, but the CERN Reports and Erice Lectures looked like manuscripts from outer space containing idiosyncratic illustrations and manners of thought. What was the origin of this graphical calculus that was difficult for others to comprehend?

Visual Learning — Johann Pestalozzi (1746–1827)

The graphical computation of the coupling given in Figure 6.17 on p. 123 can be traced back to an 18th century Swiss pedagogue and educational reformer Johann Heinrich Pestalozzi [Figure 6.8], who believed in encouraging students to visualize. "Visual understanding is the essential and only true means of teach-

Figure 6.8. Konrad Grob's painting of Pestalozzi with orphans in Stans Switzerland. (Konrad Grob (1828–1904), Public domain via Wikimedia Commons.)

ing how to judge things correctly," Pestalozzi wrote, and "the learning of numbers and language must be definitely subordinated."[22] Pestalozzi believed that the mind formed concepts by abstracting data gathered by the senses, primarily by what we saw.

Pestalozzi's educational methods were used by the cantonal school in Aarau Switzerland that Albert Einstein attended before entering university. It has been credited with fostering Einstein's visualization of problems. Einstein said of his education at Aarau: "When compared to six years schooling at a German authoritarian gymnasium, it made me clearly realize how much superior an education based on free action and personal responsibility is to one relying on outward authority" [130]. "In Aarau I made my first rather childish experiments in thinking that had a direct bearing on the Special Theory. ... If a person could run after a light wave with the same speed as light, you would have a wave arrangement which would be completely independent of time. Of course, such a thing is impossible." Here Einstein developed his idea of "gedanken," or "thought" experiments.

---

[21] Of those with a mathematical approach, only the relatively obscure George Green, now known for his "Green's function," avoided action at a distance by working with the potential function.

[22] Visual thinking evolved before linguistic thinking, supporting Pestalozzi's argument that visualization is primary. As Cormac McCarthy writes: "The pictorial is a first order representation. A picture of a deer can be understood to be a deer without further explication. But the word deer represents another category. It is a second order representation. It cannot be understood on its own. The naming of things is a wholly artificial construct" once removed from reality (McCarthy, *Nautilus*, 2017).

Manipulative Learning — Friedrich Fröbel (1782–1852)

Fröbel [Figure 6.9] was an educational reformer who attended Pestalozzi's training institute at Yverdon-les-Bains in Switzerland from 1808 to 1810, and went on to invent the kindergarten (his term) in the 1830s, and with it a method of early childhood education. He went beyond Pestalozzi. Instead of just observing children, on which Pestalozzi's methods rested, Fröbel combined observing with doing. The spread of his methods in Germany was thwarted by the Prussian government whose education ministry banned them in the "Kindergartenverbot" ["verbot" means forbidden] edict on August 7, 1851 as "atheistic and demagogic" with its "destructive tendencies in the areas of religion and politics." Teachers using Fröbel's methods fled to other countries, Austria and the United States included.

Figure 6.9. Stamps issued in 1957 by the DDR to mark the 175th anniversary of Fröbel's birth.

One of the features of Fröbel's approach to teaching young children was the use of gifts (Fröbelgaben). He developed a series of gifts that were designed to teach awareness of shapes and spatial relationships [Figure 6.10]. As Frank Lloyd Wright recalls in his autobiography [214], his mother Anna had told him that he "was to build beautiful buildings." She visited the 1876 Centennial Exhibition in Philadelphia where she saw an exhibit of Fröbel gifts, and purchased them for her son. Wright continues: "Here was something for invention to seize, and use to create. The smooth shapely maple blocks with which to build, the sense of which never afterwards leaves the fingers: form becoming feeling." And then later at age 88, "The maple wood blocks ... are in my fingers to this day."

Figure 6.10. Fröbel's 5th gift given to children that are three to four years old. (Froebel USA®.)

My Early Education

By far the greatest influence on my thinking came from my mother who was trained as a kindergarten teacher in Vienna in the 1920s. It was an exciting time for her. Pestalozzi and Fröbel's methods were taught, and Maria Montessori came from Italy to deliver lectures on early childhood education. Everyday on the way to school my mother walked past Sigmund Freud's office. She found the work of psychoanalyst Alfred Adler, who established a number of child guidance clinics, especially interesting.

Before entering college she had been largely self-taught, her childhood being disrupted by World War I and its aftermath (her well-to-do family lost everything but their lives). Fröbel's ideas must have found a welcoming home in her psyche, given her own experiences requiring independence and self-direction.

Although as a small child I was not presented with Fröbel Gifts (my parents and I were living in Detroit, refugees from Hitler's Vienna), I was given surrogates — a simple set of smooth wooden blocks without letters or numbers defiling their surfaces, and a Tinkertoy set like the one shown in Figure 6.11. Unlike Wright's mother, my mother didn't have an agenda. She just left me alone to play in peace.

Figure 6.11. A 1940s Tinkertoy set. (Courtesy Jim and Diane on eBay.)

As luck would have it, another refuge from Vienna, Annemarie Roeper, who at 19 in 1938 was planning to study with Anna Freud, ended up in Detroit where she and her husband set up a nursery school in 1941 based on the principles of Pestalozzi and Fröbel. At three and a half, I was one of her first pupils, thriving in an environment of independence, surrounded by shelves of toys from which to choose.

More than two decades later in 1963, the circular connectors in the Tinkertoy became aces, the rods the forces that bound them, at least in the fingers of my imagination. I remembered these objects from childhood and visually assembled them in many combinations to represent the known strongly interacting particles and their couplings. This was the origin of the graphical calculus, exemplified in Figure 6.17, p. 123, which was used to compute hadronic couplings. Unfortunately this calculus made the CERN

Reports and Erice Lectures difficult for others to comprehend.[23]

A Year of Independence

My four years as a graduate student at Caltech had been very stressful. At CERN with no responsibilities, I decompressed. My ex-wife and I rented a small villa in the countryside in the middle of an apple orchard in Tannay près Mies, far from CERN. Brown Swiss cows, lowing, bells swaying from their necks, grazed outside our windows. Day after day I worked at home, only going in to CERN to visit the library, talk with Rafael Armenteros (a bubble chamber experimentalists), and work on CERN's computers. I was temporarily cleansed of my professionalism, reverting back to a more primal state — my Tinkertoy days — viewing the amplitude for each decay as a sum of weighted Tinkertoy constructions. I knew what Frank Lloyd Wright meant when he said "The maple wood blocks ... are in my fingers to this day."

This was an exciting time. It was impossible to finish even the simplest calculation without jumping up, pacing back and forth for a few minutes, and rushing back to see if things were still working. I dreamed of modifying the Tinkertoy so that children could construct their own hadrons, and hadron interactions.

6.15 Politics Turned Personal

Shortly after the second CERN Report was distributed Léon Van Hove, head of CERN's Theory Division where I was visiting, came to my office, asking where I planned to publish. The *Physical Review* I replied, the most widely read and respected journal in particle physics. "That's not possible" he said. At the time CERN required that papers written by junior

---

23 Jonathan Rosner reminds me that Dmitri Mendeleev used cards to discover the periodic table. He created a card for each element, and noted how that element combined with other elements to form chemical compounds. By placing cards, whose elements formed similar compounds, in the same column, ordering them within each column by mass, he created the beginnings of the periodic table (www.youtube.com/watch?v=_yR3I8Lqx0o&ab_channel=ChemSurvival). Columns of elements are somewhat like hadron group representations — their members behave similarly.

members of the Theory Division be approved before they could be submitted for publication. Furthermore, papers had to be published in European journals in order to "build up CERN's reputation." I explained that I was a visitor with my own financial support covering my salary, $1,200 for publication costs, and also overhead expenses. Since I hadn't been hired by CERN, I quietly repeated my intention to publish in the *Physical Review*.

When I gave the second CERN Report to Tatiana Fabergé, the group secretary, to retype for publication she handed it back to me saying, with some embarrassment, that she had been instructed by Van Hove not to type *any* of my papers. Ironically, had I published in a European Journal it would have been *Il Nuovo Cimento,* a journal that no longer publishes physics papers![24]

It's customary for physicists to present their work at seminars. I had one scheduled at CERN titled "Dealer's Choice: Aces are Wild." Notices were pinned to cork boards around the Theory Division. Much to my surprise, one afternoon I saw Van Hove unpinning mine, yes, taking the announcements down. In retrospect a strange sight, the future Director-General of CERN busy with office work. I asked him what he was doing. He angrily replied: "Your seminar is canceled. You are not allowed to talk at CERN," and I didn't. Why this anger and animosity? it's hard to say, but I suspect he was a man that didn't like to be ignored.[25]

6.16   Heisenberg's Resistance

"The most radical revolutionary will become a
conservative on the day after the revolution."
- Hannah Arendt.[26]

---

[24] Getting my work on aces published in the form I wanted was so difficult that I gave up trying. Many years later, wondering where free fractionally charged particles (p. 134), if they existed, might be concentrated in the Earth, Joe Smyth and I wrote a paper titled "The Geochemical Classification of Fractionally Charged Particles" [195]. We couldn't get it published.

[25] When it came to having problems with Van Hove, I was in good company. In his historical account of physics in the CERN Theory Division, John Iliopoulos alludes to difficulties between Martinus (Tini) Veltman and his thesis advisor, Van Hove. See *Studies in CERN History*, CHS-39, p. 33. Available at: https://cds.cern.ch/record/261679/files/CERN-CH-39.pdf. Veltman, with his student Gerardus 't Hooft, received the 1999 Nobel Prize in Physics.

[26] Reflections (Civil Disobedience), p. 70, New Yorker 12 September 1970.

Heisenberg never accepted the idea of elementary particles as constituents. From his 1973 lecture at Harvard published in 1975 [120]:

> It is natural that even nowadays many experimental physicists — even some theoreticians — still look for really elementary particles. They hope for instance that the quarks, if they existed, could play this role.
>
> I think that this is an error. It is an error because even if the quarks would exist we could not say that the proton consists of three quarks. We would have to say that it may temporarily consist of three quarks, it may also temporarily consist of four quarks and one antiquark, or five quarks and two antiquarks, and so on. And all these configuration would be contained in the proton; and again one quark would be composed of two quarks and one antiquark and so on. So we cannot avoid this fundamental situation; but since we still have the questions from the old concepts, it is extremely difficult to stay away from them.

This objection is similar to one made by Gell-Mann in 1966, p. 71.

For Heisenberg particles all appeared on the same footing as different states of a single system somewhat like the stationary states of an atom or molecule. As he explained in 1976, the year of his death [121]:

> There is no difference in principle between elementary particles and compound systems. This is probably the most important experimental result of the last fifty years. ...
>
> Let us compare the so-called "elementary" particles with the stationary states of an atom or a molecule. We may think of these as various states of one single molecule or as the many different molecules of chemistry. One may therefore speak simply of the "spectrum of matter [hadrons]." ... Like the stationary states of atoms, the elementary particles can be characterized by quantum numbers — that is, by their behavior under certain transformations. The corresponding laws of conservation determine what transmutations are possible. For an excited hydrogen atom it is its behavior under rotation that determines whether it can fall into a lower state with the emission of a light quantum. In the same way *for a φ meson, it is its symmetry properties that determine whether*

it can disintegrate with the emission of a pion into a ρ meson [my italics].

This last sentence was Heisenberg's way of saying that Zweig's rule was nonsense. But understanding why the φ meson didn't "disintegrate with the emission of a pion into a ρ meson" led to the discovery of constituent quarks. Suppression had nothing to do with symmetry. The concept of "elementary" was essential.

Unsurprisingly Heisenberg also cautions:

A word should be added concerning the postulate that understanding requires a visual picture of the phenomena. ...
We will have to accept the fact that experimental data on a very large or a very small scale do not necessarily produce pictures, and we must learn to do without them. We then come to recognize that the antinomy of the smallest dimensions is solved in particle physics in a very subtle manner, of which neither Kant nor the ancient philosophers could have thought: The word "dividing" [a particle into pieces] loses its meaning.

So much for Pestalozzi, visual thinking, and the graphical calculus.

Resistance, resistance, resistance. It's not clear that Heisenberg understood that quarks with their fractional charges had been "seen" in deep-inelastic scattering experiments at SLAC. Although Heisenberg was correct that hadrons were not elementary, he never realized that quarks were not hadrons.

Other Problems With Acceptance

Although Heisenberg's argument against elementary particles wasn't based on his 1920s requirement that theory be based solely on observables, the fact that quarks had not been found at accelerators, or in cosmic rays [106], bothered others. Steve Weinberg's view of quarks was typical. In answer to the question of why he ignored quarks in his 1967 paper that unified the electromagnetic and weak interactions, he replied in his Forward to Jim Baggott's 2012 book on the Higgs particle [9]:

I did not include quarks in the theory simply because in 1967 I just did not believe in quarks. No one had ever observed a quark, and it was hard to believe that this was because quarks

are much heavier than observed particles like protons and neutrons, when these observed particles were supposed to be made of quarks. Like many other theorists, I did not fully accept the existence of quarks until the 1973 work of David Gross and Frank Wilczek, and David Politzer. They showed that in the theory of quarks and strong nuclear forces known as quantum chromodynamics, the strong force gets weaker with decreasing distance. It then occurred to some of us that in that case the strong force between quarks would counterintuitively get stronger as the quarks get farther apart, perhaps so much so as to prevent quarks from ever being separated from one another. There still is no proof of this, but it is generally believed.

Quarks also had no place in current dogma. They were incompatible with nuclear democracy, the basis of the bootstrap, and all other theories of hadron interactions. For quarks, as for Copernicus's view of the solar system, simplicity was no excuse for challenging what was "known" to be true.[27]

Talking about aces in public embarrassed me because I understood how theorists thought, and what many thought of me.[28] Working with aces even made me uncomfortable. I knew what real theories looked like. As a student I had read Julian Schwinger's paper on quantum electrodynamics. I had written "professional" papers [222]. Nothing I now did looked like that. But the methods I used, including the graphical calculus, although foreign to others, organized and predicted so much that I believed these heuristics would eventually lead to a legitimate quantum field theory.

---

[27] Wegener's theory of continental drift provides another example of discovery contradicting dogma. Evidence from geology and paleontology clearly showed that Africa and South America were connected sometime in the past, but this idea was not accepted because the Earth's crust was thought to be "frozen." A dynamical mechanism for driving continents apart was not yet known.

[28] Having worked at the Bevatron in the early 1960s, I appreciated the excitement and beauty of Berkeley and the San Francisco area. After I returned to Caltech from CERN I wanted to get a job at UC Berkeley. Gerson Goldhaber, my advocate, presented my application at a physics faculty meeting in 1966, but Geoffrey Chew blocked the appointment, passionately arguing that the ace model was the work of a "charlatan" (Gerson Goldhaber, private communication, 1966).

6.17 Leaving Physics

Constituent quarks, with their quantum numbers and mass differences, provided the *raison d'être* for the existence of the broken SU(6), SU(3), and SU(2) symmetries of the strong interactions. In 1964 nothing was known about their dynamics, except for Zweig's rule. I suspected they were real, not mathematical group-theoretic constructs.

I hated what was called the "naive" or non-relativistic quark model. The naive quark model was too naive, and subject to overfitting. Sometimes there were as many inputs as outputs, and no predictions. When I critiqued the naive quark model, complaining to Feynman, "That's not what quarks are," he replied with some annoyance, "Quarks don't belong to you anymore." He was right; I had done my part. Now constituent quarks were in the public domain. What they were was up for grabs. That was a turning point, a difficult moment of acceptance, and the third and last time Feynman showed his irritation with me.

I didn't think I could take the next step and create the quantum field theory governing quarks and the particles that bound them. All I could do was connect constituent quarks with springs and move them from one hadron to another when they interacted; that work was over after 10 months at CERN. Most discouragingly, *even if I could find the correct theory, I couldn't prove it.* The necessary calculations would be too difficult. As Steven Weinberg put it more than a dozen years later in his 1977 "The Search for Unity: Notes for a History of Quantum Field Theory" [206]:

> It wasn't that there was any difficulty in thinking of renormalizable quantum field theories that *might* account for the strong interactions — it was just that having thought of such a theory, there was no way to use it to derive reliable quantitative predictions, and to test if it were true.

Finding a quantum theory for gravity was the other great challenge, but I wasn't going to touch it having watched Feynman struggle with it for an entire year.

A Second Rejection

As a graduate student I had failed to find evidence for the violation of time-reversal (T) symmetry in K^+ decay (p. 22), but shortly thereafter Jim Cronin and Val Fitch showed that the violation of the combined symmetries of charge conjugation (C) and parity (P), so-called CP violation, occurred in neutral K decay, although at a very low level. If CPT symmetry held, they indirectly showed that time-reversal symmetry failed. Since the violation was very small it seemed to me to be impossible to predict its magnitude, but *it was possible to say something about its sign.* The abstract to the paper I wrote, titled "What determines the sign of the CP violation," summarized its conclusions:

> Observation of $K_L \rightarrow 2\pi$ allows us to assign a direction to time. The operational definition of "time direction" is discussed. A "theory" which correlates this time direction with the time directions defined by thermodynamics or cosmology must exist and take the form of a boundary condition.

This established one of the first connections between particle physics and cosmology.[29] It was presented in 1967 at a conference on CP violation organized by Cronin.

When the paper was submitted to *Physical Review Letters* shortly thereafter, it was immediately rejected. After I addressed the referee's comments, the revised paper was resubmitted, and rejected again, the Editor saying "while a report of this work deserves publication, it is not of such urgency to warrant speedy publication of a letter. On the other hand, the report as submitted appears too abbreviated to be satisfactory as an article." I set it aside, not knowing how to expand it.

Some 20 years later, while visiting the University of Chicago, I went to Cronin's office to say hello. He was expecting me, sitting behind a large oak desk, its top completely clear except for a preprint lying on the corner of the desk closest to the door. It was good to see Jim again. After we had talked a while he pointed to the preprint, asking me to take a look. I did; it was my 1967 paper! When I looked up he said: "Every time I need to be inspired I take out your paper and read it again." Although short, the paper seemed to have some merit. Jim was a nice man.

---

[29] A different connection between particle physics and cosmology was given in Andrei Sakharov's 1967 paper titled "Violation of *CP* Invariance, *C* Asymmetry, and Baryon Asymmetry of the Universe" [191].

A Door Closed: Neutral Currents

In 1970 I joined the Caltech neutrino experiment at Fermilab, putting together a proposal with Frank Scully and Charlie Peck to look for neutral currents in the weak interactions. The scheduling committee liked the proposal, but asked, "Doesn't Caltech already have an experiment at Fermilab?" They suggested we interest experimentalists at other universities, letting them take the lead. Frank, Charlie, and I turned our attention elsewhere. For me, it was hard enough doing physics without politics. Neutral currents were discovered at CERN three years later.[30]

These difficulties with publications and getting experiments approved were aberrations. Many papers were published, and I had wonderful collaborations with Fred Zacharisen, Jeff Mandula, and Jacques Weyers over a period of several years. Nevertheless, physics wasn't as interesting or exciting as it used to be.[31]

A Door Opened: Neurobiology

As an undergraduate I had noticed that scientists were most creative when they were very young. Perhaps it had something to do with their under-educated minds that didn't know better, but knew enough, confronting fields tended to by established professionals, too certain of their ways. In any case, at that time I resolved to change fields at 30.

Caltech decided to expand into neurobiology in the early 1970s when it received a very large donation from Arnold Beckman to construct the Beckman Behavioral Biology Building. To educate the faculty of the Biology Department, who would have to decide which areas of neurobiology to emphasize, an extensive series of colloquia where presented by experts in the field. I listened to all of them. Suddenly high energy physics had competition. To find out more I attended a summer school at Cold Spring Harbor designed to help young, but established, scientists switch fields to neurophysiology.

---

[30] The existence of neutral currents provided direct evidence for the electroweak theory that unified the weak and electromagnetic interactions (p. 136).

[31] With the economic strain of the war in Vietnam, and the dwindling possibilities of my graduate students getting jobs in high energy physics, I also doubted if the US would continue its commitment to the field I loved. Witness the cancellation of the Superconducting Super Collider some 20 years later.

Eventually I initiated a research program in the neurophysiology of hearing in Beckman's Building. This is how it started: While trying to understand the genetics of learning in fruit flies in Seymour Benzer's lab, but also curious about their courtship behavior, I took a small bottle of flies over to a sound booth in the Electrical Engineering Department, hoping to hear their mating calls. After listening for a long time, my ear at the mouth of the bottle, and hearing nothing, my mind began to wander. I wondered: "What does a very quiet tone sound like?" Is it just a loud tone, only much quieter? I got an HP oscillator and a pair of ear phones from the lab outside, brought them back into the booth, and started to listen to pure tones. At some frequencies a very quiet tone would be very smooth, just like a louder tone. But at other frequencies it was very rough. Sometimes I could hear *two tones*, one beating against the other. What I heard could changed dramatically as the amplitude or frequency of the tone I played was varied. The sensations were remarkable.

Then I got a second oscillator and played two tones simultaneously at frequencies f_2 and f_1, where f_2 was the higher frequency. Sometimes I heard an additional third tone, a "combination tone," whose frequency f_c shifted when either of the two primary frequencies was varied by a small amount. To determine f_c I added a third external tone, and by adjusting its frequency, got it to beat with the combination tone, whose frequency turned out to be $f_c = 2f_2 - f_1$. By changing the two primary frequencies I was able to generate about a dozen different combination tones with frequencies $3f_2 - 2f_1$, $4f_2 - 3f_1$, ... In each case the frequencies of the two primaries were adjusted to place the frequency of the combination tone at 1 kHz, about the frequency to which the ear is most sensitive. *The ear was making sounds while listening to sounds!* I thought something wonderful was happening, I just didn't know how or why.

Georg von Békésy, a physicist who started work at the Telephone System Laboratory of the Hungarian Post Office in 1923, realized that to build a better telephone it was necessary to have an understanding of how the ear works. Experimenting with human cadavers, he came to our first understanding of how the cochlea responds to sound, work for which he received the Nobel Prize in Physiology or Medicine in 1961.

The cochlear response Békésy measured was linear. The mechanics of my cochlea, whose motions

I had listened to, were nonlinear and completely unknown, probably doing remarkable things. For the first time the Mössbauer effect was being used by William Rhode to measure motion within the cochlea of living squirrel monkeys with unprecedented sensitivity, telling us how the cochlea preprocesses sound to make its information available to the central nervous system. The responses he measured were nonlinear. I realized that Békésy had studied hearing in the dead.

In cochlear mechanics there were anomalies galore, and new measurement techniques. Understanding how the cochlea functions is a very old problem in physics first addressed in detail by Hermann Helmholtz in his 1863 classic book "On the Sensations of Tone as a Physiological Basis for the Theory of Music." So with Delbrück's encouragement and Feynman's earlier example, I kissed high energy physics goodbye and switched to the neurophysiology of hearing. Rhode's Mössbauer measurements eventually became central to my research, as well as listening to, and measuring, the sounds emitted by the human cochlea as it listens to sounds. There was more to explore than high energy physics.

6.18 Looking Back

My talk on the origins of the constituent quark model at the "Baryon 1980" conference concludes with [226]:

> The intellectual history of the quark model contains an enormous number of theoretical ideas and experimental results. There are many contradictions. If asked, "In this rich environment of fact and fiction, how could you find the quark?," I would reply "By having a basic commitment to reality." Each theoretical idea was tested by experiment, and experiments tested each other. For example, Matt Roos's 1963 Reviews of Modern Physics compilation of particles and their properties referred to several hundred experimental papers. I read essentially all of them, taking care to understand how each measurement was made. Then an accurate appraisal of the results of each experiment was possible. Rational choices between conflicting experiments usually could be made. Training in experimental physics was helpful in this process. Many theories had to be judged in the face of insufficient experimental information. Theories which lacked predictive power, like Heisenberg's nonlinear spinor theory of matter, were discarded, not because they were necessarily incorrect, but because they were operationally useless. Theories of uncertain truth were compared with theories known to be either correct or incorrect. In this way it was possible to say that some theories "just didn't look right." Here training in theoretical physics was important. In addition to this combination of experimental and theoretical skills, I possessed a rather unlikely combination of personality traits that were essential to my process of discovery. Near obsession with detail coexisted with a much freer imaginative and romantic nature.

Appendices for Chapter 6

6.A   Hierarchies of Representations

This appendix expands Section 6.5: "Hadrons From Aces with Spin," p. 96, adding parity and charge conjugation to representations to flesh out ace-spin symmetry. Together with Appendix 6.B: "Evaluating Theories," p. 118, this will help you decide if you would have thought hadrons had ace constituents when their discovery was published in January 1964.

Mesons with Angular Momentum $L = 0$

Starting with three aces, each with two spin projections, $6 \times 6 = 36$ deuces with spin were created (Eq. 6.1, p. 96). By taking appropriately weighted linear combinations, 36 meson wave functions were constructed, forming a reducible representation $\mathscr{M}_{36}^{0,1}$ of SU(6), which decomposed into a sum of two irreducible representations $\mathcal{M}_{35}^{0,1} + \mathcal{M}_1^0$:

$$\text{SU(6)}: \quad 6 \otimes \bar{6} = 35 \oplus 1,$$
$$\mathscr{M}_{36}^{0,1} = \mathcal{M}_{35}^{0,1} + \mathcal{M}_1^0. \quad (6.7)$$

The italic script $\mathscr{M}$ as in $\mathscr{M}_N^{S_1,S_2}$ designates a reducible representation. A calligraphic $\mathcal{M}$ denotes an irreducible representation. The subscript N indicates the number of mesons in the representation counting their spin projections, while superscripts S_1, S_2 signify their possible spins.

While $\mathcal{M}_{35}^{0,1}$ is irreducible under SU(6), under SU(3) it becomes reducible, its 35 mesons forming a reducible representation $\mathscr{M}_{35}^{0,1}$ that decomposes into one SU(3) irreducible representation $\mathcal{M}_{8\times1}^{0-}$, and another that is reducible, $\mathscr{M}_{9\times3}^{1-}$, where the negative parity of the states in the SU(3) representations appears as a superscript to their spin:

$$\text{SU(3)}: \quad \mathscr{M}_{35}^{0,1} = \mathcal{M}_{8\times1}^{0-} + \mathscr{M}_{9\times3}^{1-}$$
$$= (8 \text{ mesons} \times 1 \text{ spin})$$
$$+ (9 \text{ mesons} \times 3 \text{ spins})$$
$$= (8 + 27) \text{ mesons with spin}$$
$$= 35 \text{ mesons, counting spin}$$
$$\text{projections,}$$

where the subscript N is written as the product of the number of mesons in the SU(3) representation

times the number of possible spin projections for each meson.

The two SU(3) representation are both reducible under SU(2), each breaking into several SU(2) irreducible representations:

$$\text{SU(2)}: \quad \mathcal{M}_{8\times1}^{0-} = {}^0\mathcal{M}_{3\times1}^{0-+} + {}^1\mathcal{M}_{2\times1}^{0-} + {}^{-1}\bar{\mathcal{M}}_{2\times1}^{0-}$$
$$+ {}^0\mathcal{M}_{1\times1}^{0-+}$$
$$= \pi + K + \bar{K} + \eta_8.$$
$$\text{SU(2)}: \quad \mathscr{M}_{9\times3}^{1-} = {}^0\mathcal{M}_{3\times3}^{1--} + {}^1\mathcal{M}_{2\times3}^{1-} + {}^{-1}\bar{\mathcal{M}}_{2\times3}^{1-}$$
$$+ {}^0\mathcal{M}_{1\times3}^{1--} + {}^0\mathcal{M}_{1\times3}^{1--}$$
$$= \rho + K^* + \bar{K}^* + \omega_8 + \omega_1, \text{ or}$$
$$+ \omega + \phi,$$
$$(6.8)$$

where the superscript $\mathbb{S}$ preceding an SU(2) representation, as in ${}^\mathbb{S}\mathcal{M}_N^{SP}$ or ${}^\mathbb{S}\mathcal{M}_N^{SPC}$, is its strangeness, and the charge conjugation is added as a second superscript to the spin of an SU(2) representation if it contains a neutral particle that is its own antiparticle, like the π^0 in ${}^0\mathcal{M}_{3\times1}^{0-+}$. The pseudoscalar meson η_8 is also called η, while ω_8, in the same SU(3) reducible representation as ω_1, mixes with it to give the observed ω and ϕ vector mesons, both members of their own SU(2) representations.

The SU(6) irreducible representation $\mathcal{M}_1^0$, found from $6 \otimes \bar{6}$ in Eq. 6.7, has only one member, a spin-0 pseudoscalar meson, ${}^0\mathcal{M}_1^{0-+}$, the η_1, which is commonly referred to as η'.

The improvement in meson classification when going from the Eightfold Way to ace-spin symmetry may be symbolically summarized by:

Eightfold Way and SU(3) $\rightarrow$ Ace-spin symmetry,

$$\left.{}^{(\mathbb{S}=?)}(\mathcal{M}=?)_{(N=?)}^{(S=?)(P=?C=?)}\right. \rightarrow {}^\mathbb{S}\mathcal{M}_{N=(2I+1)(2S+1)}^{SP=(-1)^{L+1}C=(-1)^{L+S}}. \quad (6.9)$$

Ace-spin symmetry "filled in the blanks." The expressions for parity $P = (-1)^{L+1}$ and charge conjugation $C = (-1)^{L+S}$ are discussed in Appendix A.6: "Symmetry and Quantum Numbers," p. 151.

The Lesson from Mesons

The Eightfold Way only said that hadrons were clustered in one or more SU(3) representations of unknown dimension; that's it! Ace-spin symmetry specified the combination of representations that existed at each level of the SU(6), SU(3), and SU(2) symmetries, and what combination of quantum numbers the members of those representations possessed. The predicted representations at each level of symmetry, and the quantum numbers of their hadrons, had all been observed by 1963, except for the SU(6) singlet representation consisting of the η'. Many other mesons were reported; ace-spin symmetry predicted that they did not exist, and a careful examination of the data at the time found that the evidence for them was not compelling (e.g., they were only seen in one experiment). A theory without aces that could make such detailed predictions appeared inconceivable (at least to me). *In trying to decide if mesons had ace constituents at the beginning of 1964, the unlikely possibility of a theory without aces explaining the existing observations had to be weighed against the a priori probability that aces with fractional charge existed* (Appendix 6.B: "Evaluating Theories," p. 118). Additional information helpful in making this judgement call was available in the baryon spectrum.

Baryons with Angular Momentum $L = 0$

As indicated in "Baryons from Treys with Spin," p. 97, baryons were found in the totally symmetric SU(6) irreducible representation with 56 members consisting of two irreducible representations of SU(3) with angular momentum and parity $L^P = 0^+$:

$$\text{SU(6)}: \quad \mathcal{B}_{56}^{\frac{3}{2},\frac{1}{2}}.$$

$$\text{SU(3)}: \quad \mathcal{B}_{56}^{\frac{3}{2},\frac{1}{2}} = \mathcal{B}_{8\times2}^{\frac{1}{2}^+} + \mathcal{B}_{10\times4}^{\frac{3}{2}^+}$$
$$= (8 \text{ baryons} \times 2 \text{ spins})$$
$$+ (10 \text{ baryons} \times 4 \text{ spins})$$
$$= (16 + 40) \text{ baryons with spin}$$
$$= 56 \text{ baryons, counting spin}$$
$$\text{projections}.$$

In analogy to the notation used for mesons, the $\mathcal{B}_{56}^{\frac{3}{2}^+,\frac{1}{2}^+}$ is an irreducible representation of SU(6) with 56 positive parity baryons of spin 3/2 and 1/2, counting their spin projections. $\mathcal{B}_{56}^{\frac{3}{2}^+,\frac{1}{2}^+}$ is reducible with respect to SU(3), while $\mathcal{B}_{8\times2}^{\frac{1}{2}^+}$ and $\mathcal{B}_{10\times4}^{\frac{3}{2}^+}$ are irreducible.

The two SU(3) irreducible representations each break down into several SU(2) irreducible representations:

$$\text{SU(2)}: \quad \mathcal{B}_{8\times2}^{\frac{1}{2}^+} = {}^0\mathcal{B}_{2\times2}^{\frac{1}{2}^+} + {}^{-1}\mathcal{B}_{1\times2}^{\frac{1}{2}^+} + {}^{-1}\mathcal{B}_{3\times2}^{\frac{1}{2}^+}$$
$$+ {}^{-2}\mathcal{B}_{2\times2}^{\frac{1}{2}^+}$$
$$= N + \Lambda + \Sigma + \Xi.$$
$$\text{SU(2)}: \quad \mathcal{B}_{10\times4}^{\frac{3}{2}^+} = {}^0\mathcal{B}_{4\times4}^{\frac{3}{2}^+} + {}^{-1}\mathcal{B}_{3\times4}^{\frac{3}{2}^+} + {}^{-2}\mathcal{B}_{2\times4}^{\frac{3}{2}^+}$$
$$+ {}^{-3}\mathcal{B}_{1\times4}^{\frac{3}{2}^+}$$
$$= \Delta + \Sigma^* + \Xi^* + \Omega.$$

The improvement in baryon classification when going from the Eightfold Way to ace-spin symmetry is summarized by:

Eightfold Way and SU(3) $\rightarrow$ Ace-spin symmetry,

$$\text{}^{(\mathbb{S}=?)}(\mathcal{B}=?)_{(N=?)}^{(S=?)^{P=?}} \rightarrow {}^{\mathbb{S}}\mathcal{B}_{N=(2I+1)(2S+1)}^{S^{P=(-1)^L}}. \tag{6.10}$$

As with mesons, ace-spin symmetry identified the origin of the patterns observed in the baryon spectrum, becoming the raisons d'être for the hierarchy of SU(2), SU(3), and SU(6) symmetries. It predicted which representations existed at each level of symmetry, and specified which combination of quantum numbers would be found in these representations. The predicted and observed representations, with their quantum numbers, matched perfectly in 1963, except for three baryon resonances:

1. The Ω^-, the remaining unobserved member of the $J^P = \frac{3}{2}^+$ baryon decuplet,
2. The $\Lambda(1405)$, with $J^P = \frac{1}{2}^-$ first seen in 1961, possibly an orbital excitation with $L^P = 1^-$ (Eq. 6.11),[32]
3. The "Roper resonance (1440)," with $J^P = \frac{1}{2}^+$ first seen in 1963, possibly the first radial excitation of the nucleon [28].[33]

---

[32] See arxiv.org/abs/2102.00914 for references.

[33] B. T. Feld and L. D. Roper, *Proceedings of the Siena International Conference on Elementary Particles* (Italian Physical Society, Bologna), p. 400 (1963).

Baryons with Angular Momentum $L = 1$

As for mesons, the model predicted high-mass baryons possessing rotational excitations. In addition to the 56 ground state baryons with $L^P = 0^+$, an irreducible representation of 70 states with $L^P = 1^-$ was predicted to exist. It consists of four irreducible representations of SU(3):

$$\text{SU(6)}: \quad \mathcal{B}_{70}^{\frac{3}{2}, \frac{1}{2}}.$$

$$
\begin{aligned}
\text{SU(3)}: \quad \mathscr{B}_{70}^{\frac{3}{2}, \frac{1}{2}} &= \mathcal{B}_{8\times 4}^{\frac{3}{2}^-} + \mathcal{B}_{10\times 2}^{\frac{1}{2}^-} + \mathcal{B}_{8\times 2}^{\frac{1}{2}^-} \\
&\quad + \mathcal{B}_{1\times 2}^{\frac{1}{2}^-} \\
&= (8 \times 4 \text{ spins}) + (10 \times 2 \text{ spins}) \\
&\quad + (8 \times 2 \text{ spins}) + (1 \times 2 \text{ spins}) \\
&= (32 + 20 + 16 + 2) \text{ baryons with} \\
&\quad \text{spin} \\
&= 70 \text{ baryons, counting spin} \\
&\quad \text{projections.} \quad (6.11)
\end{aligned}
$$

A Different SU(6)

The history of high energy physics reads like a mystery novel with many misleading clues. Feza Gürsey and Luigi Radicati also proposed classifying hadrons into representations of SU(6), which they justified with a theoretical argument not based on constituents [111]. Their theory combined internal with external space-time symmetries. The paper created a minor sensation among particle theorists. Sidney Coleman and Jeffrey Mandula proved that their theory could not possibly be correct, causing another minor sensation [40]. The Coleman-Mandula paper was not applicable to ace-spin symmetry because it was based on different assumptions than the Gürsey-Radicati theory. Like Coleman and Glashow (p. 101), Gürsey and Radicati got the right answer for the wrong reason!

6.B   Evaluating Theories

Bayes' theorem could have been used to decide how likely it was that aces were constituents of hadrons. With some massaging performed at the end of this section, Bayes' theorem says that the conditional probability that hadrons have ace constituents, given the experimental evidence ($\phi \not\rightarrow \rho + \pi$, mass relations, ...) is

$$P(A|E) = \frac{1}{1 + \lambda}, \text{ where } \lambda = \frac{P(E|\bar{A})}{P(E|A)} \frac{P(\bar{A})}{P(A)}. \quad (6.12)$$

Here $P(E|\bar{A})$ is the conditional probability of the experimental evidence given that aces do not exist, $P(E|A)$ is the same probability given that aces do exist, and $P(A)$ and $P(\bar{A})$ are the *a priori* probabilities ("priors") that aces do or do not exist.

Since $P(\bar{A}) \approx P(E|A) \approx 1$, the expressions for λ and $P(A|E)$ simplify to

$$\lambda \approx \frac{P(E|\bar{A})}{P(A)}, \text{ and } P(A|E) \approx \frac{1}{1 + P(E|\bar{A})/P(A)}.$$

Whether or not a rational person would have believed aces exist depends only on the ratio λ of two small numbers. They would have believed if and only if

$$\lambda \approx 0, \text{ or } P(A) \gg P(E|\bar{A}),$$

i.e., if the a priori probability of aces existing was much greater than the conditional probability of explaining the experiments if aces did not exist. When this inequality holds, even though initially $P(\bar{A}) \approx 1$, when experimental evidence is taken into account, the opposite holds, i.e., $P(A|E) \approx 1$.

Therefore both an understanding of how experiments are performed — so you know which ones to believe — and a familiarity with theories — so you know how difficult it is to create them — are both necessary to make a reasonable judgement call.

While this formulation of Bayes theorem doesn't answer the question being asked, it does makes precise the two uncertainties that must be weighed against each other to obtain an answer.

An Alternate Derivation of Eq. 6.12

Bayes' theorem says that

$$
\begin{aligned}
P(A|E) &= \frac{P(E|A)P(A)}{P(E)} \text{ and,} \\
P(\bar{A}|E) &= \frac{P(E|\bar{A})P(\bar{A})}{P(E)}. \quad (6.13)
\end{aligned}
$$

Summing both sides of these two equations and equating them,

$$
\begin{aligned}
\frac{P(E|A)P(A) + P(E|\bar{A})P(\bar{A})}{P(E)} &= P(A|E) + P(\bar{A}|E) \\
&= 1.
\end{aligned}
$$

Solving for $P(E)$,

$$P(E) = P(E|A)P(A) + P(E|\bar{A})P(\bar{A}).$$

Substituting this expression for $P(E)$ into the first form of Bayes' theorem in Eq. 6.13,

$$P(A|E) = \frac{P(E|A)P(A)}{P(E|A)P(A) + P(E|\bar{A})P(\bar{A})}$$
$$= \frac{1}{1 + \frac{P(E|\bar{A})P(\bar{A})}{P(E|A)P(A)}},$$

which is Eq. 6.12, where $\lambda = P(E|\bar{A})P(\bar{A})/P(E|A)P(A)$.

The Cardinal Question

Ace-spin symmetry suggested a broken SU(6) symmetry for hadron classification. It was not a fundamental symmetry of any field theory, but simply reflected the creation of hadrons from spin-1/2 quarks with unequal masses. It was the first of a hierarchy of representations of known dimensions that descended from SU(6) to SU(2), through SU(3) (pp. 96 and 97). *In addition to spin, each representation had all of its other quantum numbers specified for both mesons and baryons* (Eqs. 6.9 and 6.10, pp. 116 and 117).

Moreover, multiple mass and decay formulae were predicted that could not be obtained from the Eightfold Way (Sections 6.6 and 6.7, pp. 98 and 101). Weak interaction selection rules were also obtained (Section 6.8, p. 102). Where data existed they were satisfied.

But aces had fractional charge and, without modification, violated the spin-statistics theorem for some baryon states. Given the pros and cons, would you have taken aces seriously when they were first introduced, using them as a foundation for a quantum field theory governing the strong interactions? Would your *a priori* probability that they existed, which would have been quite small, have been much larger than your probability that a theory without aces could explain the data?

6.C   Graphical Calculus

Graphical thinking enabled calculations of coupling constants that would ordinarily be computed algebraically. However, the graphical calculations had a

dynamical interpretation, which made the suppression of ϕ decay both necessary and natural. SU(3) wave functions of the interacting hadrons, expressed in terms of their weighted deuces and treys, were represented graphically, and rules for connecting them were given. What would have been a contraction of an upper and lower index in tensor calculus, became an ace flowing from an initial to a final state.

Low-mass Mesons Represented Graphically

As previously described, the low-mass mesons fell into one irreducible representation of SU(6), and two irreducible representations of SU(3). One contained the pseudoscalar octet, the other the vector meson nonet (Eq. 6.8 p. 116). The graphical representation of the octet and nonet, taken from the Erice Lectures, is given in Figures 6.12 and 6.13:

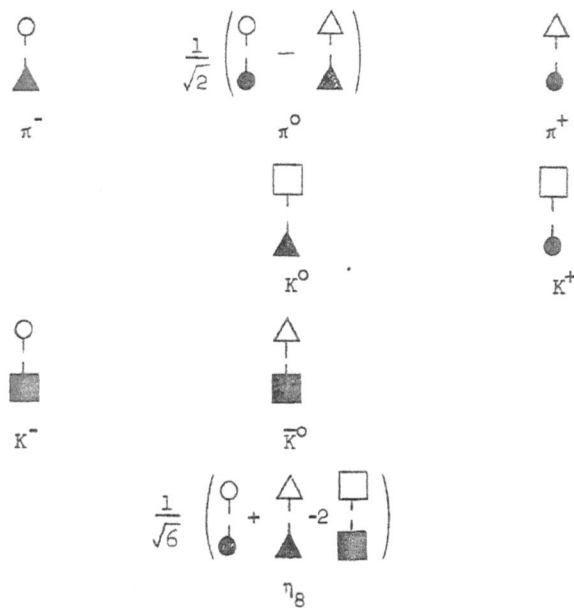

Figure 6.12. The pseudoscalar meson octet wave functions shown as a linear combinations of deuces. Ace spins have been suppressed. Recall that black and white symbols represent aces and antiaces, with the lines between them springs. The sum of the squares of the numbers multiplying the deuces in a meson sum to one. The η_8 pseudoscalar meson was also called the η.

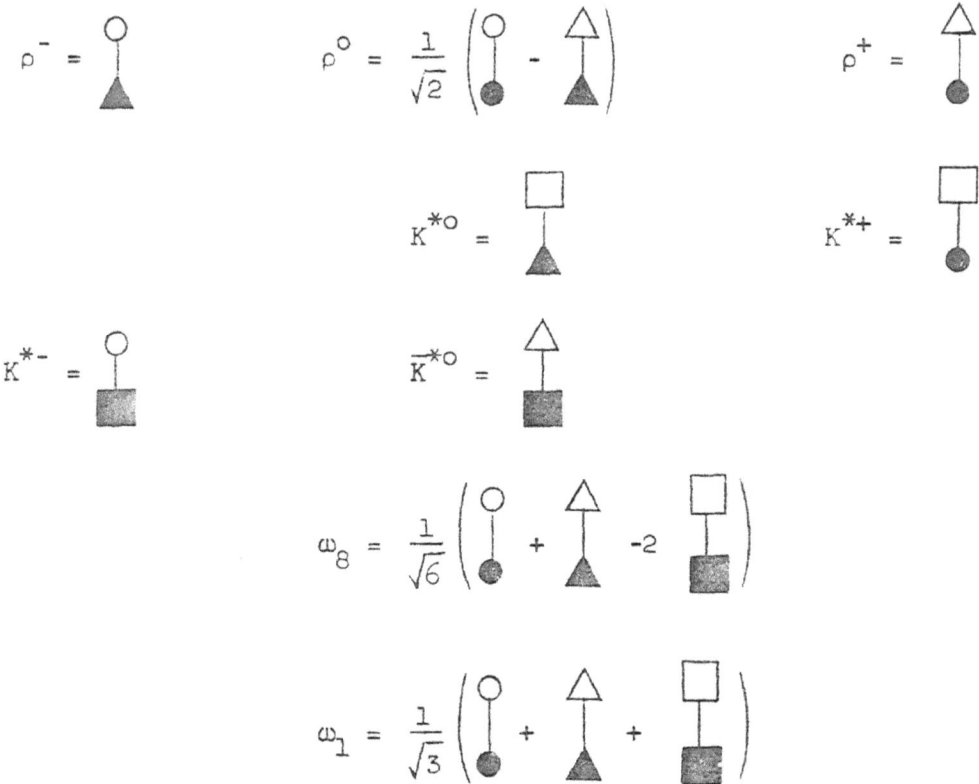

Figure 6.13. The vector meson nonet shown as a linear combinations of deuces. The octet and nonet formed an irreducible representation of SU(6).

When Λ_0 becomes distinguishable from p_0 and n_0 by increasing its mass, SU(3) symmetry is broken, and the ω_8 and ω_1 isospin singlets, with identical quantum numbers, "mix," separating the $\Lambda_0 \bar{\Lambda}_0$ from the other two lighter ace-antiace pairs, creating the observed ω and ϕ wave functions [220], shown in Figure 6.14:

$$\omega = \frac{1}{\sqrt{2}}\left(\begin{array}{c}\end{array} + \begin{array}{c}\end{array}\right)$$

$$\phi = \begin{array}{c}\end{array}$$

Figure 6.14. Graphical representation of the ω and ϕ.

The singlet SU(6) representation ($\mathcal{M}_1^0$ in Eq. 6.7, p. 116) contains the pseudoscalar meson η_1, also called the η'. Its wave function expressed as a linear combination of deuces is

$$\eta_1 = \frac{1}{\sqrt{3}}(p_0 \bar{p}_0 + n_0 \bar{n}_0 + \Lambda_0 \bar{\Lambda}_0).$$

Although the graphical representation of η_1 is the same as that of the ω_1 in Figure 6.13, the η_1 and η_8 do not mix like the ω_1 and ω_8 because they lie in different irreducible representations of SU(6).[34]

Graphical Representation of the Baryon Octet

The graphical representation of the eight spin-1/2 baryon wave functions is based on their algebraic

---

[34] Approaches to η-η' mixing are discussed in the "Eta Physics Handbook: Proceedings of the Workshop on Production, Interaction and Decay of the η Meson," held at Uppsala, October 25–27, 2001, published in *Physica Scripta,* 2002.

representation as linear combinations of treys found by antisymmetrizing the first two aces, and then adding the third ace symmetrically:

$$\begin{aligned}
D_{jk} &= A_j A_k - A_k A_j, \ j,k = \{1,2,3\}, \\
D_{ik} &= A_i A_k - A_k A_i, \ i,k = \{1,2,3\}, \\
T_{ij,k} &= A_i D_{jk} + A_j D_{ik}, \ i,j,k = \{1,2,3\} \\
&= A_i A_j A_k - A_i A_k A_j + A_j A_i A_k - A_j A_k A_i,
\end{aligned}$$

where $T_{ij,k}$ is symmetric in i and j, and $A_1 = p_0$, $A_2 = n_0$, and $A_3 = \Lambda_0$.

For example, except for a normalization constant the proton was represented by the trey $T_{11,2}$,

$$\begin{aligned}
T_{11,2} &= A_1 D_{12} + A_1 D_{12} \\
&= 2 A_1 D_{12} \\
&= 2 p_0 (p_0 n_0 - n_0 p_0) \\
&= 2 (p_0 p_0 n_0 - p_0 n_0 p_0) \\
&= 2\sqrt{2} p, \text{ or} \\
p &= \frac{1}{\sqrt{2}} (p_0 p_0 n_0 - p_0 n_0 p_0). \quad (6.14)
\end{aligned}$$

The order of aces in a trey mattered. For example, the interaction energy between the first and second ace in a trey $T_{ijk} = A_i A_j A_k$ might differ from that between the second and third, or third and first. Therefore ace order affected trey masses (Eq. 6.3, p. 99). The signs between treys in a baryon wave function, affected the baryon's coupling to other hadrons.[35]

In order of increasing mass, the SU(3) octet wave functions turned out to be [221, 223]:

$$p = \frac{1}{\sqrt{2}} (p_0 p_0 n_0 - p_0 n_0 p_0)$$

$$n = \frac{1}{\sqrt{2}} (n_0 p_0 n_0 - n_0 n_0 p_0)$$

$$\begin{aligned}
\Lambda = \frac{1}{\sqrt{12}} (&p_0 n_0 \Lambda_0 - n_0 p_0 \Lambda_0 + n_0 \Lambda_0 p_0 \\
&- p_0 \Lambda_0 n_0 + 2\Lambda_0 n_0 p_0 - 2\Lambda_0 p_0 n_0)
\end{aligned}$$

---

35 A simple example of how the order of particles in the wave function of a multiparticle system enter into its symmetry and decomposition into irreducible representations is given in "Isospin States," p. 149.

$$\Sigma^0 = \frac{1}{2} (p_0 n_0 \Lambda_0 + n_0 p_0 \Lambda_0 - p_0 \Lambda_0 n_0 - n_0 \Lambda_0 p_0)$$

$$\Sigma^+ = \frac{1}{\sqrt{2}} (p_0 \Lambda_0 p_0 - p_0 p_0 \Lambda_0)$$

$$\Sigma^- = \frac{1}{\sqrt{2}} (n_0 n_0 \Lambda_0 - n_0 \Lambda_0 n_0)$$

$$\Xi^0 = \frac{1}{\sqrt{2}} (\Lambda_0 \Lambda_0 p_0 - \Lambda_0 p_0 \Lambda_0)$$

$$\Xi^- = \frac{1}{\sqrt{2}} (\Lambda_0 n_0 \Lambda_0 - \Lambda_0 \Lambda_0 n_0). \quad (6.15)$$

Each trey in a baryon wave function was represented graphically as a scalene triangle with aces at its vertices. Baryon states were linear combination of these triangles. The graphical representation of the baryon octet defined by Eq. 6.15 is given in Figure 6.15, p. 122. This graphical representation was used to compute relations between baryon masses (Eq. 6.3, p. 99), and their couplings to mesons (Figure 6.18, p. 124).

The proton p is represented by the difference of two scalene triangles corresponding to the difference of two treys: $p = \frac{1}{\sqrt{2}} (p_0 p_0 n_0 - p_0 n_0 p_0)$. The first p_0 in the first trey corresponds to the top circle in the first triangle, the remaining p_0 and n_0 to the circle and triangle found by traveling along the first triangle's edge in a clockwise direction. In a scalene triangle the interaction energies between the different pairs of aces differ, corresponding to edges of different length.

Symmetries are apparent. For example, Ξ^0 is found from p by replacing each p_0 with a Λ_0, and n_0 with p_0, decreasing the strangeness by two and the charge by one.

Couplings Computed Graphically

Armed with the graphical representations of hadrons as weighted linear combinations of deuces or treys, couplings were computed by connecting aces to their antiaces in all possible combinations, as exemplified in Figure 6.17, p. 123.

In pseudoscalar meson-baryon coupling there are two different types of coupling conventionally called "f" and "d." Graphically it is simpler to use their sum and difference, as in the Zweig diagram for $f + d$ pion-nucleon coupling shown in Figure 6.16, p. 122.

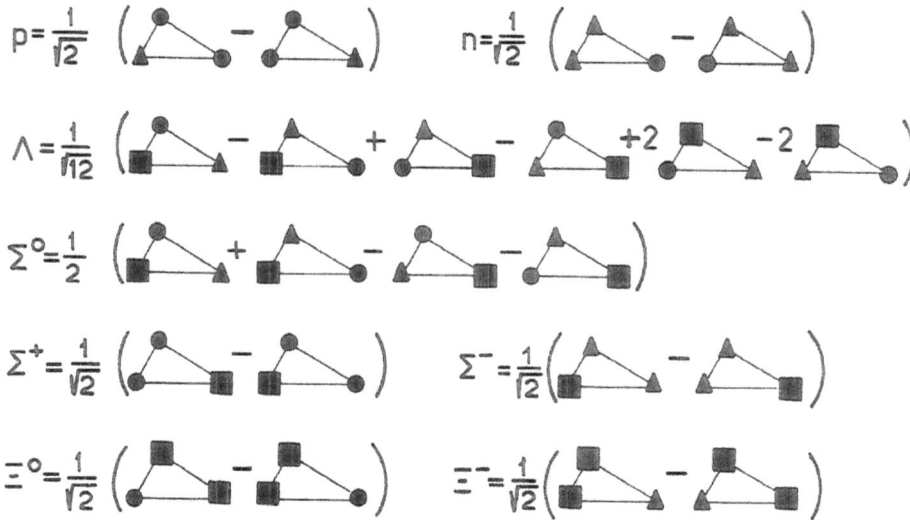

Figure 6.15. Spin-1/2 baryon wave functions represented graphically as weighted linear combinations of treys following from Eq. 6.15. Lines represent springs or pair-wise interaction energies. The probability that a trey exists inside a baryon is the square of its weight. A circle, triangle, or square represents a p_0, n_0, or Λ_0 whose strangeness is 0, 0, or -1, respectively. The larger the symbol, the heavier its corresponding ace, with the triangle only slightly larger than the circle ($M_{p_0} \lesssim M_{n_0} \ll M_{\Lambda_0}$). Therefore, as the number of Λ_0s increases in a baryon in an SU(3) representation, the strangeness of the baryon decreases while its mass increases (the less strangeness the heavier the baryon). Analogously, as the number of n_0s increases in a baryon in an isospin multiplet, the charge of the baryon decreases while its mass increases (the less charge the heavier the baryon). Similar relations held for the decuplet whose treys were equilateral triangles, corresponding to totally symmetric baryon wave functions. (Reproduced from the second CERN Report under a CC-BY license.)

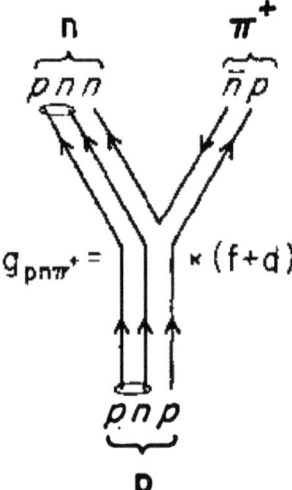

Figure 6.16. The Zweig diagram for $f + d$ meson-baryon coupling. The subscript 0 on aces has been suppressed, and the proton **p**, neutron **n**, and pion $\boldsymbol{\pi}^+$ are in bold face. The "little loop" encloses antisymmetrized aces in a proton's SU(3) wave function (the deuce D_{12} in Eq. 6.14, p. 121). In $f + d$ coupling, the antisymmetrized aces in the proton become the antisymmetrized aces in the neutron. On the other hand, $f - d$ coupling arises if the antisymmetrized aces separate in the decay, one ace terminating in the baryon, the other in the meson. See Section VII, Baryon Couplings in the second CERN Report. (Reproduction from [149].)

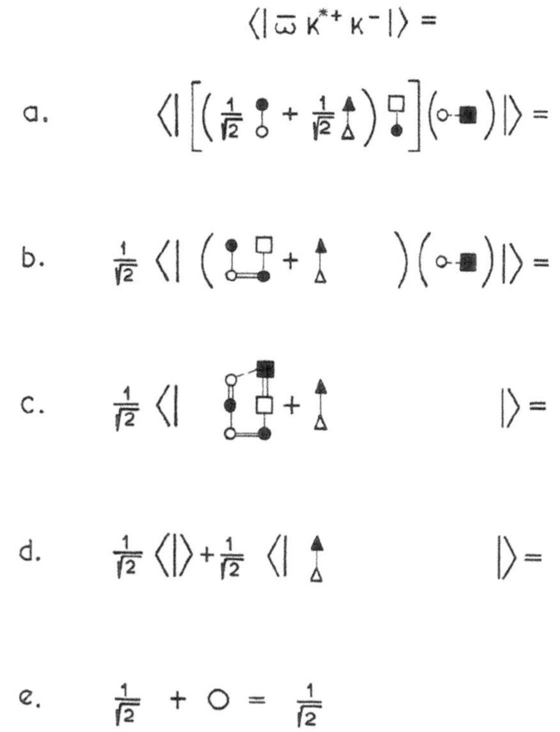

$$\langle |\, \bar{\omega}\, K^{*+}\, K^{-}\, |\rangle =$$

a.

b.

c.

d.

e. $\frac{1}{\sqrt{2}} \, + \, \bigcirc \, = \, \frac{1}{\sqrt{2}}$

Figure 6.17. The graphical "Tinkertoy" computation of the coupling constant for the virtual decay of $\omega \rightarrow K^{*+} + K^{-}$. Black circles, triangles, and squares represent p_0, n_0, and Λ_0; antiaces are white. Single lines between aces in a deuce represent their binding or interaction energies. The $\bar{\omega} = (\frac{1}{\sqrt{2}}\bar{p}_0 p_0 + \frac{1}{\sqrt{2}}\bar{n}_0 n_0)$ in line a. appears as an antiparticle (ace on top, antiace on bottom) to mark it as an initial state. The horizontal orientation of the deuce representing the K^{-} aids in matching its ace and antiace with an antiace and ace in the two other mesons. For a deuce to contribute to a coupling, both its ace and antiace must annihilate (connect) with their antiparticles in other mesons. Double lines between deuces represent those connections.

Top line: The meson coupling constant. The symbols $< |$ and $| >$ are Dirac's "bra" and "ket," dual vector and vector for the vacuum, whose product $< | >$ in line d. equals 1.

Line **a**: Deuces in the initial $\bar{\omega}$, and final K^{*+} and K^{-}, are shown.

Line **b**: The $\bar{p}_0$ in $\bar{\omega}$'s first deuce connects to p_0 in the K^{*+}.

Line **c**: The $\bar{p}_0$ and Λ_0 in K^{-} connect to the remaining p_0 in $\bar{\omega}$, and the $\bar{\Lambda}_0$ in K^{*+}. Every ace that can connect (annihilate) with its antiace has connected.

Line **d**: With all the aces and antiaces in three deuces annihilating each other, the fully-connected structure disappears, leaving the vacuum's bra projecting onto its ket.

Line **e**: That projection equals 1, which, when multiplied by the $\frac{1}{\sqrt{2}}$ weight, yields a coupling constant of $\frac{1}{\sqrt{2}}$. $\bar{\omega}$'s second deuce, whose aces can't be annihilated by the remaining aces in K^{*+} and K^{-}, does not contribute to the coupling. (Figure 10, modified, from the second CERN Report under a CC-BY license.)

The graphical Tinkertoy computation of meson-baryon couplings involved connecting deuces to pairs of treys. One such connection is shown in Figure 6.18.

The coupling shown graphically in Figure 6.18 was written algebraically as $T_{\Lambda_0 n_0 p_0} D_{p_0}^{\Lambda_0} T^{p_0 n_0 p_0}$ or $T_{321} D_1^3 T^{121}$.

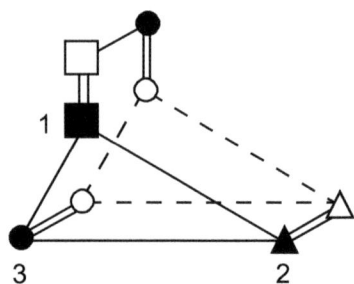

Figure 6.18. A contributor to the $\Lambda p K^-$ coupling, as seen in Figure 13 of the second CERN Report. The large triangle with three black aces at vertices 1, 2, and 3 represents a trey, one of six contributors to the graphical representation of the Λ, as seen in Figure 6.15, p. 122. The "dumbbell" is a deuce representing the K^-. Lines in the back trey with three white antiaces are dashed to aid visualization. Like single solid lines, they represent springs or interaction energies. Double solid lines connect aces with their antiaces. For a trey in a baryon to contribute to a coupling all its aces must connect to their antiaces in treys belonging to other baryons. (Figure 13 a. reproduced from the second CERN Report under a CC-BY license.)

Breaking Free

It is both remarkable and perplexing that, after spending eight years learning mathematics and theoretical physics from excellent mathematicians and the world's best theoretical physicist, I should ignore their training, start from scratch, and revert to thinking in the graphical terms depicted here. Primal influences from childhood prevailed when I was free to work alone, day after day, in a small villa in the middle of an apple orchard in Tannay près Mies ("The Origin of Graphical Calculus," p. 108). Childhood influences run deep.

Chapter 7
Epilogue

An exposed section of the 27 km-long beam tunnel of the Large Hadron Collider (LHC) at CERN. The LHC is focused on discovering new fundamental physics, testing the Standard Model, understanding the structure of exotic hadrons, and studying the quark-gluon plasma. (CERN-PHOTO-201609-210-4.)

> *Whatever happened to aces?*
> They are alive and well! In case you haven't noticed,
> quarks are aces in disguise.

7.1 Recapitulation

Quarks had two origins, definitions, purposes, and names. After his lunch with Serber in 1963, Gell-Mann used quarks as mathematical particles in a toy free field theory to abstract equal time current commutation relations [93]. Murray's quarks could not be used to create hadrons — they had no interactions. In 1966 Murray tacitly assumed that quarks had strong interactions, and used "proof by contradiction" to argue against the possibility of constituent quarks. To paraphrase: If quarks were the constituents of hadrons they would act like hadrons, but given the nature of hadron dynamics, quarks could not possibly create a hadron. Each of its quarks would just be like one of the many other hadrons contributing to its structure ("Were Murray's Quarks Real?," p. 70).

By contrast, my initial goal was to understand the suppression of ϕ decay. Since I suspected that suppression required constituents, determining their quantum numbers became my goal. Once identified, the quantum numbers of hadrons, their strong, electromagnetic, and weak couplings, together with mass relations, could be predicted and compared with experiment. These constituents — aces — couldn't act like hadrons. They obeyed unknown laws of interaction that led them, and their antiparticles, to create the hadrons we observe. If aces weren't real particles, what I was finding made no sense.

Today quarks have one definition and a single purpose. They are real particles of a well-defined quantum field theory that, together with gluons and color charge, form the basis of QCD. They are the constituents of hadrons. The idea that hadrons had constituents that weren't hadrons, and had fractional charge that couldn't be observed in isolation, made the reality of those constituents difficult to accept, and truly revolutionary.

In January 1964, after Murray submitted his paper to *Physics Letters*, he stopped working on quarks. They had served his purpose. Almost a decade later, after their reality had been established, he started up again with Harald Fritzsch and Heinrich Leutwyler as his collaborators, finally allowing quarks to interact [78].

I stopped working on quarks after the Erice Lectures in the late summer of 1964. My problem was not to expand the applicability of the quark model, finding yet another miracle of nature. Rather it was to understand one of them: the quantum field theory for quarks. Finding this field theory seemed much too difficult for me. Even if I could guess it, I wouldn't be able to prove the guess was right. I had done what I could (Section 6.17: "Leaving Physics," p. 113).

Although Niels Bohr was able to correctly calculate hydrogen's spectral lines, his model of the atom could not accurately account for the spectrum of helium, and was conceptually incomplete. The wave nature of particles had not yet been discovered by de Broglie, and was absent from Bohr's thinking. Bohr had no Lagrangian. There was no theory.

Similarly, the constituent quark model was conceptually incomplete, lacking a specification of the interactions between quarks. Like the wave-particle duality of quantum mechanics, constituent quarks had a "split personality," sometimes acting as fields in a quantum field theory for electromagnetic and weak interactions, sometimes as convenient objects for organizing hadrons into representations of SU(6), and for calculating relations between hadron masses and coupling constants. And like the Bohr atom with its stable electron orbits that were not allowed to radiate energy in violation of Maxwell's equations, the Δ^{++} spin-3/2 baryon, created from three identical quarks in the same quantum state, could not exist according to the spin-statistics theorem. I had no Lagrangian. There was no theory. But in the early 1970s quarks channeled thinking in productive directions, eventually leading to QCD.

It was generally believed by particle physicists, Feynman and Gell-Mann included, that hadrons had constituents, but those constituents were other hadrons, as quantified in the bootstrap. Some believed that field theory with elementary particles was suspect. Geoffrey Chew, the UC Berkeley bootstrap evangelist, went further when he announced in June 1961 at a La Jolla conference [34]:

> I believe the conventional association of fields
> with strongly interacting particles to be empty.

... field theory ..., like an old soldier, is destined
not to die but just to fade away.

Gell-Mann was also concerned that field theory
might not be correct, and made sure the equal time
current commutation relations he abstracted did not
depend on the toy free field theories that suggested
them. In the concluding paragraph of his February
1964 quark paper he writes [93]:

> All these [equal time current commutation]
> relations can now be abstracted from the field
> theory model and used in a dispersion theory
> treatment.

That way his work and the bootstrap could coex-
ist, even supporting each other, as Chew, Gell-Mann,
and Rosenfeld wrote in their February 1964 *Scientific
American* article [35]:

> We shall close in an optimistic spirit by men-
> tioning a fascinating possibility that would rep-
> resent the ultimate contribution of the bootstrap
> hypothesis. If the system of strongly interact-
> ing particles is in fact self-determining through
> a dynamical mechanism, perhaps the special
> strong-interaction symmetries are not arbitrar-
> ily imposed from the outside, so to speak,
> but will emerge as necessary components of
> self-consistency. It is remarkable, and puz-
> zling, that isotopic spin symmetry, strangeness
> and now the broader eightfold-way symme-
> try have never been related to other physical
> symmetries. *Perhaps their origin is destined
> to be understood at the same moment we
> understand the pattern of masses and spins
> for strongly interacting particles* [my italics].
> Both this pattern and the puzzling symme-
> tries may emerge together from the bootstrap
> dynamics.

What was "remarkable, and puzzling," was correctly
identified, as was their conjecture that the origin of
isotopic spin (isospin) symmetry, strangeness, and the
eightfold-way symmetry "is destined to be understood
at the same moment we understand the pattern of
masses and spins for strongly interacting particles." In
fact, "this pattern and the puzzling symmetries" came
from constituent quarks — aces — *precisely* crafted
through ace-spin symmetry to "understand the pattern
of masses and spins for strongly interacting particles."

Forget what was missing from the constituent quark model; the important observation was that
hadrons had constituents with baryon number 1/3, and
dynamics that suggested they were real, and therefore
should have a corresponding quantum field theory.
They do in QCD, but with a twist. Unlike other quantum fields, quark creation operators only create quarks
in confinement.[1]

The pattern of spectral lines of hydrogen that
led to the Bohr model of the atom, and so to the
eventual creation of quantum mechanics, were immediately derived from quantum mechanics. Although
ever more accurate numerical calculations of the mass
of individual low-lying hadrons are now possible with
lattice QCD, the *regularities* that exist among these
masses, regularities that led to the discovery of their
constituents, are still not understood.

How can a nonlinear theory as complicated as
QCD give rise to the simple low-energy relations of the
constituent quark model? There should be some way
to approximate QCD so that the hierarchy of representations and the existence of relations between hadron
masses becomes apparent. It's not like the physics
of water where the greater the nonlinearity, the more
complex the flow. In QCD the harder the calculation,
the simpler the result.

Recall the history of classical mechanics. Newton's laws were reformulated again and again, over
centuries, each time with a different purpose. In the
following centuries QCD will be reformulated again
and again, and with one of those reformulations the
constituent quark model may appear, *a fortiori*.

7.2 When the Fog Lifted

In 1832 Carl von Clausewitz in his book *On War*
writes:

> The factors on which action in war are based
> are wrapped in a fog of greater or lesser uncer-
> tainty. A sensitive and discriminating judgment

---

[1] In 1977 I told my 13-year-old son Geoffrey that in the beginning when the Universe was very hot, it contained a quark-gluon
plasma. As the Universe expanded and cooled it condensed into
three-quark clusters — protons, neutrons, and their antiparticles.
When I told him that current thinking held that isolated quarks
could not exist, he asked: "You mean that the number of quarks
in the Universe was divisible by three?" Put that in your pipe and
smoke it.

is called for; a skilled intelligence to scent out the truth.

So too in physics. Even the eyes of the best physicists were clouded by dense fog in the middle of the twentieth century. More than two decades of experimental discoveries and theoretical inventions — when digested — burned it off, revealing QCD, the electroweak interactions, and the Higgs field that breaks its symmetry.

The discovery and eventual acceptance of quarks would come in several steps:

1. First indirectly from the:
   (a) suppression of ϕ decay,
   (b) hadron mass relations,
   (c) regularities in hadron β-decay,
   (d) connections between hadron quantum numbers and their SU(6), SU(3), and SU(2) representations.

   These findings required constituents. There was no other way of understanding them. Here seeing was believing, but not with an optical lens.
2. Then directly from
   (a) high-momentum-transfer scattering experiments with electrons and neutrinos starting in 1968,
   (b) The discovery in 1971 of the fourth quark in Japan inside of what are now called the $D^{\pm}$ mesons.
3. Finally, indirectly from the strong suppression of J/ψ decay in 1974.

Seeing Is Believing

The spectrum of hadrons — their regularities in representations, quantum numbers, masses, and couplings — led to the discovery of aces. Confirmation was needed from independent sources. Although many experiments searched for free quarks at accelerators and in cosmic rays, experimentalists did not ask how they might look for them inside of protons and neutrons where they really resided.

Additional support for the existence of constituent quarks eventually came from high-momentum-transfer experiments [21, 25], as suggested in the second CERN Report (Section 6.9: "How to Look for Aces," p. 103), but quite by accident. Henry Kendall explains in his 1990 Nobel Lecture:

> In late 1967 the first of a long series of experiments on highly inelastic electron scattering was started at the two mile accelerator at the Stanford Linear Accelerator Center (SLAC) using liquid hydrogen and, later, liquid deuterium targets. ... *the object was to look at large energy loss scattering of electrons from the nucleon* (the generic name for the proton and neutron), a process soon to be dubbed deep inelastic scattering [my italics].

Kendall and collaborators were not looking for quarks, but their results for high-momentum-transfer electron-nucleon inelastic scattering, interpreted by Feynman and James Bjorken, showed that hadrons contained point particles [19, 72], particles Feynman called partons. However, it wasn't clear what those partons were. If partons were quarks why weren't these fractionally charged particles seen in the collision products of high-energy scattering reactions? Sidney Drell, the senior theorist at SLAC, with two collaborators, tried to solve this puzzle by making partons "bare" nucleons and pions with integer charge, hadrons without their virtual cloud of hadrons that gave them spatial extent [51].

Three years later in September 1972 at the *XVI International Conference on High Energy Physics*, using neutrinos and antineutrinos instead of electrons as projectiles, Donald Perkins reported results from CERN's "Gargamelle" bubble chamber experiment [172]:

> The preliminary data on cross sections provide an astonishing verification of the Gell-Mann/Zweig quark model of hadrons. ... In terms of constituent models, the fractionally charged (Gell-Mann/Zweig) quark model is the <u>only</u> one which fits both the neutrino and electron data.

While this conclusion doesn't say that quarks are real particles, the experiments on which it was based could only be understood that way. Protons and neutrons were probed with structureless leptons that appeared to be scattering off other structureless particles with quark quantum numbers. That meant quarks existed inside of hadrons. The high-momentum-transfer experiments responsible for this assessment led many physicists, some dragging and kicking their feet, to the acceptance of real quarks.

I had looked at nature, but through a different lens, and saw aces. That vision had now received definitive confirmation. Kendall, with his collaborators Jerome Friedman and Richard Taylor, shared the 1990 Nobel Prize in Physics "for their pioneering investigations concerning deep inelastic scattering of electrons on protons and bound neutrons, which have been of essential importance for the development of the quark model in particle physics."

The fourth quark Further confirmation of the existence of real quarks came in 1971 with Kiyoshi Niu's discovery of what would be called the D meson. Niu's particle was immediately recognized as containing the fourth charm quark (c) and the antiparticle of the down quark ($\bar{d}$), $D^+ \sim c\bar{d}$ [112]. The discovery of the fourth quark was the capstone for aces, confirming its existence, and justifying the name aces, now that there were four of them. Appendix 7.C, p. 142, details its remarkable discovery.

The J/ψ meson Three years later in November 1974 Burton Richter at SLAC and Samuel Ting at BNL discovered the J/ψ. Although it had all the attributes of an ordinary heavy strangeness $\mathbb{S}=0$ meson, its lifetime was extraordinary long, just like the lifetime of the ϕ. Soon it was realized that the J/ψ was created from the fourth quark and its antiquark ($J/\psi \sim c\bar{c}$). The decays of ϕ and J/ψ were both suppressed for the same reason: What they could decay into was severely restricted by Zweig's rule [80]. Another Nobel Prize was awarded to Richter and Ting in 1976 "for their pioneering work in the discovery of a heavy elementary particle of a new kind." The idea that hadrons contain elementary constituents with fractional charge was finally accepted by almost the entire physics community.

7.3 Asking the Right Question

In 1937 Wigner introduced SU(2) symmetry to describe the proton-neutron symmetry of the nuclear force that Heisenberg had conjectured, and Merle Tuve had then observed [202]. As the number of strongly interacting particles increased and a new quantum number was required (strangeness), theorists asked: "What symmetry group should now be used for the strong interactions?" Remarkably, the first

candidate was not SU(3), the obvious extension of SU(2), but Global Symmetry, which Gell-Mann and later Lee and Yang adopted [88, 143]. Finally Gell-Mann and Ne'eman proposed SU(3). The question then was: "What SU(3) irreducible representations — generalizations of isospin multiplets — should be used for hadron classification?" There were many possibilities besides Gell-Mann and Ne'eman's original eight-dimensional representation: Glashow and Sakurai wrote a paper titled "The 27-fold Way and Other Ways: Symmetries of Mesons-Baryon Resonances" [102]. Going in a different direction, Yoichiro Nambu and Sakurai argued that at least one hadron should not even lie in a representation of SU(3) [157]. These authors weren't lightweights. Both Glashow and Nambu went on to win separate Nobel Prizes for other work. It was a time of great confusion.

The question of which symmetry group to use for the strong interactions, while helpful in organizing data, was the wrong question to ask. The right one was: "What's the stuff of hadrons?"

7.4 The Medium Was the Message

With the discovery of the neutron, and later other hadrons, the message was that SU(2) isospin symmetry, and its SU(3) extension, were important symmetries for particle physics. The Eightfold Way was the path to follow [90, 159]. That's what we were taught in the early 1960s. But that message missed the mark. The medium — constituents — was the message.

Neither SU(2) nor SU(3) symmetry was fundamental. Instead, these approximate symmetries simply reflected the existence of three constituent quarks with unequal mass, one mass difference small, the other large.

QCD does not incorporate SU(2) or SU(3) into its structure. These approximate symmetries are an output of QCD, not an input, and therefore not vitally important as we all once thought. Evidently the Eightfold Way was not the path to follow, as beautiful as it seemed in 1961.

Intrinsic quantum numbers of hadrons were also superficial. They are derived from their constituents through the net number of aces, $\Delta(A)$, defined as the number of aces $n(A)$ minus the number of antiaces $n(\bar{A})$ in a hadronic system,

$$\Delta(A) = n(A) - n(\bar{A}), \qquad (7.1)$$

where their possible spin projections are not included. The baryon number, isospin projection, strangeness, and charge of a hadron, or system of hadrons, are found by counting the net number of their constituents, and combining them in various linear combinations:

$$B = \frac{1}{3}[\Delta(p_0) + \Delta(n_0) + \Delta(\Lambda_0)],$$
$$I_Z = \frac{1}{2}[\Delta(p_0) - \Delta(n_0)],$$
$$\mathbb{S} = -\Delta(\Lambda_0),$$
$$Q = \frac{2}{3}\Delta(p_0) - \frac{1}{3}\Delta(n_0) - \frac{1}{3}\Delta(\Lambda_0). \quad (7.2)$$

The conservation of strangeness $\mathbb{S}$ now has a simple interpretation. Rather than a biblical pronouncement or empirical fact, it holds because the net number of Λ_0s in a system of hadrons cannot be changed by the strong interactions.[2]

The charge Q of a hadron is just the sum of its constituent charges. Equation 7.2 for Q gets down to basics, a simpler alternative to the Gell-Mann–Nishijima charge formula,

$$Q = I_Z + \frac{B + \mathbb{S}}{2}, \quad (5.5 \text{ revisited})$$

which appeared to be a mysterious empirically derived linear combination of other quantum numbers.

7.5   When the Dust Settled

The CERN Reports and Erice Lectures were limited in scope, but essentially correct. Together with Gell-Mann's paper in *Physics Letters*, they initiated a long research enterprise, still ongoing, involving thousands of physicists — experimentalists and theorists alike — that eventually led to the creation of QCD, the unification of the weak and electromagnetic interactions, and the Higgs boson, i.e., the "Standard Model."

What in the two CERN Reports and Erice Lectures has survived, and what was missing?

---

[2] Since the quantization and conservation of strangeness by the strong interactions are due to a constituent carrying strangeness, one cannot help but wonder if quarks and leptons are both composites with common charged constituents whose indestructibility under all interactions is responsible for the quantization and absolute conservation of charge.

Surviving Elements

- Constituent quarks as real particles, with the three lowest-mass quarks correctly identified, and their mass differences correctly estimated.
- Spin-1/2 quarks creating hadrons with the proper quantum numbers in cascading representations of SU(6), SU(3), and SU(2).
- SU(3) and SU(2) symmetries broken by quark mass differences.
- Mass relations for mesons and baryons that continue to hold.
- Zweig's rule for coupling, but reinterpreted: As a quark and antiquark in a hadron separate, gluon-gluon interactions constrain fields with color to string-like objects called "flux tubes," which exert a constant force when stretched. As the tube grows longer, rather than increasing its length, it becomes energetically favorable to pull a quark-antiquark pair out of the vacuum, and break the tube, thereby forming a pair of hadrons [180]. This fracture prevents the isolation of quarks as free particles in QCD's explanation of Zweig's rule.
- Electromagnetic and weak interactions of hadrons created by currents constructed from quark fields.
  - The weak decays of hadrons are induced by the weak decays of their constituents.
  - The $\Delta\mathbb{S} = 0$, ± 1 and $\Delta\mathbb{S} = \Delta Q$ rules hold for the weak interactions.
- Exotic hadrons at higher mass.
- The "demotion" of hadrons and their strong interactions. Like van der Waals forces between molecules, both hadrons and their interactions are examples of emergent phenomena. In QCD the collective behavior of quarks and gluons gives rise to hadrons and their interactions, neither present in the laws of QCD.
- A correspondence between quarks and leptons.

Close, But No Cigar

The potential between two aces was assumed to grow quadratically with increasing separation (like two masses on a spring), allowing meson states to be classified according to the energy levels of a three-dimensional harmonic oscillator [224]. This quadratic-potential hypothesis was far removed from other potentials found in particle physics, like the Yukawa potential created by pion exchange, which decreases faster than exponentially with distance. We now know that the potential grows linearly at large distances, not quadratically [109].[3]

Missing Elements

- Two additional quarks with their own quantum numbers $\mathbb{B}$ and $\mathbb{T}$, analogous to strangeness $\mathbb{S}$.
- Gluons and their coupling to quarks, and to each other, in a quantum field theory.
- "Color charge" carried by quarks and gluons ("Quantum Chromodynamics," p. 134).

A Big Surprise

- While the p_0 and n_0 mass *difference* is as expected, their small masses, which are the same order as their mass difference, are a complete surprise. Therefore the mass of a nucleon is almost entirely interaction energy. From a fundamental physics perspective, those opposing 300-pound NFL linemen are essentially balls of gluons and quark-antiquark pairs crashing into one another.

A Requiem For the Past

SU(2) extended to SU(3) was really a house of cards built on Heisenberg's idea that the proton and neutron, so close in mass, should be considered the same particle when formulating a field theory of the nuclear force. This was a very useful — even beautiful — idea, eventually leading to isospin multiplets for all the strongly interacting particles. But extending SU(2) to SU(3) was quite a stretch. Now hadrons in the same SU(3) representation were widely spaced in mass, and treating the eight spin-1/2 baryons as the same particle when formulating a field theory for the strong interactions did not look promising. The group SU(2) could be extended to SU(3), but the corresponding physics didn't extend with it. The ability to classify hadrons in representations of SU(2) and SU(3) was merely a reflection of their composite nature.

The Great Yogurt Project [23]

At the last of five conferences on "Symmetry Principles at High Energy," held between 1964 and 1968, Yuval Ne'eman with great honesty proclaimed: "I am not only the summarizer of these five conferences but also the undertaker responsible for their lying in peace forever." He wistfully reminded the attendees of a story told by one of the conference organizers, Behram Kursunoglu, about Nasreddin Hodja, a fictional character in Turkish folklore:

> One morning a woodcutter saw Hodja by the edge of a lake, throwing cultured yogurt into the water. "What are you doing?" he asked. Hodja looked up and replied, "I am trying to turn the lake into yogurt." The woodcutter laughed and said, "Fool, that will never work." Hodja was silent for a while, stroking his beard. Finally he replied, "But just imagine if it worked!'

Many idiosyncratic efforts to understand the strong interactions, reminiscent of the one depicted in Figure 7.2,[4] had ended, or were about to end.

Figure 7.2. Nasreddin Hodja riding backwards on a donkey. When asked why he replied: "I do not want to be seen as a person who is following the same path as a donkey." (©Nevit Dilmen, licensed under the Creative Commons Attribution-Share Alike 3.0.)

Finally, paraphrasing Geoffrey Chew, the bootstrap, not field theory, "like an old soldier, [was] destined not to die but just to fade away."

---

[3] For small distances the coupling between two quarks approaches zero as the distance between them goes to zero; at close distances quarks act like they are almost free. This is referred to as "asymptotic freedom."

[4] Link to license: creativecommons.org/licenses/by-sa/3.0/deed.en.

7.6 Are Regularities Still Relevant?

Probably. Science has a way of forgetting the origins of its ideas. There is enough to do today without thinking about yesterday. However, there are reasons to remember the origins of QCD. The regularities that led to the discovery of constituent quarks suggest experiments and theoretical questions that are still relevant today. As artists in the Italian Renaissance understood, sometimes peering back helps in looking forward.

The internal structure of hadrons is extremely complex. Protons are not simply three quarks, but three quarks embedded in a sea of gluons, and quark-antiquark pairs, all interacting with one another. Heavier quark-antiquark pairs coming from the fourth quark (charm) also contribute, as considered by Stanley Brodsky in 1980 [27]. It took till 2022 for charm-anticharm quarks to be observed within the proton, leading Mike Williams, founder and leader of the LHCb and GlueX experimental groups at MIT, to be quoted in Quanta Magazine as saying "This is the most complicated thing [the proton] that you could possibly imagine. In fact, you can't even imagine how complicated it is."[5] Yes, but by highlighting this remark Quanta missed what's truly amazing: Given this complexity, there is no reason to expect that *the proton sometimes acts as if it were simply a composite of three fractionally charged particles.* How can this be?

The conclusion of the second CERN Report asks "Why does so simple a model yield such a good approximation to nature?" The idea that hadrons are created from constituent quarks is not as surprising as the regularities they leave on the hadron spectrum. Such simplicity is not expected from QCD, a highly nonlinear quantum field theory where perturbation theory cannot be used to understand the order observed in low-mass hadron physics. High energy theorists should be familiar with regularities in mass, long forgotten, since interpreting them provides hope for finding a deeper understanding of QCD. Feynman had the Lamb shift to inspire his creation of a renormalizable computational formulation of QED. Now theorists have the regularities of the constituent quark model. Perhaps those regularities will

inspire them to create a tractable computational framework for QCD, enabling them to *make predictions at low energies* where nonlinearities are strongest, but regularities abound. Trying to understanding this "theoretical anomaly" in QCD seems like a reasonable pursuit while waiting for the next experimental surprise.

Specifically, insight into QCD might follow from attempts to derive, from first principles, the two most remarkable constituent-quark mass relations:

$$\frac{M_\omega - M_{\rho^0}}{4} = \frac{M_\phi + M_{\rho^0}}{2} - M_{K*0}$$

$$8.5 \pm ? \qquad -\ 4 \pm ? \text{ in 1963}$$

$$1.85 \pm 0.07 \qquad 1.81 \pm 0.24 \text{ in 2023.}$$

(2.8 revisited)

and

$$(M_{\Xi^-} - M_{\Xi^0}) = (M_{\Sigma^-} - M_{\Sigma^+}) - (M_n - M_p)$$

$$5.6 \pm 1.4 \qquad 7.0 \pm 0.5 \text{ in 1963}$$

$$6.85 \pm 0.21 \qquad 6.79 \pm 0.08 \text{ in 2023,}$$

(2.9 revisited)

where masses are in MeV, and the 2023 values are from ref. [168].

Increasing the accuracy of measured mass differences in these two equations until nonzero deviations appear would challenge theorists to derive both the relations and their deviations. Increasing accuracy should not be difficult because mass measurements are old. The last measurement of the Ξ^0 mass, which is the major source of error in Eq. 2.9, was made in 2000 [54]. As the Lamb shift demonstrated, theorists are very good at finding an answer if they know it in advance ("Quantum Electrodynamics," p. 45).

So Eqs. 2.8 and 2.9 should be read in the "present perfect continuous tense" used here to describe previous findings still relevant today. Asking theorists to use QCD to understand the regularities responsible for its origin 60 years ago gives yet another meaning to William Faulkner's line from a "Requiem for a Nun:"

The past is never dead. It's not even past.

7.7 Tetra and Pentaquarks

The CERN Reports focused on hadrons created from deuces and treys, but assumed additional deuces or

---

[5] Quote from Charlie Wood, Quanta Magazine, October 19, 2022, www.quantamagazine.org/inside-the-proton-the-most-complicated-thing-imaginable-20221019/.

treys would be added to create heavier hadrons. This possibility was further explored in the 1964 Erice Summer School Lectures where the SU(6) and SU(3) irreducible representations of $A\bar{A}A\bar{A}$ ("tetraquarks") were constructed ("Exotic Hadrons," p. 98).

The first exotic hadron, the tetraquark X(3872), requiring an extra deuce was seen in 2003 by some 400 physicists and engineers in the Belle experiment in Tsukuba, Japan [37]. It was composed of $u\bar{u}c\bar{c}$ (c is the fourth quark), and had $J^{PC} = 1^{++}$, where J, P, and C are spin, parity, and charge conjugation. The 3872 referred to its mass in MeV, a little more than four times the mass of the proton. The X(3872) was seen decaying into $\pi^{+} + \pi^{-} + J/\psi$. Understanding tetraquarks and their excitations will not be easy. Even the excited states of the alpha particle remain a mystery [133].

A dozen years later in 2015 two "pentaquark" baryons composed of $uudc\bar{c}$ were accidentally discovered at CERN by the LHCb collaboration composed of approximately 1260 physicists and engineers from 74 scientific institutes, representing 16 countries. They were called $P_c^{+}(4380)$ and $P_c^{+}(4450)$, decaying into $p + J/\psi$. A second 2019 study showed that the $P_c^{+}(4450)$ actually consisted of two peaks (the $P_c^{+}(4440)$ and $P_c^{+}(4457)$), and an additional narrow $P_c^{+}(4312)$, also composed of $uudc\bar{c}$ [1].

In some exotics all quarks are tightly bound, like quarks in non-exotics. In other exotics, quarks are partitioned into two non-exotics, with quarks in each non-exotic tightly bound, the non-exotics weakly held together. The "hexaquark" (uud-udd), commonly called the deuteron, is the simplest example, but it isn't called exotic.

"Glueballs" built only from gluons, are also possible, but if they have the quantum numbers of $q\bar{q}$ states they will "mix" with them, making identification difficult. "Oddballs," glueballs with quantum numbers not found in quark-antiquark pairs (e.g., $J^{PC} = 1^{-+}$), are *new states of matter* yet to be observed. While not mixing with ordinary mesons they can, in principle, mix with exotic mesons like the tetraquarks. If their peaks are broad, discovery will be challenging.

In other exotics $q\bar{q}$ states combine with gluon states to form "hybrid" mesons.

Many particle detectors located throughout the world are searching for exotics and determining their internal dynamics. The effort is staggering. A 2020 181-page review paper of an important but restricted class of exotics referenced 977 papers, some of them lengthy reviews [24]. Still, the experimental and theoretical study of exotics is in its infancy. If ace-spin symmetry applies to exotics, as described in the 1964 Erice Summer School Lectures, exotic spectra are enormously complex.[6]

7.8   The Quark-Gluon Plasma

Squeeze ice and it turns into water. Squeeze it some more and it turns into ice with a different crystal structure called ice III. Eighteen phases of ice are known. Hadrons are tiny "crystals." What happens when you heat or squeeze a handful? Can these crystals stick together like flakes of snow? Probably not, but what are the phases of quark-gluon matter? The answer to this question is relevant for solving problems in astrophysics and cosmology.

Heavy ions colliding at the Relativistic Heavy Ion Collider and the Large Hadron Collider have created a "quark-gluon plasma," a new phase of matter. Quarks and gluons, no longer bound together as deuces and treys, formed a nearly inviscid "fluid" [113]. In the very early intensely heated universe, quarks and gluons were also unconfined. As the universe expanded and cooled, one or more phase transitions turned them into protons and neutrons.

When a star explodes into a supernova, extraordinarily dense states of matter are created in its interior. A neutron star may emerge, typically 40% more massive than the Sun with a radius of only 10 km [114]. If it were to get any smaller, two or more neutrons would have to occupy the same quantum state, not allowed by the spin-statistics theorem (p. 92). A "degeneracy pressure" would develop, preventing further collapse. However, if external pressures were great enough the neutrons would melt into a plasma of quarks and gluons. A further increase in external pressure could result in the formation of a second degeneracy pressure now designed to keep any two quarks from occupying the same quantum state.

---

[6] For example, the Erice lectures gives the SU(6) irreducible representations for the tetraquarks: $6 \otimes 6 \otimes \bar{6} \otimes \bar{6} = 2 \otimes 1 \oplus 4 \otimes 35 \oplus 189 \oplus 280 \oplus \overline{280} \oplus 405$. These representations are further decomposed into irreducible representations of SU(3) with different spins, leading to a staggering number of possibilities for the 1296 tetraquarks, which are not repeated here.

A "quark star," or a quark star with a "neutron mantle," might form. Do these structures exist, and can they be detected by the gravitational waves created by their merger with astronomical objects?

7.9 Free Fractionally Charged Particles?

The fundamental unit of charge is $Q = 1/3$. Do particles with the fundamental unit of charge exist as free particles? Quarks are confined because of their color, not their fractional charge. Therefore free fractionally charged particles, other than quarks, might exist as leftovers from the Big Bang, although the *a priori* probability is small.[7] They would be colorless and carry one or more of the charges $Q = \pm 1/3, \pm 2/3, \dots$. If they exist at least one of them, and its antiparticle, would be stable. Stability follows from the conservation of charge, which requires that any unstable fractionally charged particle decay into one or more other fractionally charged particles, each lighter than the preceding one. Eventually this sequence of decays must terminate, resulting in one or more stable particles with fractional charge. Each of these particles would have a corresponding stable antiparticle of opposite charge. Some, like negative muons, might not interact strongly and catalyze fusion at room temperature [225].

When I told Feynman about the possibility of free fractionally charged particles in 1978, wondering where they might be, he immediately replied, "In the center of the Earth," because massive stable particles, like the heavy elements, end up there. One must look and see, but in an asteroid like Psyche, which has an exposed nickel-iron core whose composition is most probably similar to the core of the Earth [Figure 7.3]. If that's not possible, looking in iron meteorites that have fallen to Earth would be a good place to start.

7.10 The Standard Model

The Standard Model of particle physics describes three of the four fundamental forces that are currently known to exist. QCD, defining the force between quarks, carried by gluons, is its first component. The electroweak theory of the electromagnetic and weak forces, carried by the photon and three intermediate

Figure 7.3. Psyche, a large M-type asteroid, thought to be the metal core of a differentiated planetesimal whose outer mantle has been blown away. If free fractionally charged particles exist they would have been created in the Big Bang. Because they would be heavy (otherwise they would have been seen at accelerators), they would eventually be concentrated in the center of planets and differentiated planetesimals, together with the heavy elements. (Screenshot courtesy of NASA.)

vector bosons, is its second component. Finally, the Standard Model contains the mass-generating Higgs field with its Higgs boson.

Quantum Chromodynamics

- *Quarks:* There are six quarks called up (u), down (d), strange (s), charm (c), bottom (b), and top (t). Their quantum numbers are given in Table 7.1.

		u	d	s	c	b	t
Q	electric charge	2/3	−1/3	−1/3	2/3	−1/3	2/3
I	isospin	1/2	1/2	0	0	0	0
I_Z	Z-component of $\vec{I}$	1/2	−1/2	0	0	0	0
$\mathbb{U}$	upness	1	0	0	0	0	0
$\mathbb{D}$	downness	0	−1	0	0	0	0
$\mathbb{S}$	strangeness	0	0	−1	0	0	0
$\mathbb{C}$	charm	0	0	0	1	0	0
$\mathbb{B}$	bottomness	0	0	0	0	−1	0
$\mathbb{T}$	topness	0	0	0	0	0	1

Table 7.1 Quark quantum numbers with columns listed in order of increasing mass. All quarks have spin 1/2 and baryon number $B = 1/3$. (Table 15.1 of the Quark Model Review [168], modified.)

---

[7] I once asked Edward Witten if string theory contains free fractionally charged particles. Amazed by the naiveté of my question he exclaimed "Of course they do!"

- *Flavor:* The generic name for the six different kinds of quarks, with their associated flavor quantum numbers: $\mathbb{S}$, $\mathbb{C}$, $\mathbb{B}$, and $\mathbb{T}$. These quantum numbers for a hadron, or system of hadrons, are defined by $-\Delta(s)$, $\Delta(c)$, $-\Delta(b)$, and $\Delta(t)$, where $\Delta(q) = n(q) - n(\bar{q})$ and $n(q)$ is the number of quarks q of a particular flavor (Eq. 7.1, p. 129). Although not conventional, an upness and downness flavor quantum number can be defined by $\mathbb{U} = \Delta(u)$ and $\mathbb{D} = -\Delta(d)$. Then

$$I_Z = \frac{\mathbb{U} + \mathbb{D}}{2}.$$

- *Charge:* Define the "hypercharge" Y of a hadron by

$$Y = B + \mathbb{S} + \mathbb{C} + \mathbb{B} + \mathbb{T},$$

where the quantum numbers in the expression for Y are its baryon number, strangeness, charm, bottomness, and topness. The average charge of all the hadrons in an isospin multiplet is half the isospin's hypercharge.

The hadron's charge Q is its isospin projection I_Z plus half its hypercharge, which may be rewritten in a simpler form:

$$\begin{aligned} Q &= I_Z + \frac{Y}{2} \\ &= \frac{\mathbb{U} + \mathbb{D}}{2} + \frac{B + \mathbb{S} + \mathbb{C} + \mathbb{B} + \mathbb{T}}{2} \\ &= \frac{B}{2} + \frac{\mathbb{U} + \mathbb{D} + \mathbb{S} + \mathbb{C} + \mathbb{B} + \mathbb{T}}{2}, \end{aligned}$$

so Q is half the hadron's baryon number plus half its total flavor quantum numbers.

- *Color:* Quarks carry a new quantum number called color that has one of three possible values (color charges): red (r), green (g), or blue (b). Antiquarks carry anticolor: cyan ($\bar{r}$), magenta ($\bar{g}$), or yellow ($\bar{b}$), complementary to the colors of quarks. A color annihilates an anticolor if, and only if, their colors are complementary.

- *Gluons:* The force between quarks is carried by a set of eight massless neutral spin-1 gluons that carry color-anticolor pairs *mathematically* analogous to the octet of pseudoscalar mesons whose wave functions contain ace-antiace pairs. The octet of gluons forms an irreducible representation of an exact SU(3) for color, while the pseudoscalar meson octet forms an irreducible representation of a broken SU(3) for hadrons. The colors and anticolors of gluons couple to the anticolors and colors of antiquarks and quarks, as well as those of other gluons, making gluons the carriers of the force responsible for the creation of hadrons.

It is remarkable that although gluons are massless, like the photon and graviton, they only get a chance to act over short distances. Creation of virtual quark-antiquark pairs, as color charges separate, restrict the gluon's range of interaction.

- *Confinement:* Particles created from quarks and gluons exist as free particles if, and only if, they are color SU(3) singlets, i.e., when adding their colors as light, the result is white. Therefore they must carry all of the three primary colors (red + green + blue = white). Since a quark carries only one color, quarks and diquarks cannot exist as free particles. Although QCD is well-established and confinement is widely accepted, the confinement of color has never been proven.

- *Resolution of the spin-statistics problem:* In creating the spin-3/2 baryons from quarks, we encountered a difficulty associated with the spin-statistics theorem, which says that no more than one quark with a particular set of quantum numbers can occupy a quantum state. This theorem forbade the existence of the three spin-3/2 baryons whose quarks were indistinguishable: the $\Delta^{++} \sim p_0 p_0 p_0$, $\Delta^- \sim n_0 n_0 n_0$, and $\Omega^- \sim \Lambda_0 \Lambda_0 \Lambda_0$. The problem is solved by attaching three different colors to the three quarks in each of these hadrons, also making them color SU(3) singlets.

- *Graphical calculus:* The same graphical calculus used to compute SU(3) couplings of hadrons can be used to compute SU(3) color couplings of gluons to quarks, and gluons to gluons. The pictorial representation of the pseudoscalar mesons, shown in Figure 6.12, p. 119, can be mapped into one for gluons by changing the black circles, triangles, and squares to circles of equal area, and coloring them red, green, and blue, respectively. The white $\bar{p}_0$, $\bar{n}_0$, and $\bar{\Lambda}_0$ also become identical circles, but with the complementary colors of cyan, magenta, and yellow. The methods of graphical calculus in Appendix 6.C will then apply to computing color coupling constants.

Zweig diagrams for hadron couplings can be used to represent color couplings. For example, while Figure 2.6, p. 25 represents the decay of meson $A\bar{A}$

into mesons $A\bar{A}'$ and $A'\bar{A}$, it can also represent the interaction of gluons.

The Electroweak Theory

A theory of the electromagnetic and weak interactions, with the photon and three massive intermediate vector bosons as their force carriers, was developed in stages independently by Sheldon Glashow, Abdus Salam, and Steven Weinberg in the 1960s. This "electroweak theory" unified the description of two of the four fundamental forces.

In the electroweak theory there are two charged $W^{\pm}$ bosons (named after the "w" in the <u>w</u>eak interactions), and a neutral Z^0 boson (named for the "z" in its <u>z</u>ero charge). Quarks interact with all three intermediate vector bosons and the photon, but only change their flavor through their interactions with the W^+ and W^-. For example, the β-decay of n_0 occurs in a two-step process,

$$n_0 \quad \rightarrow \quad p_0 + W^-, \text{ followed by}$$
$$W^- \quad \rightarrow \quad e^- + \bar{\nu}_e, \text{ together leading to}$$
$$n_0 \quad \rightarrow \quad p_0 + e^- + \bar{\nu}_e,$$

where the W^- is virtual.

The Z^0 couples to all particles in the Standard Model, and like the photon, only transfers energy, momentum, and spin.

The Higgs field The Higgs field, which permeates all of space, breaks the symmetry of the electroweak force, whose carriers are the photon and three intermediate vector bosons. The symmetry, which views the four carriers as massless, is broken when the bosons interact with the Higgs field, giving them mass. The photon does not interact, and remains massless.

The quantum excitation of the Higgs field is what we observe as the Higgs boson. Unlike other elementary particles, the Higgs boson has an integral spin of 0. Besides the three intermediate vector bosons, the Higgs field gives itself, the six quarks, and the three charged leptons their mass, but leaves the three left-handed neutrinos massless. Since these neutrinos have mass, *the Standard Model is incomplete*.[8]

---

[8] See Francois Englert's Nobel Prize Lecture: www.nobelprize.org/ prizes/physics/2013/englert/lecture\

7.11 Lessons Learned

On the occasion of Einstein's centennial in 1979, Gell-Mann noted "two lessons we can learn from Einstein's work" [99]:

> One is that while cultivating successful ideas in physical theory we must be careful to prune away any unnecessary intellectual foliage that accompanies them, assumptions that we accept out of laziness or vested interest but that we do not require for success ... The second lesson is to take very seriously ideas that work and see if they can be usefully carried much further than the original proponent suggested.

Gell-Mann certainly followed his advice. From 1957 to 1964 he continually removed "unnecessary intellectual foliage" from his toy field theories, finally getting rid of all hadron couplings and real particles, replacing them with Serber's quarks (right column, p. 79).

The lessons we learn from the history described here are many. Theorists must understand how experiments are performed so that they can test their theories. Experimentalists must understand theories to know which ones to test, and how to test them. Physicists must learn both these lessons to evaluate theories with Bayes' theorem (Appendix 6.B, p. 118). There are three additional lessons learned: Pay attention to anomalies, master the literature, and dig deeply.

Beware of Blindness

Chapters 1, 2, and Appendix 2.A highlight the importance of anomalies, especially those that contradict existing theory, and are experimentally unassailable. Too often anomalies are overlooked or dismissed out of hand — a form of mental blindness. The dismissal of left-handed electrons from β-decay is a good example. Shortly after de Broglie suggested that particles could behave like waves, Richard Threlkeld Cox, a young assistant professor at New York University, wondered if electrons could, like X-rays, become polarized by scattering off objects (X-rays and electrons from β-decay have similar wavelengths). Working with C. G. McIlwraith and B. Kurrelmeyer, he presented compelling evidence that electrons that scattered without loss of energy were polarized, but not because of scattering. It appeared that the *electrons* emitted from β-decay emerged as polarized! They moved like little left-handed screws, with their spin pointing in a direction opposite to their velocity. This discovery was described in a 1928 paper, "Apparent Evidence of Polarization in a Beam of β-rays," that was published in the *Proceedings of the National Academy of Sciences* [42].

Their finding was unintelligible. How would the electron "know" which way to turn? Carl Chase, Cox's graduate student, carefully repeated the experiment multiple times, definitively confirming their discovery. In 1930 Chase published his results in the *Physical Review* [32]. Puzzling as it was, β-rays turned with a left-handed screw.

Despite the strength of these experiments, the anomaly was simply ignored. The idea of a weak interaction responsible for particle decay did not exist, let alone a theory that would require electrons to spin in opposition to their direction of motion. By the time theory had advanced enough to describe a weak decay with a left-handed electron, Cox and Chase's measurements were forgotten [110].

Like the left-handed electron, the long lifetime of the ϕ meson was an anomaly that had no place in theory, and could easily have been completely ignored. Sakurai was aware of its lifetime measurement, but didn't appreciate its significance (Section 2.6: "An Overlooked Anomaly," p. 23). Decoding this anomaly led to the discovery of quarks.

Turn Every Page

Mastering the literature is essential. Even though parity violation was in clear view, Tsung-Dao Lee and Chen-Ning Yang missed it when they went looking for it. In their 1956 paper "Question of Parity Conservation in Weak Interactions," which led to their Nobel Prize, Lee and Yang write [141]:

> It will become clear that existing experiments do indicate parity conservation in strong and electromagnetic interactions to a high degree of accuracy, but that for the weak interactions (i.e., decay interactions for the mesons and hyperons, and various Fermi interactions) *parity conservation is so far only an extrapolated hypothesis unsupported by experimental evidence* [my italics].

That's what I learned as a graduate student. But Lee and Yang didn't comb the literature.[9] It took almost 30 years for Chien-Shiung Wu to confirm Cox and Chase's result that parity was violated in β-decay [215].

So listen to Mr. Hathway: When Robert Caro, Lyndon Johnson's legendary biographer, first went to work at *Newsday* as an investigative reporter, he was admonished by his editor Alan Bonnell Hathway:[10]

> "Just remember," he said. "Turn every page. Never assume anything. Turn every goddamn page."

Dig Deeply

The organization of data related to physical phenomena without theories of causal explanation is called phenomenology. It is based entirely on observation, and is a tool for discovering fundamental laws and their particles.

Practicing phenomenology requires mastering the literature and "turning every page." The development of quantum mechanics provides a wonderful example of the transition from phenomenology to fundamental physics. Bohr integrated Rutherford's model of the atom with Planck's quantum hypothesis. Using the empirical formulae for the frequencies of spectral lines discovered by Balmer, Rydberg, and Ritz, Bohr developed his phenomenological model of the atom. Although influenced by Bohr, Heisenberg took a different tack, because Bohr's phenomenological formulation of quantum mechanics "contain[ed], as basic elements, relationships between quantities that are apparently unobservable in principle" [115]. Heisenberg's formulation of the quantum-mechanical atom only contained relations between observables — the intensity and polarization of the radiation atoms emit — that could be used to *compute* their values. Now that's fundamental — that's deep!

To discover quarks, reading the voluminous literature on all strongly interacting particles was

---

[9] One of the authors (Bernard Kurrelmeyer) of the 1928 National Academy paper had been in the same department at Columbia where Lee worked, and was at Brooklyn College in 1956.

[10] Robert A. Caro "The Secrets of Lyndon Johnson's Archives," *The New Yorker*, January 28, 2019: www.newyorker.com/magazine/2019/01/28/the-secrets-of-lyndon-johnsons-archives.

necessary. Connections had to be made, errors uncovered — again and again — until a coherent picture emerged. Data from new technologies — bubble chambers and real-time data analysis — were often misinterpreted; papers could not be taken at face value ("Problems with Experiment," p. 105 and Appendix 7.A: "Bumps Along the Way," p. 139). Here some knowledge of experimental physics was essential, especially if you were a theorist and had no idea of what experimental physics was all about.

Regularities in the hadron spectrum, mass formulae, and Zweig's rule were phenomenology. The quarks they reflect, and the deeper layer of reality they inhabit, are fundamental, and will be with us forever.

7.12 Remaining Questions

For almost a hundred years two layers of realities have coexisted — the classical and quantum worlds — one responsible for the other, the other more easily observed. In the first 60 years, nine independent ways of formulating quantum mechanics were created, each connected differently to classical mechanics [197]. As previously noted, another pair of realities have been identified — the world of hadrons, and that of their constituents. Their connection is poorly understood, even after 50 years. The tension between them is apparent in the dual structure of the nucleon, which was determined in two very different ways, one with low-momentum-transfer elastic scattering in 1956 (p. 31), the other with high-momentum-transfer inelastic scattering beginning in 1968 (p. 128). Both types of experiments increased our understanding of the nucleon. The first, with low spatial resolution, established the existence and size of the hadron cloud that surrounds the nucleon. The second, with higher resolution, confirmed the existence of point-like constituent quarks inside the nucleon. These two views of the same object are disjoint. Can one, when passing through intermediate-momentum-transfers, go "smoothly" from cloud to point? And how can one understand the successes of the bootstrap, which, while limited, were still substantial?

In QCD mass-energy in hadrons has several spatial distributions due to gluons, antigluons, quarks, and antiquarks. What are these distributions, and how do they compare to the distribution of hadrons?

Zweig's rules for hadron couplings and masses are analogous to Bohr's rule for spectral lines. Bohr asserted that these lines of light arose when electrons jumped from one quantized orbit to another. Bohr's rule was understood to follow from the theory of quantum mechanics soon after it was developed. Fifty years have passed since QCD was created, but theorists still cannot derive Zweig's rules from first principles, *quantitatively* comparing theory with experiment in the process.

At a more fundamental level, are quarks and leptons really related? If so, how? Why should one have fractional, the other integral charge? Quarks stand beside leptons in equal number, both without internal structure at our energy scale, both with seemingly random values of mass. Their connection was obscure in 1963, and remains so.

As previously related, when the muon was discovered, identical to the electron except for mass, Rabi at a Chinese restaurant famously asked, "Who ordered that?" Rabi's question, in its modern form, might be "Why don't we live in a world with two quarks and two leptons, or four quarks and four leptons?" Who needs all those other quarks and leptons? Would life be possible without them? Although Newton taught us to ask "How" not "Why," we will always ask: Why are things the way they are? Why is there necessity?

7.13 What About the Other Stuff?

Quarks and QCD — with their associated unanswered questions — loom large in the minds of high energy physicists. They lie at the end of a long road that started in 1896 with a handful of photographic plates that Becquerel placed in his bureau drawer with phosphorescent rocks inherited from his father.

But we now know that quarks, gluons, and electrons — the stuff we are made of — comprise only some 5% of the mass-energy of the Universe. Dark energy and dark matter, today's most striking anomalies, constitute the rest. It's time to figure out what's going on with that other 95%, but it's your turn to try.

Appendices for Chapter 7

"Courage to those who would carry on!"
From Chris Quigg's summary talk at IP2I: "Double
Charm Tetraquark and Other Exotics," Lyon,
November 23, 2021.

7.A   Bumps Along the Way

Figuring out which experiments were incorrect was a major difficulty in discovering constituent quarks, as it had been for developing the V–A theory of parity violation. Knowing that, the path to discovery was relatively straightforward. Unfortunately issues with data continued to plague particle physics well past 1963, casting doubt for more than six years on the existence of constituent quarks.

Starting in 1965, several different experiments presented overwhelming statistically significant evidence that the first orbitally excited state of the ρ meson, the spin-2 "A2" in which its quark and antiquark orbited around each other with angular momentum $L = 1$, was really two closely spaced resonances, *not the single resonance predicted* (Appendix, A.7: "Orbital and Vibrational Excitations," p. 152). The A2 meson was "split."

The data from three "missing mass" experiments shown in Figure 7.4 were production experiments,

$$\pi + p \rightarrow \text{A2} + p,$$

where the momentum, energy, and angle of the recoil proton were measured to determine the mass of the A2.

In 1970, five years after the A2 was first reported, Peter Schübelin summarized the situation [192]:

> Instead of the single peak of a normal resonance, the A2 has two closely spaced peaks with the separation between the peaks roughly equal to their width. This double peak implies that the A2 is really two particles with nearly the same mass, or perhaps a single object of an entirely new type, a double resonance or "dipole." The very existence of the splitting of the A2 has been a topic of controversy for five years. Now, with the splitting all but certain, the question remains: What does this structure imply for the rest of high energy physics? The

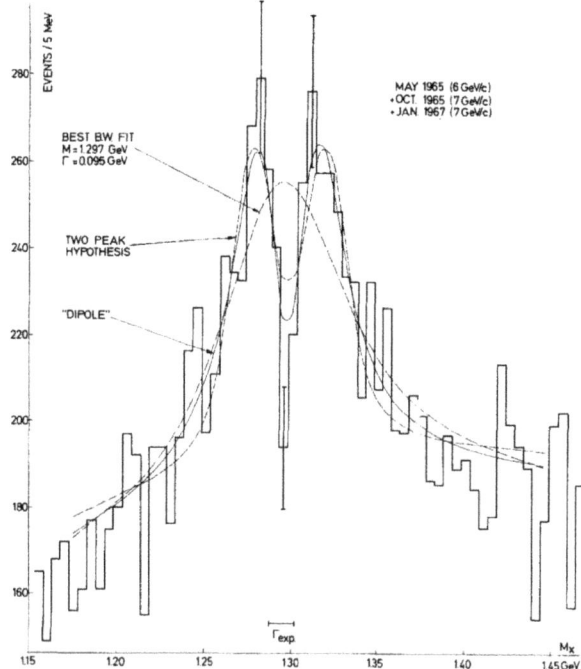

Figure 7.4. A 1968 histogram presenting evidence from three CERN missing mass experiments that the A2 meson was split into two. Several additional experiments confirmed the splitting, including an almost identical experiment across the ocean at Brookhaven National Laboratory. (From W. Kienzle, " Compiled Evidence for a splitting of the A2 meson" [137].)

> most recent work indicates that the splitting is independent of how the A2 is produced and how it decays; furthermore both halves appear to have the same intrinsic quantum numbers.

The view that hadrons were created from constituent quarks was now untenable, and those that believed in the constituent quark model were "ostriches," as illustrated in Figure 7.5.

However, there was a potential problem with the missing mass experiments. In each of them the histogram, like the one shown in Figure 7.4, was built up in real time inside the experimenters' hut on the accelerator floor while the experiment was running. This was a new exciting way to see how the experiment was progressing, alerting the experimenters to problems in time to fix them. But this procedure was fraught with danger: Suppose there was only one peak, but a statistical fluctuation split it in two? The dip

Figure 7.5. A cartoon shown in April, 1969 by Harry Lipkin when speaking about "The spectrum of hadrons" at the Inaugural Conference of the European Physical Society in Florence Italy [146]. (Reproduced with permission from CERN.)

between peaks would invariably start filling in. Seeing this, and perhaps thinking a split peak more interesting than a single one, an experimenter might wonder if the energy of the proton beam circulating in the accelerator had started to drift. Going to the control room he would ask the machine operator to "tune the beam," i.e., check that the machine energy was correct. The operator would retune, probably ending up with a slightly different energy. The final state proton momentum and energy, and therefore the missing mass, would shift accordingly. This, or something like it, probably split the peak.

I thought the splitting was an artifact, and ignored the problem. It eventually went away when the splitting wasn't seen in bubble chamber experiments, leaving egg on many faces, including those of theorists who invented theories to explain it.[11]

7.B   Moving in the Right Direction

The Baryon Number of SU(3) representations

Haim Goldberg and Yuval Ne'eman addressed a shortcoming of SU(3) symmetry shortly after it was introduced by Gell-Mann and Ne'eman. In a paper titled "Baryon Charge and R-Inversion in the Octet Model," published January 1, 1963, they write [105]:

---

[11] When it was all over I remember Murray coming into my office to congratulating me for not writing a paper proposing a theory with a split A2.

The unitary symmetry approach to the strongly interacting particles makes use of irreducible representations of SU(3). In the octet model as it now stands, one has to attach the *same* representation, namely (1 1), whether it be to the baryons, to the antibaryons, or to the mesons. Thus, the model does not reflect baryon charge, which has to be introduced as an external feature. To overcome this defect, it is suggested that the group U(3) be used instead of SU(3) (as in the Sakata model). This enables us to attach to the 3 octets three different irreducible representations, distinguishable exactly by the one extra infinitesimal operator.

They then define a baryon operator,

$$B = \begin{pmatrix} \frac{1}{3} & 0 & 0 \\ 0 & \frac{1}{3} & 0 \\ 0 & 0 & \frac{1}{3} \end{pmatrix},$$

allowing them to assign the proper baryon numbers of 1, −1, and 0 to baryons, antibaryons and mesons. While they mention in passing "the tendency to consider the mesons and the new baryons (N^*, Y^*) as composite particles," they are composites of other baryons. *Since they don't model hadrons as composites of elementary point particles,* their paper is not related to quarks. They solve the problem of assigning baryon number to hadrons in a different way, continuing to use group theory.

The Case of André Petermann

In *March 1965* Nuclear Physics published a short paper of little more than three pages, written in French, by André Petermann titled "Properties of Strangeness and a Mass Formula for Vector Mesons" [174]. The paper had been received *December 30, 1963*.

The paper is something of a puzzle because of the disconnect between the title and body of the paper, and two sentences in its conclusion. Those two sentences have generated some comment in recent years.

Petermann's paper begins with:

Despite the success of the Gell-Mann-Okubo formula in giving mass differences of the ordinary baryons (N, Λ, Σ, Ξ), the latter proved powerless to shed any light on the differences in mass of vector mesons. The purpose of this note is to make up for this deficiency. and to give, at least tentatively, a dynamic explanation of the role of strangeness in these mass formulas.[12]

Then comes the original contribution of his paper:

> Consider for this purpose two spinors s and s' (their antispinors being denoted by $\bar{s}$ and $\bar{s}'$), with s' strange ($|S| = 1$) while s is not. Electromagnetic and weak interactions are decoupled so that only the world of interactions stronger than they are will be considered. It should be noted that if the electromagnetic interaction were present, we would be obliged to consider 3 spinors s, $\hat{s}$, and $\hat{s}'$, i.e., the isospinor $(\hat{s}, s)$ with $S = 0$ and the isoscalar s' with $|S| = 1$. So the case considered here simplifies from the degeneracy $s = \hat{s}$ which manifests itself in the absence of the electromagnetic field.

This is similar to the Sakata model, but the three spinors are not identified with p, n, and Λ. That way Petermann is not saddled with the problems of Sakata's model. However, Petermann is not sure how the baryons should be constructed. In particular, he doesn't know how many spinors to use, only stating that there must be "at least three."

To get at the mass difference between s and s', he notes that "strangeness is conserved like charge: We call it the 'strange charge'" and proceeds to treat strangeness perturbatively in the strong interactions, somewhat as charge is handled in the electromagnetic interactions. In a toy field theory with a gauge invariance he computes the difference in self energy between s' and s to first order, showing it is positive. From this he concludes that s' is heavier than s.

He then notes that masses of baryons in the octet increase approximately linearly with the absolute value of strangeness, as expected from the positive spinor mass difference.

To get the vector meson mass formula his paper is centered on, and its relation to strangeness, he assigns his spinors to the vector mesons exactly as they might have been assigned in the Sakata model using s' for Λ and s for N. This makes the masses of the ρ, K^*, and ϕ increase linearly, because the number of strange spinors they contain increases linearly. Petermann expresses this as:

$$m_\phi = 2m_{K^*} - m_\rho.$$

He then shows that this equation is satisfied within experimental error.

This mass relation was well known, but Petermann does not reference its prior derivation. It was first published six months before Petemann's submission date by Okubo using other assumptions [166], as discussed here in "The Mesons," p. 24.

In his conclusions, after stating his model is necessarily quite crude since it cannot account for differences in spinor binding energies, Petermann makes this puzzling assertion:

> When the electromagnetic interaction is present, difficulties concerning the electric charge present themselves either in the form of a non-conservation of charge when the particles bind to form the particles that we observe in the physical world. Or, if one wants to preserve the conservation of the charge, which is highly desirable, then the s particles must have non-integer values.

This appears out of the blue and is unintelligible. Charge conservation is not mentioned elsewhere in the paper, and there is no explanation given here for its violation. Neither is there an explanation of why assigning non-integral charges to the spinors would preserve charge conservation, or what those non-integral values would be.

That charge is not conserved in Petermann's theory, and that it can be restored by giving spinors fractional charges, are remarkable claims calling for detailed explanation, certainly warranting more than off-the-cuff assertions in the conclusion to his paper. The puzzle is why an otherwise tightly organized, coherent paper concludes with radical statements unsubstantiated by the results it presents, and is even unrelated to its putative topic.

Attribution of credit and priority of results can be gnarly. Petermann has his supporters (see "Who Invented Quarks?," a letter by Álvaro De Rújula submitted to the CERN Courier in 2014,[13] "The Idea of Quarks: Towards Restoring of Historical Justice," 2020 by Vladimir Petrov,[14] and "The Yang-Mills Model" by Glashow[15]).

---

[12] My translation.

[13] home.cern/fr/node/2420.

[14] arxiv.org/abs/2001.04843.

[15] Published online in *Inference*, **5** (2), May 2020, inference-review .com/article/the-yang-mills-model .

On the other hand, in light of the incongruities described above, and the timeline of submission and publication of Petermann's paper — with the quark papers by Gell-Mann and Zweig appearing more than a year earlier than Petermann's publication — it has been suggested that the conclusions were edited while the paper was being corrected in proof. I leave it to you to decide. A link to a complete translation into English of Petermann's paper is given in Footnote 14.

In any case Petermann's spinors differ from quarks in several ways:

1. The number of spinors in a baryon is not specified.
2. The spinors' charges are unknown.
3. The spinors are not associated with SU(2), SU(3), or SU(6), and all that these connections imply for the prediction of hadron multiplets, their quantum numbers, and couplings.

7.C   Discovering the Fourth Quark

While reading the November 2021 issue of *Physics Today* I stumbled across a sentence in an obituary for Toshihide Maskawa, who shared the 2008 Nobel Prize in physics with Makoto Kobayashi "for the discovery of the origin of the broken [CP] symmetry which predicts the existence of at least three families of quarks in nature" [138]. The obituary said [125]:

> *In 1971* Kiyoshi Niu and colleagues discovered a new type of event in an emulsion chamber experiment, and Shuzo Ogawa suggested it was evidence of the fourth quark [my italics].

I thought evidence for the fourth quark was first seen *three years later* inside the J/ψ meson! How could this be?

After its defeat in World War II, when accelerator physics was rapidly developing in the US, Japanese physics was left with cosmic radiation, nuclear emulsions, and little money. Niu and collaborators at Nagoya University, working for more than 15 years, tailored and perfected the "emulsion chamber" to observe both charged particles and neutral gamma rays created by cosmic rays to study strong interactions in the 10 TeV range. A typical 50 kg chamber, shown in Figure 7.6, consisted of many alternating layers of lucite and photographic emulsion for producing cosmic-ray interactions, and visualizing the charged particles that were created. Further down the chamber the emulsion was layered with thicker lead plates to convert neutral gamma rays, created by $\pi^0 \to \gamma + \gamma$, into electron-positron showers.

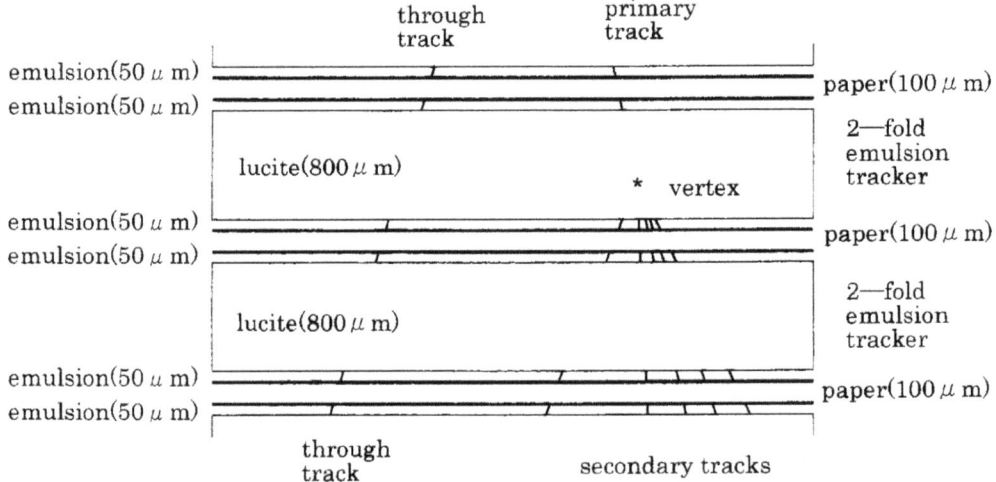

Figure 7.6. Lucite-emulsion sandwiches in the upper section of an emulsion chamber. Most interactions were initiated in the lucite. The resulting charged particles created tracks in the emulsion that became visible when the emulsion was developed. Neutral π^0s created in the interaction became visible when they decayed into photons that were converted into electron-positron showers in lead-emulsion sandwiches near the bottom of the chamber (not shown). (From [164].)

The plates, double-coated with emulsion, were specially manufactured by Fuji Photo Film, and a special optical system, with a large depth of field for viewing tracks in the emulsion at 1μ resolution, was built by Chiyoda Optical Co. This resolving power, and the large dimensions of the chamber, were essential to directly observe *both* production and decay vertices of particles with long lifetimes of approximately 10^{-12} seconds. Such hadrons had never been seen.

Between August and December 1969, 12 compact emulsion chambers were flown in Japan Air Lines cargo jets, each exposed to cosmic radiation for 500 hours [164]. This, and the resulting data analysis, were a massive undertaking. Developing all layers of emulsion while keeping everything in register was a real nightmare. The required microscopic scanning of the emulsion, micron-by-micron, is unimaginable. This method of exploration led Niu to muse [164]: "What is its strength [resolution] is also its weakness."

After almost two years of analysis, *one event* with a new type of particle, denoted by "X^+," together with its antiparticle X^-, was discovered by Niu and collaborators [163, 164]. Since the X (now called D), was produced by the strong interactions, and presumably was massive, its long lifetime was unprecedented, much longer than those of any other hadron that decayed into hadrons. The event, observed by Niu, et al., is shown in Figure 7.7.

Immediately Shuzo Ogawa, and collaborators, proposed that the X particle contained the fourth quark, which they called "p'" [112]:

> We use the ace-quark assignment for the composite system and assign the charge 2/3 to p'.

The name p' was a natural choice because they adopted my notation for aces, but dropped the subscript 0 (the charges on p' and p_0 are identical). In Ogawa's notation, $X^{\pm}$ was written as $p'\bar{n}$ and np'. Today the fourth quark p', and the $X^{\pm}$, are called the charm quark c [16], and the $D^{\pm}$ meson. The quantum number charm carried by c, which is responsible for D's long lifetime, is "a second type of strangeness."

Unlike the single "Barkus event" ("Weak Interactions of Hadrons — Advice Not Taken," p. 49), the "Niu event" was not a statistical fluctuation. After redesigning the emulsion chamber to improve the analysis of X decays, additional data were taken for three more years. By 1974, the year J/ψ was discovered, 20 charm events had been observed, most of them with charmed particle-antiparticle pairs. The lifetimes of the charged and neutral X were $1.25 \pm 0.25 \times 10^{-12}$ and $0.35 \pm 0.05) \times 10^{-12}$ seconds, the charged X living longer [Figure 11 in [164]]. The lifetimes of the charged and neutral D are $1.040 \pm 0.007 \times 10^{-12}$ and $0.4101 \pm 0.0015 \times 10^{-12}$ seconds [168], close to the original estimates. The branching ratio of the D^+ meson into $\pi^+ + \pi^0$ is only about 1.2×10^{-3}. All told, the discovery of the $X^{\pm}$ with its fourth quark was a remarkable achievement.

CP Violation

While the discoveries made with emulsion chambers had no impact in Western countries, they were important in Japan. The number of quarks grew from four to six in 1973, when Kobayashi and Maskawa noted that the experimental observation of CP violation could be incorporated into the recently created theory of electro-weak interactions [205], if another pair of quarks existed. Already believing that there were four quarks, not three, helped them jump to six. As noted in Kobayashi's Nobel Lecture,

> These works [with four elementary particles] were revived in 1971, when Niu and his collaborators found new kind of events in emulsion chambers exposed to cosmic rays. One of the events they found is shown in Figure 3 [Figure 7.7] above. In this event, we see kinks on two tracks, which indicate the decay of new particles produced in pairs.

As Kiyoshi Higashijima states in Maskawa's obituary:

> The KM theory was a beautiful Nagoya flower that bloomed in Kyoto. I asked Kobayashi why they came up with six quarks when only three quarks were known. His answer was simple: "There were four quarks in Nagoya."

Both Kobayashi and Maskawa were educated at Nagoya University in Sakata's group, and then moved to Kyoto University.

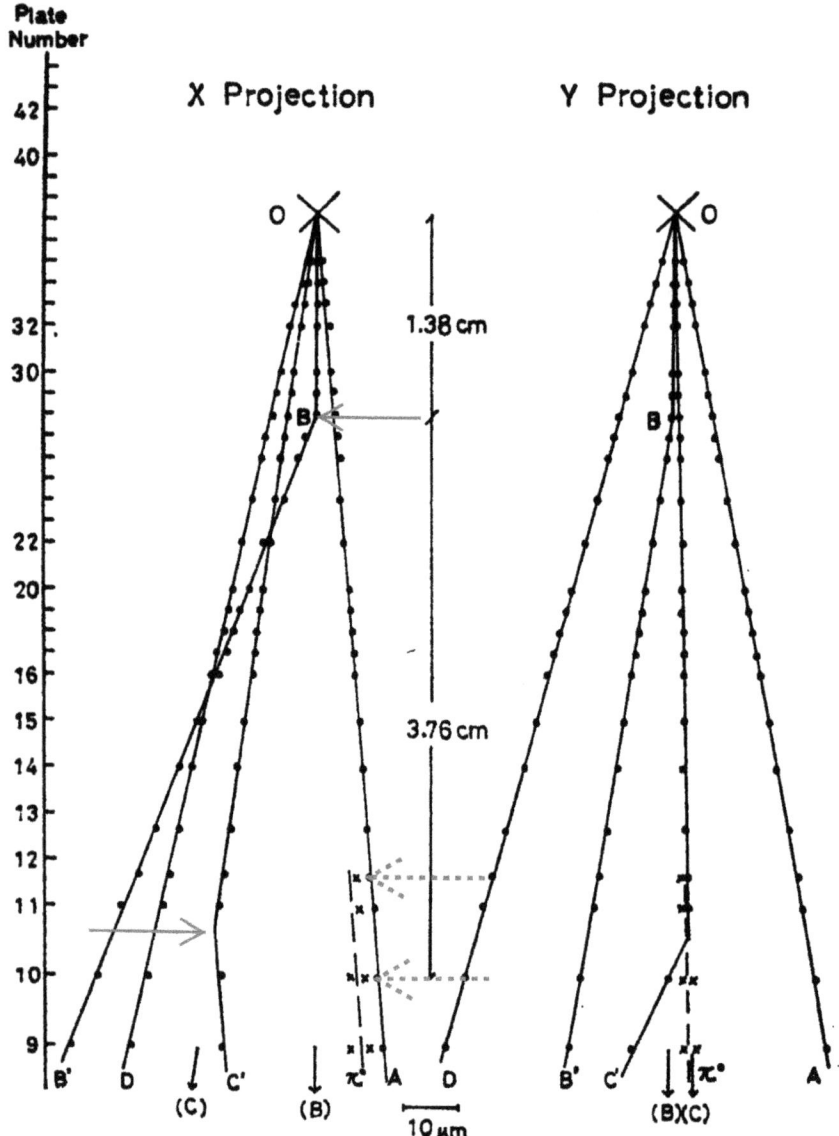

Figure 7.7. Two projections, labelled "X" and "Y," in an emulsion chamber of an interaction between a neutral cosmic ray and a nucleus, occurring at the × near the top of the figure. The interaction takes place at the bottom of the 37th lucite plate, where plate numbers are listed on the left (also see Figure 7.6). Four charged particles, labeled A through D, emerge from the interaction. Particles A and D fly straight through the chamber. Particle B decayed at B into B′ and at least one π^0 at the point marked by the top solid arrow 1.38 cm below the interaction point in the left X Projection. Two γ rays, daughters of the π^0, initiated electron showers at plates 12 and 10 marked by the two dashed arrows. Particle C, with a lifetime similar to B's, decayed into C′ and one or more invisible neutral hadrons at the point marked by the bottom solid arrow. B's decay occurs at the bottom of plate 28, C's in the middle of plate 10. While B's decay is 1.38 cm below the interaction point, C's is lower by another 3.50 cm. The upper *dashed* arrow points to the start of an electron-positron shower created by the first of two gamma rays coming from a decaying π^0 created by B's decay. The lower dashed arrow points to the top of the shower created by the second gamma ray, 3.76 cm below the B decay. (From [163].)

Niu's Contributions

The interpretation of Niu's 20 events as containing hadrons with a new quantum number attributable to the fourth quark was much more straightforward than the analysis of the J/ψ discovered three years later by Richter and Ting. The long lifetime of the X gave it away: Charm was conserved by the strong interactions, so the X, the lightest charmed particle, decayed weakly (unlike the long-lived ϕ). Charm was directly observed. As the Japanese said, it lay "naked" inside the X, immediately "visible," and therefore immediately interpretable [112]. In contrast, charm was "hidden" inside the J/ψ. Zweig's rule was necessary to interpret its unexpectedly long lifetime.

Niu's experiments encouraged Kobayashi and Maskawa to propose the existence of two more quarks in order to incorporate CP violation into the theory of electro-weak interactions. These emulsion experiments had important consequences, more so than the Richter–Ting experiments, which were more like "frosting on the cake."

Niu's earlier discovery of the fourth quark, and its quantum number $\mathbb{C}$, was not properly recognized. Here was another instance of a quantum number discovered in Japan, but this time not championed in the West, as Gell-Mann had with Nishijima's discovery of strangeness. Niu should have received the Nobel Prize together with Richter and Ting for "for their pioneering work in the discovery of a heavy elementary particle of a new kind."

The emulsion chamber was not a dead-end technology. It is a novel method for observing particles. After Kimio Niwa developed a fully automatized scanning device for nuclear emulsion, the emulsion chamber was used in 2000 at Fermilab to discover the τ neutrino [139], and later to directly observe $\nu_\mu - \nu_\tau$ oscillations at the Gran Sasso National Laboratory in Italy, with muon neutrinos coming from CERN [4]. A new emulsion chamber (FASER) consisting of 1,000 emulsion films interspersed with 1.2 tons of tungsten plates, now running at CERN, has ushered in a new era of neutrino physics at colliders [3], and the search for new physics unobtainable with other types of detectors.

When credit is concerned, sometimes it doesn't matter who discovered what, and when, or where ideas originated. The assignment of credit grows through social interactions, sometimes tweaked by its participants. Credit and prizes are the imperfect reflections of real contributions.

Appendix A
A Primer on Quarks

A.1 Prologue

The 1957 article "Elementary Particles," which Murray Gell-Mann wrote with Edward Rosenbaum (Section 2.2, p. 16), was viewed as a great success by Dennis Flanagan, editor of the *Scientific American*. In 1971 he flew Francis (Frank) Bello, an associate editor, to Pasadena to help Murray and me write an article on quarks, the new "elementary particles." Murray and I were to write two separate sections. Frank and I worked together for a week before he returned to New York to combine what we had written, with his and Murray's introduction, but without Murray's promised section. Almost one year later, after showing a draft to his editor Dennis Flanagan, Frank wrote me a letter, shown in Figure A.1, together with a draft of the article.

Shortly after I received the draft, Murray popped his head into my office saying that we should abandon the article. I reluctantly agreed. By then I had already switched fields to Neurobiology.

"A Primer on Quarks" is based in part on what Frank and I created. Quarks in the Primer mean real particles, the constituents of hadrons, i.e., aces.

A.2 Forces

Underlying all the laws of physical science are the microscopic laws of the particles out of which all matter is composed. The particles and the laws governing them have been, so far as we can tell, always the same through out the universe. One would like to find a simple description of these laws but that goal has proved elusive. Here we simply start by classify particles into categories according to their properties, and describe the several kinds of forces to which they respond. Two are familiar to everyone: the gravitational and

February 25/72

Dear George,

Remember quarks?

This is the version I turned in to Flanagan last week. And, to my immense surprise, he thinks it's publishable about as it stands.

As you will see, I made a few small changes in the last draft we worked on, principally in the introduction.

I'm sending a copy to Murray and have proposed to Flanagan that if Murray has no objection we simply forget the section he was supposed to write and tell the reader this is a primer on the quark and nothing else.

If you think we have a presentable article we should get busy on illustrations more or less immediately.

Am anxious to have your reaction.

Cordially,

Frank

Figure A.1. Letter from Frank Bello informing me that *Scientific American* would like to publish an article we had written with Murray.

electromagnetic force. Both operate over an infinite distance. Although it is not obvious from everyday experience, the electromagnetic force greatly exceeds the gravitational force; for example, the magnitude of the electrical force between two protons is greater than the gravitational force by a factor of about 10^{36}. We are keenly aware of the extraordinarily weak force of gravity only because as Earth-dwellers our many atoms are attracted to the many more atoms in the Earth. The number of atoms in two magnets that we might try to hold apart is far less.

For practical purposes, gravitation has been understood since Newton and in a more refined way since Einstein. Nevertheless, a quantum theory of gravity near black holes where gravitational fields are enormous still eludes us. The quantum theory of electromagnetism is much better understood. The two other forces or interactions are the strong interaction, which holds atomic nuclei together, and the weak interaction, which leads to the radioactivity of certain atomic nuclei and slow particle decay. The strong interaction, in turn, is describable in terms of a more fundamental interaction that acts on elementary particles, called quarks, that live in a deeper layer of reality. The proton, neutron, and other strongly interacting particles are created from quarks.

The strong interactions are ineffective at distances much beyond the diameter of a proton, 10^{-13} centimeters, while the range of the weak interaction is much smaller. If the strong force is assigned a strength of 1, the weak force has a strength of around 10^{-7}. On the same scale, the electromagnetic force has a strength of 10^{-2}.

It is a peculiar property of particles that not all of them feel every force. For example, electrons in the shells of an atom respond to the electromagnetic force of protons inside the atomic nucleus, but electrons can enter a nucleus without being aware of the strong, or nuclear, force. The neutrino feels neither the electromagnetic nor strong force. Responsive to the weak force, it can pass through the Earth, or even the sun, with only a small probability of interacting. The only force felt by all particles is gravity, even the massless, but energetic, photon.

The Creation of Forces

According to quantum theory, forces are transmitted between the objects they affect by means of one or more particles, or quanta that serve as carriers. Thus the massless photon is the quantum of electromagnetism, and the heavy intermediate vector

bosons are the quanta of the weak interaction. The quantum of gravity is the graviton, which presumably will never be detected.

Particle physicists once believed that the quantum or carrier of the strong force was a single particle — the pion — and that protons and neutrons were the elementary building blocks of the cosmos. The proliferation of strongly interacting particles, the hadrons, changed all that.

It appeared that no hadron was more fundamental than any other and that all of them served to greater or lesser degree as carriers of the strong force. On this view, the hadron spectrum is a democracy: One hadron is as fundamental as any other. All hadrons can be viewed as bound states of hadrons held together by the attractive forces generated by the exchange of hadrons. This idea, which asks hadrons to create themselves through their strong interactions, is known as the bootstrap hypothesis. Although it seemed very promising, it was difficult to implement and has yielded few quantitative predictions about hadronic behavior.

Because hadrons interact strongly, any process that is not forbidden by a symmetry or conservation law occurs rapidly with the maximal strength allowed by the conservation of probability. Thus at sufficiently high energies the collision of two hadrons leads to the production of many other hadrons.

Notwithstanding the complexity of hadronic systems and the certainty that none of the hadrons is fundamental, this Primer outlines an approach that makes hadronic systems extremely simple. In this view hadrons come from quarks. It deals primarily with the symmetries that characterize the pattern of hadronic states and the selection rules that tell which hadrons can be produced when hadrons collide or decay. It does not address the problem of how quarks interact with each other to create the hadrons.

If we were developing this approach in an historical fashion, we would begin by describing the scores of hadron states known in 1963 and point out that the proliferation of particles, discovered with the help of bubble chambers at particle accelerators, did not imply an increasing complexity of subnuclear physics. Although the spectrum of particle states goes on and on with increasing energy, much as atomic or nuclear spectra, we would find an underlying pattern governing the properties of hadrons that is repeated at higher and higher mass. The pattern would not be unlike that observed in the periodic table of elements.

We would continue by describing how hadrons react and decay into one another, how they scatter on impact, or produce other hadrons. This would allow us to identify conservation laws and symmetries that govern these processes. Only then, if we followed history, would we introduce quarks and show how they explained a great many observations. Instead, because starting with an answer and finding its consequences is much easier than reversing the process, especially if there are experimental errors and statistical fluctuations, we shall postulate the properties of quarks and show how they can be used to account for the hadrons and their properties.

A.3 Quarks

Quarks q are a family of six (originally three) spin-1/2 point particles, and their antiparticles $\bar{q}$, from which all hadrons are created. Quark-antiquark pairs $q\bar{q}$ form the basis of the lighter mesons that have baryon number $B = 0$ and integral spin. They can rotate around each other with angular momentum $L = 0, 1, \ldots$. Quark triplets qqq create the lightest baryons with $B = 1$ and half-integral spin. As a whole they too can also rotate with $L = 0, 1, \ldots$.

Heavier mesons and baryons are created by adding quark-antiquark pairs, or three-quark triplets, to quarks in the lighter hadrons. Such additions were expected, and are now called exotics.

Quarks say nothing about the photon, electron, muon, or their neutrinos, which are oblivious to the strong interactions.

A quark is the carrier of all the indestructible or quasi-indestructible quantities that are embodied in various combinations inside the hadrons. These quantities, or quantum numbers, are strictly conserved in strong interactions, but they need not be conserved in electromagnetic or weak interactions. Quantum numbers are the modern expression of the primitive notion that in the ceaseless flux of things certain essential features of matter endure. It is this enduring property of quantum numbers that gives them their importance.

Quantum numbers of quarks determine the quantum numbers of their hadrons. In turn, a hadron's quantum numbers can be measured. The two must

match, testing the idea that hadrons have quark constituents.

A.4 Quantum Numbers

Before we can describe in an adequate manner how quarks combine to create the known hadrons, we must describe the quantum numbers used to label particles and collections of particles. The quantum numbers of a system of particles can be obtained by combining the quantum numbers of the individual particles composing that system, and possibly the quantum numbers of the angular momentum those particles posses. In spite of the fact that interactions between particles can be complex — often involving the destruction of the original particles and the creation of a whole set of new ones — the quantum numbers of the system survive unchanged. The set of quantum numbers that remain unchanged depends on the interactions involved.

In 1963 when the discovery of quarks was underway, all hadrons and systems of hadrons could be completely specified by the assignment of nine quantum numbers, that will be enumerated and then individually explained below. They are the electric charge Q, baryon number B, parity P, charge conjugation C, the total angular momentum vector $\vec{J}$ and its projection J_z along some axis "z" in ordinary space, the total isospin vector $\vec{I}$ and its projection I_Z along some axis "Z" in a three-dimensional mathematical "isospin space," and strangeness $\mathbb{S}$. Now three additional quantum numbers similar to strangeness are required that label three additional quarks.

The most familiar quantum number is the electric charge Q. Particles and systems of particles always carry discrete values of Q. It was previously believed that the smallest observed quantum of electric charge is the positive charge of the proton ($Q = 1$) and the negative charge of the electron ($Q = -1$). Quarks, as we shall see, carry fractional charges: either $Q = 2/3$ or $Q = -1/3$.

The baryon number B is a label so assigned that it divides the total family of hadrons into two types: baryons and mesons. Baryons, such as the proton and neutron, are assigned a baryon number B equal to 1; their antiparticles have $B = -1$. Mesons are the class of hadrons with $B = 0$. Baryons have half-integral spin, 1/2 and 3/2 at low mass, meson spins are integral, starting with 0 and 1. Mesons and baryons are the simplest varieties of nuclear species that extend all the way in nature from the nucleus of hydrogen with $B = 1$ to the nucleus of oganesson with $B = 294$. The total baryon number is conserved in all reactions. A baryon and an antibaryon can annihilate, producing one or more mesons with $B = 0$, together with any number of baryon-antibaryon pairs permitted by the conservation of energy.

Mesons and baryons, in turn, are divided into subclasses according to their strangeness $\mathbb{S}$. One finds that in certain high energy reactions, some mesons, such as pions, can be produced in an arbitrary number whereas other mesons, such as kaons, are only produced in particle-antiparticle pairs or together with baryons like the Λ and Σ, but not with others. This can be accounted for by assuming that strangeness is conserved in the strong interactions, and assigning the pions a strangeness of 0, kaons a strangeness of 1, while the Λ and Σ are given a strangeness of -1. The proton and neutron are assigned strangeness $\mathbb{S} = 0$. The strangeness of antiparticles is reversed in sign.

Although the assignment of isospin $\vec{I}$ and its projection I_Z can likewise be made on the basis of allowed and forbidden reactions, there is an easier way to make the assignment. Hadrons come in isospin multiplets — clustered in mass — having identical quantum numbers except for charge. The number of hadrons in each multiplet is $N = 2I + 1$, where I can be any nonnegative half-integer or integer starting from 0. Hadrons in a multiplet are labeled by I_Z, the particle with the largest charge having $I_Z = I$, with both I_Z and charge Q on the remaining particles decreasing in steps of 1 until $I_Z = -I$ is reached. The classic example is that of the proton and neutron for which N, of course, is 2; thus $I = (N - 1)/2 = (2 - 1)/2$, which reduces to $I = 1/2$.

A.5 Addition of Isospin, Spin, or Angular Momentum

Up to this point we have tacitly assumed that one combines quantum numbers by the ordinary laws of addition. Because isospin $\vec{I}$ is a vector, the laws of combination used to find the total isospin of a system from the isospin vectors of its constituents, which can point in a variety of directions, is more complex.

On the other hand, if all the vectors are projected onto a single axis Z, the ordinary rules of addition

again apply. The rule for combining isospins says that if one combines I_1 and I_2, (where I_2 is greater than or equal to I_1), one obtains the ordinary sum $(I_1 + I_2)$, the ordinary-difference $(I_2 - I_1)$, and all values between the sum and the difference obtainable by counting down by one from the larger value. This says, for example, that if we start with a system of two particles of isospin 1/2, then the isospin of the combined system can be either the sum, $1/2 + 1/2 = 1$, or the difference, $1/2 - 1/2 = 0$. Schematically, this is written:

$$1/2 \otimes 1/2 = 1 \oplus 0. \tag{A.1}$$

The system with $I = 1$ has $2I + 1 = 2 \times 1 + 1 = 3$ physical manifestations while the system with $I = 0$ has only one $(2 \times 0 + 1 = 1)$. Paying attention to the number of particles, rather than their isospin, one writes:

$$2 \otimes 2 = 4,$$
$$= 3 \oplus 1.$$

The rules for combining ordinary spin are the same as those for isospin: Just replace I by spin S everywhere.

Isospin States

If the two isospin-1/2 particles above are assumed to be the nucleons $\mathcal{N}$ — the proton p or neutron n — then the possible combined systems would be pp, pn, np, and nn. These would be organized into an $I = 1$ multiplet with three members $\mathcal{N}_1 = (pp, \, pn + np, \, nn)$, and an $I = 0$ multiplet with one member $\mathcal{N}_0 = (pn - np)$. Note that $pn \neq np$; the order in which p and n are written matters when itemizing the possible composite particles found in nature.

This partition of four elements into three plus one is dictated by the symmetry properties these two sets of elements have under the simultaneous interchange of p to n and n to p, a transformation that may be written as $\left(\begin{smallmatrix} p \, n \\ n \, p \end{smallmatrix}\right)$. Here the first column says replace p by n, the second n by p. With this notation the identity transformation is $\left(\begin{smallmatrix} p \, n \\ p \, n \end{smallmatrix}\right)$.

The operator $\left(\begin{smallmatrix} p \, n \\ n \, p \end{smallmatrix}\right)$ transforms the $I = 0$ state $pn - np$ into itself except for a change of sign: $pn - np \to np - pn = -(pn - np)$. Under the same transformation the three $I = 1$ states transform into each other: $pp \to nn$, $pn + np \to np + pn = pn + np$, and $nn \to pp$. This is written as

$$\begin{pmatrix} p\ n \\ n\ p \end{pmatrix} \mathcal{N}_0 = -\mathcal{N}_0 \ \text{ and } \ \begin{pmatrix} p\ n \\ n\ p \end{pmatrix} \mathcal{N}_1 = \mathcal{N}_1.$$

Therefore exchanging proton and neutron multiplies the isospin part of the two-nucleon state by a factor of $(-1)^{I+1}$,

$$\begin{pmatrix} p\ n \\ n\ p \end{pmatrix} \mathcal{N}_I = (-1)^{I+1} \mathcal{N}_I.$$

The two transformations $\left[\begin{pmatrix} p\ n \\ p\ n \end{pmatrix}, \begin{pmatrix} p\ n \\ n\ p \end{pmatrix} \right]$ form the "permutation group," or "symmetric group" S_2, of the two elements (p, n).

The two permutations form a group because:

1. There is an identity transformation that takes the original elements back into themselves.
2. Each permutation has an inverse, so the product of a permutation with its inverse is the identity transformation.
3. The product of any two permutations is a permutation.
4. The product of any three permutations is associative, i.e., the product of the first two, times the third, is the same as the product of the first, times the product of the last two.

The four $I = 0$ and $I = 1$ states together are called a "reducible representation" of S_2, while the $I = 0$ and $I = 1$ states form two "irreducible representations"; they cannot be divided into smaller representations whose elements transform entirely among themselves.

Total Angular Momentum

The total angular momentum $\vec{J}$ of a system of particles differs fundamentally from the quantum numbers so far described in that it is not simply a combination of the spins $\vec{S}$ associated with the individual particles (their intrinsic angular momentum), but also includes the angular momenta $\vec{L}$ produced by the relative motion of all the particles within the system (the extrinsic angular momentum). The relations among the different types of angular momenta is given by the equation

$$\vec{J} = \vec{L} + \vec{S}.$$

The rules for combining the individual angular momentum vectors and individual spin vectors to obtain L and S as well as the rules for combining L and S to give J are the same as the rules we have given for combining isospin vectors. The quantization of angular momentum requires that L, S and J take on discrete values. L can equal 0, 1, 2, 3 ... , while S can be 0, 1/2, 1, 3/2, 2, ... Of course, J can assume any of the values permitted to S. The quantum number J can have either any integer or half-integer value between $L + S$ and $L - S$ or $S - L$, whichever is not negative.

Let us take, for example, a system consisting of two particles with spin $S = 1/2$, the proton and neutron, spinning around one another with angular momentum L. To find the possible values of total angular momentum J, we first combine the spins of the two particles ($\vec{S}_p$ and $\vec{S}_n$) to obtain $\vec{S} = \vec{S}_p + \vec{S}_n$. Using Eq. A.1 above, $S = 1$ or 0. When $S = 0$, the total angular momentum J simply equals L. When $S = 1$, J can have by the rules of vector addition any one of three possible values: $L + 1$, L, or $L - 1$.

The concept of an orbital angular momentum in isospin space does not exist, although it was tried unsuccessfully as an alternative to strangeness in 1953.

Spin States

We would like the reader to appreciate the analogy between the way in which the spins of the proton and neutron are added here to give their total spin S and the way the individual isospins of the proton and neutron were combined in the previous discussion of isospin. Let us take a particle with spin $S = 1/2$. Its spin projection $S_z = +1/2$ can be represented by an arrow pointing up, $\uparrow$. Similarly, its spin projection $S_z = -1/2$ can be represented by an arrow pointing down, $\downarrow$. Using this notation the three projections $S_z = +1$, 0 and -1, corresponding to the state $S = 1$ of the proton and neutron, can be represented by $\mathcal{S}_1 = (\uparrow\uparrow, \ \uparrow\downarrow + \downarrow\uparrow, \ \downarrow\downarrow)$. The state $S = 0$ can be represented by $\mathcal{S}_0 = (\uparrow\downarrow - \downarrow\uparrow)$. The analogy in first case is with $\mathcal{N}_1 = (pp, \ pn + np, \ nn)$. The analogy in the second case is with $\mathcal{N}_0 = (pn - np)$.

Instead of using the symmetric group $S_2 = \left[\begin{pmatrix} p\ n \\ p\ n \end{pmatrix}, \begin{pmatrix} p\ n \\ n\ p \end{pmatrix} \right]$ of the two elements $\mathcal{N} = (p, n)$, here $S_2 = \left[\begin{pmatrix} \uparrow\ \downarrow \\ \uparrow\ \downarrow \end{pmatrix}, \begin{pmatrix} \uparrow\ \downarrow \\ \downarrow\ \uparrow \end{pmatrix} \right]$ of the two elements $\mathcal{S} = (\uparrow, \downarrow)$. Interchanging the first and second spins of an $S = 1$ state with the operator $\begin{pmatrix} \uparrow\ \downarrow \\ \downarrow\ \uparrow \end{pmatrix}$ takes you to the same $S = 1$ state. Operating on the $S = 0$ state changes its sign. Therefore swapping spins multiplies the spin part of the two-spin state by a factor of $(-1)^{S+1}$,

$$\begin{pmatrix} \uparrow & \downarrow \\ \downarrow & \uparrow \end{pmatrix} \mathscr{S}_S = (-1)^{S+1} \mathscr{S}_S. \qquad (A.2)$$

Measuring Spin Projections

When discussing the $2I + 1$ states of the same isospin I, but different isospin projections, we saw that these states could be distinguished by their differences in charge, and slight differences in mass. To demonstrate that a particle with spin S actually has $2S + 1$ spin components one can repeat the experiment first performed in 1921 by Otto Stern and Walter Gerlach. If one passes a beam of these particles through a non-uniform magnetic field one finds that the emerging beam is split into $2S + 1$ components. If $S = 1/2$ the beam will be split into two components, one consisting entirely of particles with spin "up" (equivalent to a projection of $+1/2$ on the z axis), the other entirely of particles with spin "down" (equivalent to a projection of $-1/2$ on the z axis).

A.6   Symmetry and Quantum Numbers

There is a deep connection between the conservation of a quantum number and some symmetry of the world. This is readily exemplified by the conservation of angular momentum and its relationship to rotational symmetry. Thus there is no way to tell from a movie of two colliding objects in space, for example, whether one is seeing their original orientation or whether the image has been rotated by tilting the camera. This symmetry, corresponding to the conservation of angular momentum, corresponds to the rotation group SO(3). One other symmetry of nature is that of time reversal T. If it holds, one cannot tell whether a movie of colliding objects is being run forward or backward. This is the only symmetry that usually has no quantum number associated with it. Weak interactions break this symmetry.

Parity and Charge Conjugation

The last two quantum numbers we will discuss are related to symmetry under mirror reflection and the symmetry involved in exchanging particles for antiparticles. The first is called parity P, the second charge

conjugation C. Both of these quantum numbers can assume only one of two values, either 1 or -1. These quantum numbers are conserved under the strong and electromagnetic interactions.

The conservation of parity is illustrated by one's inability to determine whether a picture of our two colliding objects was taken directly, or through a mirror. Conservation of charge conjugation reflect one's inability to tell whether the colliding objects are composed of particles or their antiparticles.

Only two possible states of parity or charge conjugation exist because applying either of these operations twice takes you back to the initial states. Alternatively, there are only two possible configurations: In the case of parity, the mirror is there or it is not, whereas in the case of charge conjugation, the objects can consist of particles or their antiparticles. In the case of angular momentum, on the other hand, there can be an infinite number of values (orientations) of $\vec{J}$ because a system can be rotated continuously.

Previously we have seen that quantum numbers could be combined in two ways: by simple addition and by vector addition. Parity and charge conjugation are combined in yet a third way: by multiplication. This insures that their values are limited to 1 and -1, and that the operation performed twice returns a system to its original state.

Like angular momentum, parity and charge conjugation can have both intrinsic and extrinsic components. The total parity P or charge conjugation C of a system consists of the product P_I of the intrinsic parities of the constituent particles times an extrinsic value P_E provided by the relative motion of the particles or how their spins combine,

$$P = P_I P_E.$$

The intrinsic parity of a quark q is set to 1. The intrinsic parity of an antiparticle is the negative of the particle's parity, so the intrinsic parity of a quark-antiquark pair is $P_I = (1)(-1) = -1$. For a quark-antiquark system $(q\bar{q})_L$ with angular momentum L, the extrinsic parity P_E is equal to -1 multiplied L times, that is,

$$P_E(q\bar{q})_L = (-1)^L (q\bar{q})_L.$$

Therefore the parity of a quark-antiquark pair with angular momentum L is

$$P = P_I P_E,$$
$$= (-1)(-1)^L,$$
$$= (-1)^{L+1}.$$

Since $L = 0$ for the pion its parity is -1. Had its parity been 1 it would have been called a "scalar" rather than "pseudoscalar" meson.

Charge conjugation interchanges particles with antiparticles. For a quark-antiquark system, their angular momentum L contributes a factor $(-1)^L$, as it did to parity. Two additional factors appear, one from their spin, the second from the exchange of a fermion (a particle with spin $S = 1/2, 3/2, ...$) with its antifermion. Exchanging spin supplies a factor $(-1)^{S+1}$, as given in Eq. A.2. Exchanging quark and antiquark adds another factor of (-1) since quarks are fermions, so that

$$C = (-1)^L(-1)^{S+1}(-1) = (-1)^{L+S}.$$

The only particles that can be assigned a charge-conjugation quantum number are their own antiparticles, that is, they must have zero baryon number, charge, and strangeness. A familiar examples is the neutral pion, which has as charge conjugation $C = 1$ because its quark-antiquark have $L = S = 0$,

$$C\pi^0 = (-1)^{(L=0)+(S=0)}\pi^0 = \pi^0.$$

A.7   Orbital and Vibrational Excitations

When one looks at the extensive spectrum of hadrons, one observes that it contains recurrences of particles with similar sets of quantum numbers, with increasing spin. A given set of quantum numbers embodied in a particle makes its first (lowest-mass) appearance in conjunction with its lowest value of internal angular momentum L. These low-lying ground states are represented by three quarks or by a quark-antiquark pair that exhibit no relative motion. The more massive states are produced by introducing energy into a quark system, either in the form of rotation, in which case particles of ever higher spin and alternating parity result, or through vibration where both spin and parity remain the same.

As noted in the section above, changes in angular momentum L produce sequential changes in parity P and charge conjugation C for mesons: $P = (-1)^{L+1}$ and $C = (-1)^{L+S}$. Meson spins $\vec{J} = \vec{L} + \vec{S}$ come from combining $S = 0, 1$ with $L = 0, 1, 2, ...$ in all possible

combinations (Primer's A.5, p. 149). From the states with $S = 0$, $J = L$ results. Their J^{PC} is given in the top row of Table A.1 for $L = 0, 1,$ and 2. From states with $S = 1$, $J = L + 1$, $J = L$, and $J = L - 1$ are possible. Their J^{PC} is given in the rows below the top row of Table A.1. The larger the L, the higher the mass expected.

L	S	J^{PC}	L	S	J^{PC}	L	S	J^{PC}
0	0	0^{-+}	1	0	1^{+-}	2	0	2^{-+}
0	1	1^{--}	1	1	0^{++}	2	1	1^{--}
			1	1	1^{++}	2	1	2^{--}
			1	1	2^{++}	2	1	3^{--}

Table A.1 Quantum numbers of mesons with spin J, parity P, and charge conjugation C created from quark-antiquark pairs whose angular momentum L is less than or equal to two. C refers to the charge-conjugation quantum number of the neutral member of an isospin multiplet with strangeness 0. The quark-antiquark pairs can only have a total spin S of 0 or 1 for each L value. The three major columns correspond to mesons with $L = 0, 1,$ and 2. For example, when $L = 0$, only two types of mesons are allowed, the pseudoscalar mesons with $J^{PC} = 0^{-+}$ and the vector meson with $J^{PC} = 1^{--}$. (Adapted from [168].)

Note an entire series of mesons with

$$J^{PC} = 0^{+-}, \ 1^{-+}, \ 2^{+-}, 3^{-+}, \ ...$$

is absent, as is the meson with $J^{PC} = 0^{--}$. These mesons, which could not be created from quark-antiquark pairs, were predicted not to exist at low mass, and haven't been positively identified, even at higher mass.

A.8   The Limits of Conservation

Most quantum numbers are not absolutely conserved. For example, particles within an isospin multiplet, like the proton and neutron, do differ slightly in mass, which means that isospin conservation is violated. In 1963 the origin of this violation had not been firmly established but was presumed to be due to the electromagnetic interactions. We now know that the mass differences between particles in the same isospin multiplet originate in part with the mass differences of their constituent quarks.

Only total angular momentum $\vec{J}$, charge Q, and baryon number B are absolutely conserved. The impact of interactions on conservation laws is summarized in Table A.2.

	STRONG	ELECTROMAGNETIC	WEAK
Angular momentum $\vec{J}$	yes	yes	yes
Charge Q	yes	yes	yes
Baryon number B	yes	yes	yes
Isospin $\vec{I}$	yes	no	no
Isospin projection I_Z	yes	yes	no
Strangeness $\mathbb{S}$	yes	yes	no
Parity P	yes	yes	no
Charge conjugation C	yes	yes	no
CP	yes	yes	no

Table A.2 Conservation laws that are, or are not, respected by different interactions. While experimentally conserved, baryon number is not conserved in unproven grand unified theories (GUTs).

A.9 Reverse Engineering Quarks

Quarks were introduced not only to account for the many new hadrons being discovered, but also for the absence of the still larger number of hadrons that were not observed.

The most familiar restriction is associated with electric charge. For one class of hadrons (mesons), one finds only charges of 1, 0, and −1. For the other class (baryons), charges are limited to 2, 1, 0, and −1. For strangeness the restrictions are also stringent: for mesons $\mathbb{S} = 1, 0,$ or -1 and for baryons $\mathbb{S} = 0, -1, -2,$ or -3, antibaryons having their charge and strangeness reversed in sign.

In addition to such limitations on the intrinsic quantum numbers of hadrons, there are also restrictions on quantum numbers where extrinsic factors are involved. As we have seen, an entire series of mesons symbolized by $J^{PC} = 0^{-+}, 1^{+-}, 2^{-+}, \ldots$ are absent, as is the 0^{--}.

These strong constraints suggested that hadrons might be thought of as being created from a very few constituents — quarks — who carry, in different proportions, the known quantum numbers of the hadrons. Let us try to reconstruct how such a hypothesis might have been developed in 1963:

- Since particles with spin 1/2 exist, it's simplest to give all quarks spin $S = 1/2$.
- Create each meson from a quark and antiquark to give mesons $B = 0$.
- Since pairs of hadrons with isospin 1/2 exist, like the kaon and nucleon, it's simplest to give one pair of quarks isospin $I = 1/2$. Call them a and b with $I_Z = 1/2$ and $I_Z = -1/2$.

While mesons can easily be created from quark-antiquark pairs, the construction of baryons is not so obvious. The baryon with the most extreme charge is the Δ^{++} with charge $Q = 2, I = 3/2, I_Z = 3/2$, and strangeness $\mathbb{S} = 0$. Since quark a has $I_Z = 1/2$, it is natural to assume that Δ^{++} is created from three of them, $\Delta^{++} \sim aaa$, which implies that a has baryon number $B = 1/3$, and strangeness $\mathbb{S} = 0$. Quark b must have the same baryon number and strangeness because particles in the same isospin multiplet have identical quantum numbers, except for charge.

Δ^{++}'s charge of two implies that a's charge is $Q = 2/3$, so b's charge, in the same isospin multiplet as a, is reduced by one to $Q = -1/3$.

Since hadrons with nonzero integral strangeness exist, a third quark c with nonzero strangeness must exist. Although it had not been observed till early 1964, a spin-3/2 baryon, the Ω^- with strangeness $\mathbb{S} = -3$, charge $Q = -1$ and isospin $I = 0$ was expected to exist on the basis of SU(3) arguments (The Ω^- was in the same SU(3) representation as the Δ^{+++}). Therefore it would have been natural to assume that the Ω^- is created from ccc, where c has strangeness $\mathbb{S} = -1$, baryon number $B = 1/3$, charge $Q = -1/3$, and isospin $I = 0$.

These quantum numbers assigned to a, b, c are consistent with those found from the Gell-Mann–Nishijima charge formula,

$$Q = I_Z + \frac{B + \mathbb{S}}{2}, \qquad \text{(5.5 revisited)}$$

as they must be, since the quantum numbers in this equation are additive.

Therefore $a, b,$ and c are just $u, d,$ and s (or p_0, n_0, and Λ_0). It seems so simple. What was all the fuss about?

Appendix B
The First CERN Report

AN SU$_3$ MODEL FOR STRONG INTERACTION SYMMETRY AND ITS BREAKING

G. Zweig [*]

CERN — Geneva

A B S T R A C T

Both mesons and baryons are constructed from a set of three fundamental particles called aces. The aces break up into an isospin doublet and singlet. Each ace carries baryon number $\frac{1}{3}$ and is consequently fractionally charged. SU$_3$ (but not the Eightfold Way) is adopted as a higher symmetry for the strong interactions. The breaking of this symmetry is assumed to be universal, being due to mass differences among the aces. Extensive space—time and group theoretic structure is then predicted for both mesons and baryons, in agreement with existing experimental information. An experimental search for the aces is suggested.

[*] This work was supported by the Air Force Office of Scientific Research and the National Academy of Sciences — National Research Council, U.S.A.

8182/TH.401
17 January 1964

1. INTRODUCTION

We wish to consider a higher symmetry scheme for the strongly interacting particles based on the group SU_3. The way in which this symmetry
is broken will also concern us. Motivation, other than aesthetic, comes
from an attempt to understand certain regularities, described below, in
the spectra of particles and resonances. Since we deal with the same
underlying group as that of the Eightfold Way [1], particle classification
will be similar in the two models. However, we will find restrictions on
the representations that may be used to classify particles, restrictions
that are not contained in the Eightfold Way. The $(N, \Lambda, \Sigma, \Xi)$ and the
pseudoscalar mesons will fall into octets; the vector mesons will be
grouped into an octet and singlet, where the two representations will mix
by a predictable amount when unitary symmetry is broken; while the
$(N^*_{\frac{3}{2}}(1238), Y^*_1(1385), \Xi^*_{\frac{1}{2}}(1530), Z^-_0(1675?))$ will form a decuplet in the
usual manner. The restriction of representations will allow us to under-
stand certain features concerning the organization of these particles.
We will also be able to obtain a deeper understanding of both the meson
and baryon mass spectrum by relating one to the other.

The two symmetry schemes differ in the way particles or resonances
are constructed. In the Eightfold Way, the 8 pseudoscalar mesons may be
thought of as bound states of a fundamental triplet (p, n, Λ). For example,
the π^+ would be represented by $\bar{n}p$, the K^- by $\bar{p}\Lambda$, etc. In the
language of group theory, the 8-dimensional representation of SU_3
containing the mesons is included in the 9-dimensional baryon $\otimes$ antibaryon cross product space, i.e., $\bar{3} \otimes 3 = 8 \oplus 1$. However, if as in
the Sakata model [2] we attempt to construct the baryons out of this triplet
(for example $n \sim \bar{p}pn$, $\Xi^- \sim \bar{p}\Lambda\Lambda$, etc.) we are no longer able to
classify them into the familiar group of 8 particles. The difficulty stems
from the fact that the eight-dimensional representation describing the
baryons is not contained in the 27-dimensional antibaryon $\otimes$ baryon $\otimes$
baryon cross product space, $\bar{3} \otimes 3 \otimes 3$. In the decomposition

2.

$\overline{3} \otimes 3 \otimes 3 = 3 \oplus 3 \oplus \overline{6} \oplus 15$, only the 15-dimensional representation can accommodate all 8 baryons. Unfortunately this representation contains other particles whose masses may be predicted by the Gell-Mann – Okubo mass formula [3)]

$$m = m_o \left\{ 1 + a\, Y + b \left[I(I+1) - 1/4\, Y^2 \right] \right\} \tag{1.1}$$

Since these particles or resonances do not seem to be present in nature, we must abandon the Sakata model and work with the 8 baryons themselves as "fundamental" units.

There is, however, another possibility based on a genuine desire to keep certain elements of the Sakata model. If we build the baryons from a triplet of particles (p_o, n_o, Λ_o), (p_o, n_o) being a strangeness zero isospin doublet and Λ_o a strangeness -1 singlet, using $3 \otimes 3 \otimes 3$ instead of $\overline{3} \otimes 3 \otimes 3$ we find that classification of baryons into a set of 8 is possible since $3 \otimes 3 \otimes 3 = 1 \oplus 8 \oplus 8 \oplus 10$. We note that the 10-dimensional representation is present so that the $N_{\frac{3}{2}}^*$ decuplet may also be constructed from our three fundamental units. The 27-dimensional representation which occurs naturally in the Eightfold Way and which does not seem to be used by nature is suggestively absent. The only difficulty is that now the baryons seem to have baryon number 3. This we get around by assigning baryon number 1/3 to each member of the basic triplet, which leads via the Gell-Mann – Nishijima charge formula, $Q = e\left[I_z + \frac{1}{2}(B+S)\right]$, to non-integral charges for (p_o, n_o, Λ_o) [4)]. The isospin doublet (p_o, n_o) contains charges $(\frac{2}{3}, -\frac{1}{3})$ while the isospin singlet Λ_o has charge $-\frac{1}{3}$. We shall call p_o, n_o, or Λ_o an "ace". Note that the charges of the aces are just those of (p, n, Λ), but shifted by a unit of $-\frac{1}{3}$. The isospin and strangeness content, along with space-time properties, remain the same. We will work with these aces as fundamental units from which all mesons and baryons are to be constructed. It is quite possible that aces are completely fictitious, merely providing a convenient way of expressing a symmetry not present in the Eightfold Way. On the other hand, as we shall see, an experimental search for aces would definitely seem worthwhile.

2.   THE BARYON OCTET

For convenience, let us designate the accs (p_o, n_o, Λ_o) by (A_1, A_2, A_3). In order to construct the states representing the eight baryons we consider the reduction of the 27-dimensional cross product space of "treys" $A_\alpha A_\beta A_\gamma$ [5] $(\alpha, \beta, \gamma = 1,2,3)$ into irreducible representations [6].

$$A_\alpha \otimes A_\beta \otimes A_\gamma \sim T_{\alpha\beta\gamma,} \oplus T_{\alpha\beta,\gamma} \oplus T_{\alpha\gamma,\beta} \oplus T_{\alpha,\beta,\gamma}$$

$$3 \otimes 3 \otimes 3 \sim 10 \oplus 8 \oplus 8 \oplus 1 \tag{2.1}$$

Here $T_{\alpha\beta\gamma,}$ is totally symmetric in its indices and will represent members of the $N^*_{\frac{3}{2}}(1238)$ decuplet, while $T_{\alpha\beta,\gamma}$ is symmetric in α, β; being explicitly given by

$$T_{\alpha\beta,\gamma} = \frac{1}{2\sqrt{2}} \left\{ T_{\alpha\beta\gamma} - T_{\alpha\gamma\beta} + T_{\beta\alpha\gamma} - T_{\beta\gamma\alpha} \right\}$$

and will be taken to represent the nucleon octet ($T_{\alpha\gamma,\beta}$ could of course be used just as well). $T_{\alpha,\beta,\gamma}$ is totally antisymmetric in α, β, γ and allows for the existence of an $I = 0$, $S = -1$ singlet to be identified with the $Y^*_0(1405)$. The fact that the $N^*_{\frac{3}{2}}$ does not seem to belong to the 27-dimensional representation [7] of SU_3 may be taken as a prediction of this model.

We now list the baryon states :

$$n = -T_{22,1} \qquad\qquad p = T_{11,2}$$

$$\Lambda = \sqrt{2/3} \left\{ T_{23,1} - T_{13,2} \right\}$$

$$\Sigma^- = T_{22,3} \qquad \Sigma^0 = \sqrt{2}\, T_{12,3} \qquad \Sigma^+ = -T_{11,3} \tag{2.2}$$

$$\Xi^- = -T_{33,2} \qquad \Xi^0 = T_{33,1}$$

For example, by inspection of the subscripts, $T_{22,3}$ has

4.

$I_z = (-\frac{1}{2}) + (-\frac{1}{2}) + 0$ and strangeness $S = 0+0+(-1)$. In the limit of unitary symmetry the 3 aces are indistinguishable and all baryon states have the same structure and mass. This is represented in Fig. 1a. The mass of a baryon $T_{\alpha\beta,\gamma}$ may be thought of as $m_\alpha + m_\beta + m_\gamma - E_{\alpha\beta} - E_{\alpha,\gamma} - E_{\beta,\gamma}$ where, for example, m_α represents the mass of the ace α and the E's are binding energies. ($E_{\alpha\beta}$ is the binding energy between the two aces α and β when they are connected by a solid line. Binding energies for dashed line connections are given by $E_{\alpha,\beta}$.) In the unitary symmetric limit we have $m_\alpha = m_\beta = m_\gamma$, $E_{\mu\nu} = E_{\sigma\delta}$, $E_{\mu,\nu} = E_{\sigma,\delta}$ so that the masses of all the baryons are identical. We now assume that unitary symmetry is broken due to the fact that the singlet A_3 is heavier than the doublet (A_1, A_2) [8], in analogy to the Sakata model where the Λ was assumed heavier than the (p,n). The baryons now break up into distinguishable groups, so that instead of Fig. 1a, we have Fig. 2a. As a first approximation, neglecting differences in binding energies, we immediately find [9]:

$$m(\Lambda) = m(\Sigma) \,, \quad [m(\Sigma) + m(\Lambda)]/2 = [m(\Xi) + m(N)]/2 \qquad (2.3)$$
$$(1115) \quad (1193) \qquad\qquad (1154) \qquad\qquad (1130)$$

The Σ and Λ masses are expected to differ, however, because the ace A_3 is bound differently in the two cases. To obtain more accurate results one would have to say something about the binding energy between aces.

It is interesting to note that if one assumes that the breaking of unitary symmetry by electromagnetism takes place by virtue of the fact that the $A_2 A_1$ mass difference is not zero, then independent of the values of the binding energies we have the mass difference equation [10]:

$$m(\Xi^-) - m(\Xi^0) = m(\Sigma^-) - m(\Sigma^+) - [m(n) - m(p)] \qquad (2.4)$$
$$(5.6 \pm 1.4) \qquad\qquad (7.0 \pm 0.5)$$

Assuming that A_2 (the more negative member of the doublet) is heavier than A_1 and neglecting shifts in binding energies due to the electromagnetic breaking of the symmetry we find the qualitatively correct result that within any charge multiplet, the more negative the mass, the heavier the particle.

8182

3.   THE BARYON DECUPLET

Figure 1b represents the decuplet $T_{\alpha\beta\gamma}$, in the limit of unitary symmetry. A_1, A_2, and A_3 are indistinguishable and all binding energies are equal. The 10 members are completely degenerate. As for the baryon octet, we assume that unitary symmetry is mainly broken by virtue of the fact that A_3 is heavier than A_1 and A_2. The objects of the decuplet will no longer be identical but appear as in Fig. 2b. Neglecting shifts in binding energies due to the breaking of unitary symmetry it is clear that the decuplet resonances increase their masses linearly with strangeness, i.e.,

$$m(Y_1^*) - m(N_{3/2}^*) = m(\Xi_{1/2}^*) - m(Y_1^*) = m(Z_0^-) - m(\Xi_{1/2}^*) \qquad (3.1)$$
$$\quad (147) \qquad\qquad\qquad (145) \qquad\qquad\qquad (\overset{?}{.})$$

Since the decuplet and octet are constructed from the same set of particles we may try to obtain a formula relating the masses of the two different representations. The $\Xi_{\frac{1}{2}}^*$ Y_1^* mass difference is given by $m_3 - m_2 - E_{33} + E_{ab}$ (a,b = 1,2 depending on the charges we take) [11]. $m_\Lambda - m_N$, $m_\Sigma - m_N$, $m_\Xi - m_\Lambda$, $m_\Xi - m_\Sigma$ all contain the difference $m_3 - m_2$ and are of roughly the right order of magnitude. If we pick $\Xi - \Sigma$, the only mass difference whose binding energy term is $-E_{33} + E_{ab}$ we find :

$$m(\Xi_{1/2}^*) - m(Y_1^*) = m(\Xi) - m(\Sigma) \qquad (3.2)$$
$$\quad (145) \qquad\qquad\qquad (130)$$

Note that we do not expect this equation to hold exactly, even in the limit of unitary symmetry, because the spins, and hence the ace dynamics or binding energies, differ for the two representations.

8182

6.

The Baryon Singlet

The $Y_o^*(1405)$, in the limit of unitary symmetry, is shown in Fig. 1c. Figure 2c indicates the Y_o^* when the symmetry is broken by increasing the A_3 mass. Since the Y_o^* is a unitary singlet nothing quantitative can be said about its mass.

4. THE VECTOR MESON OCTET AND SINGLET

Meson states are built from the same units (A_1, A_2, A_3) as the baryons. They are contained in the anti-ace $\otimes$ ace cross product space :

$$A^\alpha \otimes A_\beta \sim \left(D^\alpha_\beta - 1/3\, \delta^\alpha_\beta\, D^\gamma_\gamma \right) \oplus \delta^\alpha_\beta\, D^\gamma_\gamma$$
$$\bar{3} \otimes 3 \sim \quad\quad 8 \quad\quad \oplus \quad 1 \tag{4.1}$$

where A^α stands for the anti-ace of A_α. Because of the nature of the decomposition of $\bar{3} \otimes 3$, mesons can only fall into groups of 8 or 1. The Eightfold Way would allow, in addition, groups of 10 and 27, possibilities which nature does not seem to take advantage of. We have pictorially represented in Fig. 1d, 1e the two possible meson representations in the limit of unitary symmetry.

The vector meson states are given by :

$$\rho^- = D^1_2 \qquad\qquad \rho^0 = 1/\sqrt{2}\left(D^1_1 - D^2_2 \right) \qquad\qquad \rho^+ = D^2_1$$

$$K^{*0} = D^3_2 \qquad\qquad\qquad K^{*+} = D^3_1$$

$$K^{*-} = D^1_3 \qquad\qquad \bar{K}^{*0} = D^2_3$$

$$\omega_8 = 1/\sqrt{6}\left(D^1_1 + D^2_2 - 2 D^3_3 \right) \tag{4.2}$$

for the octet, and

$$\omega_0 = 1/\sqrt{3}\, D^\gamma_\gamma = 1/\sqrt{3}\left(D^1_1 + D^2_2 + D^3_3 \right) \tag{4.2a}$$

for the unitary singlet. In the limit of unitary symmetry the masses of the singlet and octet must be the same because the binding is identical in both representations and all aces are degenerate. In fact, if the

8182

8.

forces are such as to bind the aces into an octet, they must also bind the aces into a singlet. It is important to note that this is not the case for baryons where the singlet, octet, and decuplet bindings all differ, even in the unitary symmetric limit.

Unitary symmetry must be broken for the mesons in exactly the same way as it was broken for the baryons, that is, the isospin singlet A_3 (or its anti-ace A^3) must become heavier than the isospin doublet (A_1, A_2). Breaking the symmetry by giving A_3 a larger mass not only splits the masses of the eight vector mesons, but it also mixes the singlet ω_0 with the $I = 0$ member, ω_8, of the octet. As a result of mixing the physically observable particles ω and φ are formed. Since A_3 becomes distinguishable from A_1 and A_2, ω_0 and ω_8 must mix in such a way as to separate (A_1, A_2) from A_3. This immediately leads to

$$\varphi = D_3^3 \qquad (4.3)$$
$$\omega = 1/\sqrt{2}\,(D_1^1 + D_2^2)$$

The plus and not the minus sign that appears in the "deuce" expression for ω distinguishes the ω from the ρ^0. Figure 2d shows the vector meson states after unitary symmetry has been broken. Using the empirical fact that when dealing with mesons one must always work with squares of masses, and neglecting changes in the bindings due to the breaking of unitary symmetry we immediately have [12]

$$m^2(\omega) = m^2(\rho) \qquad (4.4)$$
$$(784)^2 \quad (750)^2$$

$$m^2(\varphi) = 2\,m^2(K^*) - m^2(\rho) \qquad (4.5)$$
$$(1018)^2 \qquad (1007)^2$$

8182

9.

Mixing has made the φ as heavy as possible. The mixing angle Θ defined by

$$\varphi = \omega_o \sin\Theta - \omega_8 \cos\Theta$$

$$\omega = \omega_o \cos\Theta + \omega_8 \sin\Theta \tag{4.6}$$

comes out to be

$$\sin\Theta = \sqrt{1/3} \; , \; \cos\Theta = \sqrt{2/3} \; , \; \text{or} \; \Theta = 35.3° \tag{4.7}$$

as compared with the empirical value of $\Theta \approx 38°$ [13].

Only now has the real power of dealing with three basic objects become apparent. When working with the baryons, one could easily say, for example, that the more strangeness a particle carries, the heavier it is. But by using the basic triplet of aces we are able to say, after inspecting the baryons, that for an octet and singlet of mesons it is a non-strange particle that is heaviest of all; for it contains more A_3 than the strangeness carrying meson does.

Interestingly enough, we are able to improve equations (4.4) and (4.5). If we define the traceless matrix V of the vector meson octet in the conventional way :

$$V = \begin{pmatrix} \omega_8/\sqrt{6} + \rho^o/\sqrt{2} & \rho^+ & K^{*+} \\ \rho^- & \omega_8/\sqrt{6} - \rho^o/\sqrt{2} & K^{*o} \\ K^{*-} & \bar{K}^{*o} & -2\omega_8/\sqrt{6} \end{pmatrix}$$

10.

and let the matrix G be given by

$$G_{\alpha\beta} = V_{\alpha\beta} + \delta_{\alpha\beta}\omega_o/\sqrt{3} \qquad (4.8)$$

then the mass formulae (4.4), (4.5) may be alternatively derived by assuming that

$$H^2 \approx H_1^2 = m_1^2 \, \text{Tr}\,\bar{G}G - m_2^2 \, \text{Tr}\,\bar{G}[G\lambda_8 + \lambda_8 G] \qquad (4.9)$$

for the mass terms in the square of the Hamiltonian H [14]. Here Tr stands for trace while $m_1^2 = (2m^2(K^*)+m^2(\rho))/3$, $m_2^2 = (m^2(K^*)-m^2(\rho))/\sqrt{3}$, and

$$\lambda_8 = \frac{1}{\sqrt{3}}\begin{pmatrix} 1 & 0 & 0 \\ 0 & 1 & 0 \\ 0 & 0 & -2 \end{pmatrix}$$

Note that we have suppressed all terms involving $\text{Tr}\,G = \sqrt{3}\,\omega_o$.

More generally, however, we may write for the mass terms in the square of the Hamiltonian,

$$H^2 = H_1^2 + m_3^2 \, \text{Tr}\,\bar{G}\,\text{Tr}\,G + m_4^2\left[\text{Tr}\,\bar{G}\,\text{Tr}\,G\lambda_8 + \text{Tr}\,G\,\text{Tr}\,\bar{G}\lambda_8\right] +$$
$$m_5^2 \, \text{Tr}\,\bar{G}\lambda_8 \,\text{Tr}\,G\lambda_8 + m_6^2 \, \text{Tr}\,\bar{G}\lambda_8 G\lambda_8, \qquad (4.10)$$

where we treat the terms in m_3^2 to m_6^2 as perturbations to H_1^2. Since the term $m_3^2 \, \text{Tr}\,\bar{G}\,\text{Tr}\,G$ is invariant under SU_3 while the terms multiplying m_4^2, m_5^2, and m_6^2 are not, we might expect that to a good approximation we only need keep the perturbation $\text{Tr}\,\bar{G}\,\text{Tr}\,G$, i.e.,

$$H^2 = H_1^2 + m_3^2 \, \text{Tr}\,\bar{G}\,\text{Tr}\,G \qquad (4.11)$$

Doing this we immediately arrive at

$$[m^2(\omega)-m^2(\rho)]/2 = m^2(\varphi)+m^2(\rho)-2m^2(K^*) \qquad (4.12)$$

which is correct to the known accuracy of the masses.

8182

5. THE PSEUDOSCALAR MESONS

In the limit of unitary symmetry we have nine pseudoscalar mesons of equal mass, just like the vector meson case. The members of the octet we call (π, K, η_8) while the singlet is denoted by η_0. Breaking the symmetry by increasing the A_3 mass yields relations analogous to (4.4) and (4.5), i.e.,

$$m^2(\pi_0^0) = m^2(\pi) \tag{5.1}$$

$$m^2(\eta) = 2m^2(K) - m^2(\pi) \tag{5.2}$$
$$(690)^2$$

where η and π_0^0 are the physically observable particles that result from mixing η_8 and η_0, just as φ and ω are mixtures of ω_8 and ω_0. Furthermore, by using arguments identical to those given in the vector meson case we obtain the analogue of the mass relation (4.12)

$$\left[m^2(\pi_0^0) - m^2(\pi) \right]/2 = m^2(\eta) + m^2(\pi) - 2m^2(K) \tag{5.3}$$

Substituting the physical masses for π, K, and η we see that $m^2(\pi_0^0)$ comes out negative !

Fortunately we have an argument that alleviates the difficulties. After increasing the A_3 mass we found $m^2(\pi_0^0) = m^2(\pi)$. Therefore in this approximation, and this is the crucial point, $m^2(\pi_0^0)$ is very small compared to the mass square differences that exist among the pseudoscalar mesons. A small perturbation (one which changes mass squares by an amount small compared to changes initiated by the A_3 mass increase) may be enough to shift the mass square of the π_0^0 down to zero or even negative values. We might say that the π_0^0 is formed from two very massive objects that are extremely tightly bound. Energy conservation leaves the π_0^0 with a small positive energy or mass. If we introduce

12.

a perturbation that decreases the mass of the fundamental objects or increases the binding strength then the π_0^0 may no longer possess a net positive energy and cannot correspond to a physical particle. This, or something like it is evidently the situation in the pseudoscalar meson case.

It is interesting to note that we would not expect the removal of the ω in analogy to the elimination of the π_0^0. The perturbation given by (4.11) is expected to shift $m^2(\omega)$ by an amount small compared to the mass square splittings induced by the increase of the A_3 mass. Since $m^2(\omega)$ is larger than the vector meson mass square splittings there is no danger of the ω's disappearing through the introduction of a perturbation.

With the removal of the π_0^0 we expect that the pseudoscalar mesons behave as an isolated octet. This is indicated in Fig. 3. Neglecting changes in the binding energies due to the breaking of unitary symmetry we immediately obtain, by counting squares, the celebrated Gell-Mann — Okubo formula :

$$m^2(K) = 3/4\, m^2(\eta) + 1/4\, m^2(\pi) \tag{5.4}$$

Neglecting differences of binding energies within octets it is clear that we have the relation

$$m^2(K^*) - m^2(\rho) = m^2(K) - m^2(\pi) \tag{5.5}$$
$$(.22\,GeV^2) \qquad (.22\,GeV^2)$$

6. COMMENTS

The degree to which unitary symmetry is violated seems precarious: it appears to change from one representation to another. For the pseudoscalar mesons, for example, the violation seems enormous. Unitary symmetry gives $m^2(\pi) = m^2(K) = m^2(\eta)$, yet, for physical particles $m^2(\pi) \ll m^2(K) \approx \approx m^2(\eta)$. For the baryons, on the other hand, unitary symmetry works reasonably well, predicting $m(N) = m(\Lambda) = m(\Sigma) = m(\Xi)$. In spite of these differences, our model suggests that the strength of unitary symmetry violation is the same in both cases: for the breaking of unitary symmetry is measured by ace mass splittings, i.e.,

$$\left[m(A_3) - m(A_1)\right]/m(A_1)$$

and not by $(m^2(K) - m^2(\pi))/m^2(\pi)$ or $(m(\Lambda) - m(N))/m(N)$. The amount of unitary symmetry breaking is universal, it is the same for mesons as baryons, it is identical for octets and decuplets. This accounts for roughly the same mass differences within the meson octets, the baryon octet, and the baryon decuplet, irrespective of the masses of the members of these representations.

Although our aces p_0, n_0, Λ_0 have "peculiar" baryon number and charge, their space-time properties should be identical to p, n, Λ (in this respect we may think of them as p, n, Λ with charge translated by a unit of $-\frac{1}{3}$). This places a restriction on the quantum numbers that a meson may possess. For example, for spin 0 or spin 1 non-strange mesons, the following J^{PG} are excluded :

1) 0^{+-}, 0^{--}, 1^{-+} for isospin 0 states;
2) 0^{++}, 0^{-+}, 1^{--} for isospin 1 states.

Up to now no resonances have been found with these quantum numbers.

14.

   It is natural to associate the baryons with the lowest energy state of the trey system that represents them. This presumably means that the 3 aces are all in orbital angular momentum S states with the spin of one pair summing to O. Similarly, the pseudoscalar mesons would correspond to an ace and anti-ace whose orbital angular momentum and total spin are both O (i.e., $^1S_0$ state). Since the parity of a nucleon (ace) and antinucleon (anti-ace) state are opposite, we see that the intrinsic parity of the pion should be odd while that of the nucleon should be even.

   We have obtained the result that Λ_0 is heavier than (p_0, n_0) by an amount characteristic of the gross mass splittings within an octet, i.e., ~ 200 MeV. We therefore expect that Λ_0, if it exists, would undergo the β decays

$$\Lambda_0^{-1/3} \rightarrow p_0^{+2/3} + e^- + \nu$$
$$\rightarrow p_0^{+2/3} + \mu^- + \nu$$

just as

$$\Lambda \rightarrow p + e^- + \nu$$
$$\rightarrow p + \mu^- + \nu$$

On the basis of the electromagnetic mass splittings within a given isotopic spin multiplet we are also tempted to conjecture that n_0 is heavier than p_0, making p_0 completely stable (like p) but allowing the decay

$$n_0^{-1/3} \rightarrow p_0^{+2/3} + e^- + \nu$$

just as

$$n \rightarrow p + e^- + \nu$$

An experimental search for the p_0, n_0, Λ_0 might prove interesting.

7. CONCLUSIONS

The scheme we have outlined has given, in addition to what we already know from the Eightfold Way, a rather loose but unified structure to the mesons and baryons. In view of the extremely crude manner in which we have approached the problem, the results we have obtained seem somewhat miraculous.

A universality principle for the breaking of unitary symmetry has been suggested. From this followed a qualitative understanding of the meson mass splittings in terms of the baryon mass spectrum, e.g., $m(\Lambda) > m(N)$ implies that $m(\varphi) > m(K^*) > m(\omega) \approx m(\rho)$. The proportionately larger mass splittings within the pseudoscalar meson octet have been explained. Mass formulae relating members of different representations have been suggested, e.g., $\left[m^2(\omega) - m^2(\rho)\right]/2 = m^2(\varphi) + m^2(\rho) - 2m^2(K^*)$, $m^2(K^*) - m^2(\rho) = m^2(K) - m^2(\pi)$.

Nature's seeming choice of 1, 8, and 10-dimensional representations for baryons along with 1 and 8-dimensional representations for the mesons has been accounted for without dynamical or "bootstrap" considerations. The existence of a unitary singlet ω_0 which mixes with the octet of vector mesons has been predicted (along with the amount of mixing), while the absence of a unitary singlet for pseudoscalar mesons has been made plausible. For the baryons the model predicted that there were no analogues of $\omega - \varphi$ mixing for either the octet or decuplet, even though there might be singlet baryon states.

The quantum numbers available to a meson have been restricted to those which may be formed from the p, n, Λ and their antiparticles. Finally, the odd intrinsic parity of the pion and opposite nucleon parity fit naturally into the model.

There are, however, a number of unanswered questions. Do aces bind to form only deuces and treys ? What is the particle (or particles) that is responsible for binding the aces ? Why must one work with masses for the baryons and mass squares for the mesons ? And more generally, why does so simple a model yield such a good approximation to nature ?

16.

Our results may be viewed in several different ways. We might say :

1) The relationships we have established are accidents and our model is completely wrong. The formula $m(\Xi) = \left[3m(\Sigma) - m(N)\right]/2$ is correct to electromagnetic mass splittings and yet seems entirely "accidental". It certainly would be no great surprise if our mass formulae were accidents too.

2) There is a certain simplicity present, additional to that supplied by the Eightfold Way, but this simplicity has nothing to do with our model [15]. For example, the Gell-Mann – Okubo mass formula may be written for any SU_3 representation as :

$$m^2 = m_0^2 \left\{ 1 + b'(m_0^2)\left[I(I+1) - 1/4\,Y^2\right] \right\}$$

for mesons,

$$m = m_0 \left\{ 1 + a(m_0)\,Y + b(m_0)\left[I(I+1) - 1/4\,Y^2\right] \right\}$$

for baryons, where m_0, b', a, b vary from one representation to another. The quantities b', a, and b may be considered functions of m_0 or m_0^2. Equation (5.5) may be "explained" by postulating that $b'(m_0^2)$ goes like : $b'(m_0^2) \sim 1 / m_0^2$. Equation (3.2) would follow if $a(m_0)$ and $b(m_0)$ were any slowly varying function of m_0, going for instance like $1/m_0$. Relations of this type could undoubtedly result from many different theories.

3) Perhaps the model is valid inasmuch as it supplies a crude qualitative understanding of certain features pertaining to mesons and baryons. In a sense, it could be a rather elaborate mnemonic device.

4) There is also the outside chance that the model is a closer approximation to nature than we may think, and that fractionally charged aces abound within us.

17.

ACKNOWLEDGEMENTS

We would like to thank Dr. J. Prentki and Dr. L. Van Hove. Their comments and criticisms have been very useful.

8182

R E F E R E N C E S

1) M. Gell-Mann, Phys.Rev. 125, 1067 (1962).

2) S. Sakata, Prog.Theor.Phys. 16, 686 (1956).

3) S. Okubo, Prog.Theor.Phys. 27, 949 (1962).

4) Dr. Gell-Mann in a recent preprint, Physics Letters, to be published, has independently speculated about the possible existence of these particles. His primary motivation for introducing them differs from ours in many respects.

5) In general, we would expect that baryons are built not only from the product of three aces, AAA, but also from $\overline{A}AAAA$, $\overline{A}AAAAAA$, etc., where $\overline{A}$ denotes an anti-ace. Similarly, mesons could be formed from $\overline{A}A$, $\overline{A}AAA$ etc. For the low mass mesons and baryons we will assume the simplest possibilities, $\overline{A}A$ and AAA, that is, "deuces and treys".

6) R.E. Behrends, J. Dreitlein, C. Fronsdal and W. Lee, Rev.Mod.Phys., 34, 1 (1962).

7) S.L. Glashow and A.H. Rosenfeld, Phys.Rev. Letters 10, 192 (1963).

8) Since A_1 and A_2 form an isospin doublet their mass difference must be electromagnetic in origin and hence negligible in first approximation.

9) These formulae are obtained by counting the number of shaded squares (the number of A_3's) that are present in each baryon. We have averaged the masses of the Λ and Σ somewhat arbitrarily in Eq. (2.3).

10) This formula was first derived with other techniques by :
S. Coleman and S.L. Glashow, Phys.Rev. Letters 6, 423 (1961).

8182

20.

11) Charge independence would require $E_{11} = E_{22} = E_{12}$ and
$E_{1,1} = E_{2,2} = E_{1,2}$, but not $E_{11} = E_{1,1}$, etc.

12) For example, we would write $m^2(\rho^+) = m_1^2 + m_2^2 - (E_1^2)^2$ where E_1^2 is
the binding between ace 1 and anti-ace 2. The author has no
explanation for why squares of masses or binding energies should
appear when working with mesons. This is especially mysterious
in any model, like ours, where particles are treated as composite.

13) J.J. Sakurai, Phys.Rev. <u>132</u>, 434 (1963).

14) For a discussion of the G matrix see :
S. Okubo, Physics Letters <u>5</u>, 165 (1963).

15) In a recent preprint S. Coleman and S.L. Glashow have considered
another mass formulae producing model.

FIGURE CAPTIONS

Figure 1   These deuces and treys correspond to all known particle
representations in the limit of unitary symmetry.

    a)   This trey stands for a member of a baryon octet. The
shaded circles at the vertices are aces, while the
solid and dashed lines denote two different types of
binding.   The trey is symbolically given by $T_{\alpha\beta,\gamma}$
while the binding energies are $E_{\alpha\beta}$ (solid line):
$E_{\alpha,\gamma}$; and $E_{\beta,\gamma}$ (dashed lines). In the unitary
symmetry limit the three aces are indistinguishable.

    b)   This trey represents a member of a baryon decuplet and
is written as $T_{\alpha\beta\gamma,}$.

    c)   The trey stands for $T_{\alpha,\beta,\gamma}$, a unitary singlet.

d)—e)   The deuces shown correspond to members of meson octets
$D^{\alpha}_{\beta} - (\delta^{\alpha}_{\beta}/3) D^{\gamma}_{\gamma}$: and singlets, $(1/\sqrt{3}) D^{\gamma}_{\gamma}$. The open
circles are anti-aces.

Figure 2   We view the particle representations with unitary symmetry
broken. One of the three aces has now become distinguishable
from the other two. It is pictured as a shaded square. The
open squares are anti-aces. The mass splittings within
representations are induced by making the squares heavier
than the circles. Since the same set of aces are used to
construct all particles, mass relations connecting mesons and
baryons may be obtained.

Figure 3   The isolated octet of pseudoscalar mesons is represented
after unitary symmetry has been broken and the π^{0}_{0} has
been removed.

Figure 1 176

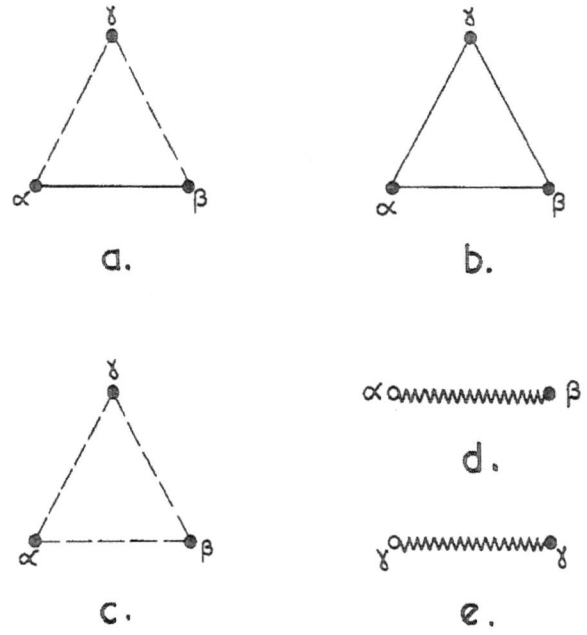

FIG. 1

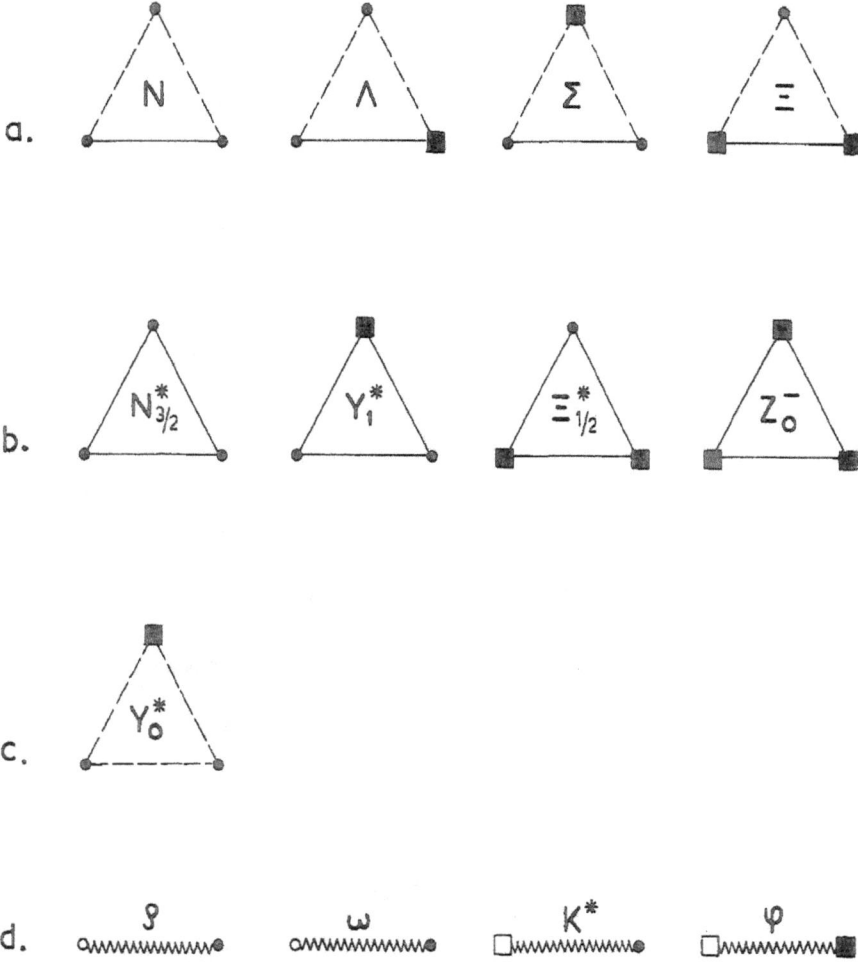

FIG. 2

Figure 3 178

π

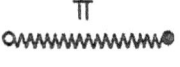

K

η

⅓ ○wwwwwwwww● + ⅔ □wwwwwwwww■

FIG. 3

Glossary
A Guide for the Perplexed

aces Elementary spin-1/2 particles from which hadrons are created. The first three, $\mathcal{A} = (p_0, n_0, \Lambda_0)$ or (A_1, A_2, A_3), have baryon number $B = 1/3$, fractional charge $Q = (2/3, -1/3, -1/3)$, and strangeness $\mathbb{S} = (0, 0, -1)$. When their discovery was published in January 1964, a fourth ace was expected; there were four elementary spin-1/2 particles for the weak interactions, the leptons, $\mathcal{L} = (e^-, \mu^-, \bar{\nu}_e, \bar{\nu}_\mu)$, so four for the strong interactions seemed natural. Subsequently, a fourth ace, c (for "charm"), with $B = 1/3$, $Q = 2/3$, strangeness $\mathbb{S} = 0$, and charm $\mathbb{C} = 1$ was found. The name "aces" has been superseded by "constituent quarks," or simply "quarks," with $u = p_0$, $d = n_0$, and $s = \Lambda_0$. Currently there are six quarks and six leptons. Quarks and leptons are building blocks for the Standard Model.

additive quantum number When a quantum number of a collection of particles is the sum of their individual quantum numbers, that quantum number is said to be additive. Additive quantum numbers are primarily related to quantities we can count, like charge, baryon number, strangeness, and charm. The projection of hadron isospin vectors along any fixed axis in isospin space is also additive.

alpha particle – α Along with the beta particle (an electron), one of two types of charged particles discovered by Rutherford in 1899 as decay products of radioactivity. An alpha particle consists of two protons and two neutrons, tightly bound, identical to a helium nucleus. It was distinguished from a beta particle by its much shorter range when traveling through matter.

antiparticle A particle having the same mass, spin, and lifetime as its corresponding particle, but with all its quantum numbers reversed in sign; a particle-antiparticle pair has the quantum numbers of the vacuum. Every particle has an antiparticle, denoted by a bar above its symbol; thus $\bar{p}$ is the antiproton. The photon, Z^0 intermediate vector boson, graviton, and Higgs boson are their own antiparticles, as are some hadrons like the neutral pion.

asymptotic freedom The force between quarks becomes vanishingly small as the distance between them approaches zero. Small separations occur with large momentum transfers. In this regime, calculations are possible with perturbation theory, providing tests of QCD.

baryons Strongly interacting particles with half-integral spin, like the spin-1/2 proton and neutron. Low-mass baryons are created from three quarks, but exotic baryons with additional quark-antiquark pairs – pentaquarks – exist at higher mass. All baryons, except for the proton, are unstable, decaying either through the strong, electromagnetic, or weak interactions.

baryon number – B The difference between the number of baryons and antibaryons in a system of particles. Baryon number is conserved by all interactions, leading to the stability of the proton.

beta particle or beta ray – β Like an alpha particle, this particle or "ray" is a charged decay product of radioactivity. It was distinguished from an alpha particle in 1899 by Rutherford who noticed that It traveled much further in matter. Beta particles were responsible for the images Becquerel found on his photographic plates, since alpha particles were absorbed by the plates' paper wrappings. [Figure 1.2, p. 2]. In 1900, Becquerel showed that β's charge-to-mass ratio was the same as the electron's, correctly leading him to conclude that beta particles were actually electrons.

beta decay A type of radioactive decay in which a beta particle is emitted from a neutron, usually in a nucleus. The underlying reaction is $n \rightarrow p + e^- + \bar{\nu}_e$.

Bevatron An early proton accelerator completed in 1955 at the Lawrence Berkeley National Laboratory. An innumerable number of resonances were discovered there by Luis Alvarez's bubble chamber group, starting in the early 1960s. The Betatron was built to find the antiproton, where it was discovered in 1955. The antineutron was discovered there a year later in proton-antiproton collisions.

blackbody An idealized object that absorbs all electromagnetic radiation that falls upon it. No radiation is reflected, or can pass through it. It is also a perfect emitter, converting all its thermal energy into electromagnetic radiation.

blackbody radiation Electromagnetic radiation emitted from a blackbody by virtue of its heat. The radiation emitted at room temperature is mostly infrared light, but as the temperature increases it starts to emit visible wavelengths. Its unexpected temperature-dependent frequency spectrum led Max Planck to treat its radiant energy as quantized in packets (photons) whose energy is proportional to the radiation frequency v, $E = hv$, where h is Planck's constant.

bosons Particles with integral spin ($S = 0, 1, 2, ...$) named after Satyendra Nath Bose. They can be elementary or composite. The carriers of fundamental forces (the photon, intermediate vector bosons, gluon, and graviton) and the Higgs boson are elementary. Mesons and nuclei with integral spin are composite. Except for the graviton, the elementary bosons are part of the Standard Model. Bosons obey the spin-statistics theorem, e.g., the wave function of two identical bosons is symmetric under their interchange. Unlike the fermions, an unrestricted number of bosons can occupy the same quantum state.

bremsstrahlung The electromagnetic radiation created by the acceleration (positive or negative) of a charged particle. The greater the acceleration, the higher the photon energies in its spectrum. The word comes from the German *bremsen* (to brake) and *strahlung* (radiation), i.e., "braking radiation."

broken strong interaction symmetries The strong-interaction symmetries of SU(2) and SU(3) are broken by the mass differences of the first three quarks, leading to the observed mass splittings of hadrons in SU(2) and SU(3) irreducible representations.

Brookhaven National Laboratory – BNL Established in 1947 as a nuclear research institution located on Long Island in New York. From 1960 to 1968 its Alternating Gradient Synchrotron (AGS) was the highest energy particle accelerator. The suppression of ϕ decay was discovered there in 1963, but not recognized as such by the experimentalists who observed it. Additional discoveries include the muon neutrino v_μ, CP violation, the $J \sim c\bar{c}$ and $Y \sim b\bar{b}$ (Upsilon) mesons. The discovery of the Y in 1977 confirmed the existence of the bottom quark and its corresponding antiquark. Research in particle physics at BNL now takes place at the Relativistic Heavy Ion Collider (RHIC), where the properties of the quark-gluon plasma are being established. An electron ion collider (EIC) is under active development.

bubble chamber A particle detector invented by Donald Glaser at the University of Michigan, but whose potential was first fully realized by Luis Alvarez at the Bevatron. Charged particles were identified by the trail of bubbles they left behind in a superheated liquid, typically liquid hydrogen. Many hadrons responsible for the discovery of quarks were found in bubble chambers at the Bevatron and the AGS at Brookhaven National Laboratory. Glaser was one of Carl Anderson's graduate students at Caltech.

CERN The Centre Européen de Recherche Nucléaire near Geneva Switzerland, established in 1954. CERN is the home of the Large Hadron Collider (LHC), the highest energy particle accelerator. LHC has a ring circumference of approximately 27 kilometers and is located about 100 meters underground, crossing the border between Switzerland and France. The $W^\pm$ and Z^0 intermediate vector bosons were discovered there in 1983, confirming the electroweak theory. The discovery of the Higgs boson in 2012 confirmed the existence of the last elementary particle in the Standard Model. The exotic pentaquark was also discovered at the LHC in 2015. Aces (constituent quarks) were discovered in CERN's Theory Division in January 1964.

charge conjugation – C The transformation that exchanges particles and antiparticles. Only particles that are their own antiparticles can have a charge-conjugation quantum number. Like parity, charge conjugation is a multiplicative quantum number with values of $C = 1$ or -1, that is conserved by the strong and electromagnetic interactions, but not by the weak interactions. The charge-conjugation operator acts both on the particles and their spatial coordinates. The charge conjugation of a neutral $\mathbb{S} = \mathbb{C} = 0$ meson created from a quark-antiquark pair with spin S and angular momentum L is $C = (-1)^{L+S}$. Since $L = S = 0$ for the π^0, its charge conjugation is 1, in agreement with experiment.

charge independence Strong interactions between a proton or neutron and a nucleus are the same, a symmetry called charge independence, or isospin symmetry. This symmetry extends to all hadrons in the same isospin multiplet. The group describing this symmetry is SU(2). Charge independence is broken by the electromagnetic interactions.

***charm quantum number* – $\mathbb{C}$** The quantum number associated with the fourth quark c. Like strangeness $\mathbb{S}$, it is an additive quantum number for hadrons that is conserved by the strong, electromagnetic, but not weak interactions. It is the difference between the number of charm and anticharm quarks in a system of hadrons.

cloud chamber An early particle detector. Charged particles were identified by the trail of ionized atoms they left behind in supersaturated water vapor. These ions acted as condensation centers around which small droplets of water were formed. An alpha-particle track was thick and straight, while the track of a much lighter beta particle was wispy from deflections caused by collisions with the vapor. The positron, muon, and kaon were discovered in cloud chambers exposed to cosmic rays.

color charge A quantum number for quark and gluon interactions with three different "values," analogues of the positive electric charge of the electromagnetic interactions: "red (r), green (g), and blue (b). These charges, like the first three quarks, form a three-dimensional irreducible representation of an SU(3) symmetry, although this SU(3) symmetry of color is exact. Its exactness confines quarks.

Anticolors, analogues of negative electric charge, have anticolor values: "cyan ($\bar{r}$), magenta ($\bar{g}$), and yellow ($\bar{b}$)." Each of the six quarks and antiquarks carry colors and anticolors, respectfully. Gluons, the mediators of the force between quarks, and the force between gluons, carry color-anticolor pairs, analogues

of the pseudoscalar mesons that carry quark-antiquark pairs. Both gluons and pseudoscalar mesons form eight-dimensional irreducible representations of their respective SU(3).

confinement According to QCD, particles with color cannot exist as free particles, and they have never been seen in isolation. Hadrons have the right combination of colors and anticolors to make them color neutral, and therefore free. The energy required to separate a (colored) quark and an (anticolored) antiquark in a meson increases with increasing separation, eventually leading to the creation of an antiquark-quark pair from the vacuum when the separation becomes large enough. The quark and antiquark from the pair combine with the antiquark and quark from the vacuum to form two meson decay products (Zweig's rule). This guarantees that colored quarks will never be found as free particles.

constituent quarks Real particles (aces), the constituents of hadrons, as opposed to Gell-Mann's mathematical quarks, or "current quarks" he used to define a field theory without strong interactions from which to abstract "current commutation relations." See "aces."

cosmic rays Energetic particles coming from outer space that produce many secondary particles in Earth's atmosphere. The term "ray" is a misnomer. These particles arrive individually (primarily as protons), not in the form of rays or beams of particles.

coupling constant A number indicating how strongly particles interact with one another. Its square, for a decaying particle, is inversely proportional to the particle's lifetime.

D meson A meson containing a charm quark and a light antiquark: $D^+ = c\bar{d}$ and $D^0 = c\bar{u}$. The D^+ and its antiparticle D^- were discovered in 1971 by Kiyoshi Niu and colleagues, who named them $X^{\pm}$. A charged X meson decayed weakly, giving it a lifetime long enough to be seen both produced, and decaying, in the same emulsion chamber. Its observation was immediately recognized in Japan as the discovery of the fourth quark, but went unnoticed in the West.

deep inelastic scattering The large-momentum-transfer inelastic scattering of a high-energy electron or neutrino off a proton or neutron, where the wavelength of the virtual photon emitted by the electron is short enough to resolve small structures within the nucleon. At its heart, deep inelastic scattering is a high-momentum-transfer *elastic* scattering of an electron off a quark in the nucleon, where the quark undergoes final state interactions with other quarks and gluons as it leaves the scattering region, creating many hadrons in its wake. Deep inelastic scattering experiments confirmed the physical reality of quarks, and catalyzed the development of QCD.

Dirac equation The relativistically invariant quantum-mechanical equation of an "idealized electron" that doesn't take into account the electron's interaction with the vacuum, which is described by quantum field theory. The Dirac equation, when properly interpreted, required the existence of an antiparticle that Carl Anderson called the positron when he discovered it in 1932.

elastic scattering In the elastic scattering of two particles, the initial and final particles remain the same, their total kinetic energy is conserved, but, in the center of mass frame, their direction of travel may rotate.

electron – e^- The lightest of the three charged leptons all having identical quantum numbers and interactions. With a mass of 0.511 MeV, it is about 2,000 times lighter than the proton. Unlike the two other charged leptons, the electron is stable.

electron volt – eV A unit of energy defined as the amount of kinetic energy gained by an electron as it accelerates across an electric potential difference of one volt.

elementary particle A particle without internal structure, not composed of other particles. All particles in the Standard Model are elementary: quarks, leptons, gauge bosons (the photon, $W^\pm, Z^0$, and gluon), and the Higgs boson, which is not a gauge boson because it has no associated gauge symmetry. The remaining elementary particle, the hypothetical graviton, is a neutral spin-2 gauge boson whose coupling is so small that it will never be detected, although its effects are obvious. Every elementary particle is a quantum fluctuation of a corresponding quantum field that can affect other particles passing through it..

exotic hadron A heavy hadron created by the addition of a quark-antiquark pair, a triplet of quarks, or a glueball, to an ordinary hadron. Tetraquarks ($q\bar{q}q\bar{q}$) and pentaquarks ($qqqq\bar{q}$) are two examples. They were expected to exist when quarks (aces) were first discovered, and the SU(6) representations of tetraquarks were described in 1964. The tetraquark X(3872) was the first exotic discovered in 2003 by the Belle experiment at KEK, an electron-positron collider in Japan.

Fermilab Fermi National Accelerator Laboratory, near Chicago Illinois, where the Tevatron was located. The bottom and top quarks, as well as the tau neutrino, were discovered at Fermilab in 1977, 1995, and 2000. The Tevatron ceased operations in 2011 due to budget cuts, and the completion of the Large Hadron Collider at CERN. Fermilab is now involved in neutrino-oscillation experiments at "DUNE" and "NOvA."

fermions Particles with half-integer spin ($S = 1/2, 3/2, ...$). They can be elementary (the six quarks and six leptons), or composite (the baryons and nuclei with half-integral spin). The elementary fermions are part of the Standard Model. Fermions obey the spin-statistics theorem, e.g., the wave function of two identical fermions is antisymmetric under their interchange. Therefore two identical fermions cannot occupy the same quantum state.

flavor The generic name for the six different kinds of quarks — up, down, strange, charm, bottom, and top (u, d, s, c, b, t), with their associated flavor quantum numbers. The flavor charges $\mathbb{S}, \mathbb{C}, \mathbb{B},$ and $\mathbb{T}$, of a hadron or system of hadrons, are defined by $-\Delta(s)$, $\Delta(c), -\Delta(b)$, and $\Delta(t)$, where $\Delta(q) = n(q) - n(\bar{q})$, and $n(q)$ is the number of quarks q of a particular flavor. Although not conventional, an upness and downness flavor quantum number can be defined by $\mathbb{U} = \Delta(u)$ and $\mathbb{D} = -\Delta(d)$. Then

$$I_Z = \frac{\mathbb{U} + \mathbb{D}}{2}.$$

The charge Q of a hadron is half its baryon number plus half its total flavor quantum numbers,

$$Q = \frac{B}{2} + \frac{\mathbb{U} + \mathbb{D} + \mathbb{S} + \mathbb{C} + \mathbb{B} + \mathbb{T}}{2}.$$

Quark flavors can only be changed in weak interactions by the $W^\pm$ intermediate vector bosons.

forces In quantum field theory particles interact through the exchange of particles that create forces between them. See "interactions."

free fractional charge Although particles with color are confined, nothing prevents free fractionally charged particles with the fundamental charges of $(\pm\frac{1}{3}, \pm\frac{2}{3})$ from existing, but none have been observed. If they do exist, at least one would be stable, because charge is conserved in all interactions.

fundamental physical laws Laws of physics that cannot be derived, even in principle, from any other physical laws. Fundamental physical laws are analogous to the axioms of mathematics, except that physicists are not allowed to make them up. If these laws correctly describe reality, they cannot contradict one another. Some mathematicians reassure themselves that their axioms are self-consistent if they are embodied in a physical system. The three laws of thermodynamics are not fundamental because they follow from other laws.

glueballs Hypothetical particles composed entirely of gluons, without any quarks. Oddballs are glueballs whose quantum numbers cannot be created from quark-antiquark pairs, e.g., their spin J, parity. and charge conjugation are $J^{PC} = 0^{-+}, 1^{+-}, 2^{-+}, \ldots$ or 0^{--}.

gamma ray – γ-ray An energetic photon emitted in a radioactive decay of a nucleus. Gamma-rays are generally more energetic than X-rays. Being neutral they travel further in matter than either alpha or beta particles.

gluons – g Eight massless neutral spin-1 particles responsible for the interactions between quarks, antiquarks, and gluons. They carry color-anticolor pairs analogous to the quark-antiquark pairs that create the pseudoscalar meson octet. Gluons form an eight-dimensional irreducible representation of an exact SU(3) used to represent the symmetry of color couplings.

gluon interactions One of the four fundamental interactions created by the exchange of gluons between quarks, antiquarks, and gluons. Gluons carry color-anticolor pairs that couple to the colors of quarks, the anticolors of antiquarks, and the color-anticolor pairs of gluons.

graviton The hypothetical massless spin-2 particle responsible for the gravitational force. It couples to both mass and energy (the stress-energy tensor), and so couples to itself. Since its coupling is small, perturbation theory may be used to compute its effects everywhere except in the vicinity of black holes where gravitational fields are enormous and perturbation theory fails.

hadrons Strongly interacting particles, either baryons or mesons, having half-integral or integral spin. They are composite particles created from three quarks, quark-antiquark pairs, and possibly balls of glue, or their combinations (exotics). In a hadron color charges of its constituents combine to form a color-neutral (color-singlet) particle. For baryons, three colored quarks combine to form a color-neutral state, while for mesons, colored quark-antiquarks or gluons create the colorless state. They decay through strong, electromagnetic, or weak interactions.

hadronization The process by which quarks and gluons transform into hadrons in high-energy inelastic scattering reactions, guaranteeing that the color they carry will not go free.

Heisenberg's uncertainty principle For pairs of "conjugate variables," like position and momentum, or energy and time, Heisenberg's uncertainty principle states that the accuracy with which both such variables can be measured simultaneously is subject to the restriction that the product of their uncertainties is at least as large as Planck's constant h divided by 4π.[1]

Higgs field or boson The quantum field responsible for creating the mass of all elementary massive particles except for the three neutrinos that are distinguished from the other particles by their left-handedness. The Higgs field breaks electroweak symmetry, giving mass to the intermediate vector bosons, while leaving the photon massless. The electrically

---

[1] Unfortunately the translated word "uncertainty" is misleading. It leaves open the possibility that precise values of both conjugate variables exist, and if we were clever enough we could measure them. It isn't just that we are uncertain of their values, we can never be certain. I prefer the term Heisenberg used [116]: *ungenauigkeit,* which indicates an inability to be exact or precise (uncertainty is *unsicherheit* in German). Conjugate variables are intrinsically "fuzzy." The lack of precision required by the uncertainty principle requires statistical relationships in quantum mechanics. Much as Einstein didn't like it, God does play dice with the Universe.

neutral massive spin-0 Higgs boson is the quantum excitation of the Higgs field.

***hypercharge** – Y* The hypercharge of a hadron is defined by

$$Y = B + \mathbb{S} + \mathbb{C} + \mathbb{B} + \mathbb{T},$$

where the quantum numbers in the expression for Y are the baryon number, strangeness, charm, bottomness, and topness of the hadron. Its charge Q is the sum

$$Q = I_Z + \frac{Y}{2},$$

where I_Z is the hadron's isospin projection. Half the hypercharge of a hadron is the average charge of all the hadrons in its isospin multiplet.

hyperons Any baryon composed of three quarks that contains at least one strange quark but does not include any charm, bottom, or top quarks. The hyperons with the lowest mass are the Λ, Σ, and Ξ. Hyperons might exist inside the cores of some neutron stars whose extremely high densities and pressures make it energetically favorable for neutrons to convert into hyperons via weak interactions through reactions like $n + e^- \rightarrow \Sigma^- + \nu_e$.

interactions There are four types of fundamental interactions: strong (gluonic), electromagnetic, weak, and gravitational. The electromagnetic and weak interactions are unified in the electroweak theory. Fundamental interactions are due to the exchange of four different types of virtual bosons: gluons for the strong, photons and intermediate vector bosons for the electroweak, and gravitons for gravity. There is a fifth non-fundamental strong interaction responsible for the nuclear force. It is a "hadronic interaction," whose force between hadrons is created by the exchange of hadrons.

***intermediate vector bosons** – $(W^\pm, Z^0)$* These massive spin-1 particles are responsible for the weak interactions. Together with the massless photon γ, they are the quanta of the electroweak force. The $W^\pm$, but not the Z^0, are responsible for changing the flavor of quarks.

ion An atom or molecule that has a net electric charge due to the loss or gain of one or more electrons.

isospin multiplet A set of strongly interacting particles, very close in mass, with members differing in

charge by one. Other quantum numbers are the same. Each multiplet is labeled by a nonnegative integer or half-integer I, where there are $2I+1$ particles in the set. Members of a multiplet have identical strong interactions, and form an irreducible representation of SU(2). The proton p and neutron n form an isospin doublet, the nucleon $\mathcal{N}$, with $I = 1/2$. Since protons and neutrons respond identically to the nuclear force, theories of the nucleus are formulate in terms of $\mathcal{N}$, where $\mathcal{N}$ refers to either p or n.

***isospin (isotopic spin)** – $\vec{I}$* Each isospin multiplet with $2I+1$ members is assigned a vector $\vec{I}$ in a mathematical three-dimensional "isospin space," analogous to $\vec{S}$, the spin vector for particles, that exists in ordinary space. The length of $\vec{I}$ is $\sqrt{I(I+1)}$. $\vec{I}$, and the cone on which it lies, starts at the origin of a coordinate system in isospin space. The Z axis defines the central axis of the cone, whose opening angle is quantized, so that the projections I_Z of $\vec{I}$ onto the Z-axis can only have values that start at $I_Z = I$ and decrease in steps of one to $I_Z = -I$.

Each value of I_Z corresponds to a hadron whose charge also decreases in steps of one, with decreasing I_Z. For example, $\vec{I}$ for the nucleon $\mathcal{N}$ may be thought of as pointing in either the proton ($I_Z = 1/2$) or neutron ($I_Z = -1/2$) direction with charges 1 and 0. $\vec{I}$ is only conserved by the strong interactions, but I_Z is conserved both by the strong and electromagnetic interactions. Isospin allows one to describe the strong interactions in terms of isospin multiplets without regard to the electric charge of its members.

***J or J/ψ** meson* The $J \sim c\bar{c}$ meson is created from a charm and anticharm quark, the analogue of $\phi \sim s\bar{s}$, and a lighter version of the Upsilon ($\Upsilon \sim b\bar{b}$). Its discovery in 1974 by two different groups confirmed the existence of the fourth ace, discovered three years earlier by Kiyoshi Niu and colleagues in what is now called the $D^\pm$ mesons.

jets A narrow cone of hadrons produced in high-energy collisions by the hadronization of a quark or gluon carrying large momentum.

***kaon** – K meson* The lightest hadron containing a strange antiquark $\bar{s}$ (or $\bar{\Lambda}_0$), giving it strangeness $\mathbb{S} = 1$. The discovery of the kaon in cosmic ray interactions by George Rochester and Clifford Butler in

1947 was a complete surprise. The $K = (K^+, K^0)$ is an isospin doublet with $I = 1/2$ and negative parity. Together with its antiparticle $\bar{K}$, and the π and η mesons, it forms the pseudoscalar meson octet. The weak decay of the K^+ (originally the θ^+/τ^+) violates reflection symmetry (parity P is not conserved).

leptons Six elementary particles with spin 1/2 that do not feel the strong (gluonic) interactions felt by quarks. Three leptons are charged (e^-, μ^-, τ^-) and three, with only left-handed spin, are neutral (ν_e, ν_μ, ν_τ). Like the six quarks they interact weakly through the intermediate vector bosons. Only the charged leptons get their mass by interacting with the Higgs field. The origin of the three neutrino masses is unknown.

mesons Strongly interacting particles with nonnegative integral spin, like the spin-0 pion and kaon. Low-mass mesons are created from quark-antiquark pairs, but exotic mesons with additional quark-antiquark pairs — tetraquarks — exist at higher mass. All mesons are unstable, decaying either through the strong, electromagnetic, or weak interactions.

MeV A unit of energy equal to 10^6 electron volts. An electron weighs about 1/2 MeV.

muon – μ The second heaviest of the three charged leptons, all having identical quantum numbers and interactions. With a mass of 105.7 MeV, it is about 200 times heavier than the electron, and about 10 times lighter than the proton.

multiplicative quantum number When a quantum number of a collection of particles is the product of their individual quantum numbers, that quantum number is said to be multiplicative. Multiplicative quantum numbers are either 1 or -1, since applying their associated operation twice leaves the system unchanged. Parity and charge conjugation are two examples.

neutrino – ν One of three light electrically neutral leptons (ν_e, ν_μ, and ν_τ) associated through the weak interactions with their charged massive lepton partners. The antineutrino $\bar{\nu}_e$ is produced in neutron β-decay: $n \rightarrow p + e^- + \bar{\nu}_e$. Since neutrinos are unaffected by both the strong and electromagnetic interactions, neutrinos pass through matter almost undisturbed, and so are difficult to detect. Their existence was predicted by Wolfgang Pauli in 1930 in order to preserve the conservation of energy in nuclear β-decay. Neutrinos

are the lightest of all known elementary particles with mass. Since they don't have right-handed spin, they are unaffected by the Higgs field, leaving them massless in the Standard Model.

nucleon – $\mathcal{N}$ The generic name used to describe the two constituents of an atomic nucleus, the proton and neutron. Although protons and neutrons have different charge, and therefore different electromagnetic interactions, there is no analogous quantity that allows them to be distinguished by the strong interactions, which treats them equally. If the nucleon is represented by its two-components, $\mathcal{N} = (p, n)$, the proton and neutron are said to form an isospin multiplet, and the nucleon forms a two-dimensional irreducible representation of the symmetry group SU(2).

orbital (rotational) angular momentum – $\vec{L}$ Orbital angular momentum is the rotational analogue of linear momentum. The orbital angular momentum of a particle in quantum mechanics is a quantized version of its classical counterpart. The quantum number L associated with the vector $\vec{L}$ is quantized at the nonnegative integers. If the projection L_z of $\vec{L}$ on an axis "z" in space is measured, only one of $2L + 1$ possible values will be found, starting with $L_z = L$, and decreasing in steps of 1 to $L_z = -L$. L is conventionally measured in units of $\hbar = h/2\pi$, where h is Planck's constant.

OZI rule A weaker version of Zweig's rule, since it does not specify the relative amplitudes of the hadron couplings that are allowed.

parity – P Like charge conjugation, parity is a multiplicative quantum number with values of $P = 1$ or -1, that is conserved by the strong and electromagnetic interactions, but not by the weak interactions. A parity transformation acts both on the intrinsic parity of a particle and its spatial coordinates. By convention the intrinsic parity of a proton is 1, so that the intrinsic parity of a quark must also be set to 1. The parity of a meson created from a quark-antiquark pair with angular momentum L is $P = [(1)(-1)](-1)^L = (-1)^{L+1}$. The factor of $[(1)(-1)]$ is the contribution to the meson's parity from the intrinsic parities of the quark and antiquark, while the $(-1)^L$ comes from their angular momentum. Since $L = 0$ for the pion, its parity is -1, in agreement with experiment. Spin-0 mesons with negative parity are called pseudoscalar mesons.

Pauli exclusion principle Pauli's principle, which he applied to electrons in atoms, states that two or more identical electrons cannot occupy the same quantum state. Therefore, two electrons in the same atomic orbit must have opposite spin directions. This principle, when extended to include all particles with half-integral spin in any quantum system, is called the "spin-statistics theorem."

phi – ϕ meson The $\phi \sim s\bar{s}$ meson (originally $\Lambda_0 \bar{\Lambda}_0$) is created from a strange and antistrange quark. It decays into $K + \bar{K}$, but not $\rho + \pi$, as was expected in 1963. This anomaly led to the discovery of aces in 1964.

photon – γ The massless neutral spin-1 particle responsible for the electromagnetic force. The photon and the massive spin-1 $W^{\pm}$ and Z^0 intermediate vector bosons are carriers of the electroweak force. The parity and charge conjugation of the photon are both -1.

pion – π meson The pion is the lightest hadron and carrier of the long-range component of the nuclear force. Its existence was predicted by Hideki Yukawa in 1935 and was first observed in cosmic ray interactions by Cecil Powell and collaborators in 1947. Pions have negative parity; neutral pions have positive charge conjugation. With the K, $\bar{K}$, and η mesons, they form the pseudoscalar meson octet. The pion is an $I = 1$ isospin triplet consisting of the π^+, π^0, and π^-, all treated equally by the strong interactions.

Planck's constant – h The proportionality constant relating the energy E of a photon to the frequency ν of its associated electromagnetic wave: $E_\gamma = h\nu$. Planck's constant divided by 2π is denoted by $\hbar$, the unit used to measure quantized spin and orbital angular momentum. In these units the electron spin is 1/2.

positron – e^+ The electron's antiparticle.

positronium An electron-positron atom.

psi – ψ meson See "J meson."

quantum chromodynamics – QCD The quantum field theory of quarks, gluons, and color that is responsible for the creation of hadrons and a description of their strong interactions. It was completed in the mid-1970s. Since gluons interact with each other, as well as with quarks, QCD is computationally challenging,

making it difficult to work out its consequences, particularly at low energy where the gluon's color couplings are large.

quantum electrodynamics – QED The quantum field theory of charged particles and photons that describes the electromagnetic interactions responsible for the creation of atoms and molecules. It was completed in the late 1940s. Since the coupling between photons and charged particles is small, and photons do not couple to other photons, QED is computationally tractable, and well tested.

quantum numbers Numerical attributes of particles, or systems of particles, that remain constant in time when one or more types of interactions are present. They are linked to symmetries. For example, the conservation of the total angular momentum of a set of particles is related to the rotational symmetry of three-dimensional space. Total angular momentum is conserved by all interactions, and is described by the symmetry group SO(3). Each quantum number has only certain values, and together, they help characterize the properties of a particle or system of particles. Quantum numbers also provide a systematic way to categorize and predict the behavior of particles.

quarks Initially three mathematical entities invented by Robert Serber early in 1963. After Serber told Murray Gell-Mann about them, Gell-Mann used them in February 1964 to construct a toy field theory, without strong interactions, from which he abstracted the equal-time commutation relations of the electromagnetic and weak currents. George Zweig discovered quarks in January 1964, but called them aces, thinking that there would eventually be four of them. Aces acted like real particles with their own laws of interaction responsible for the creation of hadrons. They also explained weak-interaction selection rules. Quarks as mathematical entities were eventually forgotten, and real particles that are the constituents of hadrons — aces — were called quarks. Quarks increased in number from three to six. They carry the fractional electric charges of 2/3 and $-1/3$. Quarks and gluons, together with the color they carry, form the building blocks of quantum chromodynamics (QCD), the fundamental theory of the strong interactions.

quark-gluon plasma – QGP A pressurized mixture of quarks and gluons created at accelerators in which colliding heavy nuclei are squeezed into a nearly invis-cid fluid in which protons and neutrons have melted into their constituents. A QGP filled the entire Universe shortly after the Big Bang, before it expanded, cooled, and condensed into nucleons.

radioactivity – radioactive decay The spontaneous disintegration of an atomic nucleus into a lighter nucleus, accompanied by the emission of radiation (alpha particles, beta particles, or gamma-rays).

resonance A highly unstable strongly interacting particle that decays rapidly into other hadrons in about the time it takes light to travel across its diameter, approximately 10^{-23} seconds. It may be observed as one of the products of a scattering reaction, or in isolation as a peak in the total cross section of two-particle scattering. A resonance can either be a baryon or meson. Resonances are typically discovered as peaks in invariant mass plots of their decay products. The peak is characterized by the mass and width of the resonance. The width of the peak is inversely proportional to the resonance's lifetime, as dictated by the uncertainty principle.

spin – $\vec{S}$ The angular momentum "carried" by a particle or set of particles. The associated quantum number S is either integral or half-integral. Fermions have half-integral spin ($S = 1/2, 3/2, ...$), while bosons have nonnegative integral values. If the projection S_z of $\vec{S}$ on an axis "z" in space is measured, only one of $2S + 1$ possible values will be found, starting with $S_z = S$, and decreasing in steps of 1 to $S_z = -S$. S is conventionally measured in units of $\hbar = h/2\pi$, where h is Planck's constant. Spin is part of a particle's total angular momentum, which includes both orbital angular momentum and intrinsic spin.

Elementary particles are point-like, without structure, so they have noting to rotate to create their spin. Instead, they carry "packets" (quanta) of angular momentum. Spin angular momentum can be converted into orbital angular momentum, and conversely, in interactions possibly involving the creation or destruction of other particles. However, the total angular momentum $\vec{J}$ of a system can not change in the process. The spin of a hadron is formed by combining the total orbital angular momenta $\vec{L}$ of its quark constituents with their total spin $\vec{S}$, $\vec{J} = \vec{L} + \vec{S}$.

spin-statistics theorem The wave function of a system of identical particles with integral spin (bosons) remains unchanged when the positions of any two particles are exchanged. If the identical particles have half-integral spin (fermions), the wave function changes sign. Therefore, two or more identical fermions cannot occupy the same quantum state.

Standard Model QCD and the electroweak theory, with the Higgs field and its quantum fluctuation — the Higgs boson — form the Standard Model. It is based on quarks, leptons, and the carriers of the forces that act between them: gluons, photons, and the $W^{\pm}$ and Z^0 intermediate vector bosons. While phenomenally successful, the Standard Model is incomplete. It predicts massless neutrinos, but neutrinos have mass, and it doesn't include gravity, dark energy, or dark matter.

Stanford Linear Accelerator Center – SLAC A two-mile-long linear electron accelerator completed in 1966, site of the deep inelastic scattering experiments, started in !968, that confirmed the existence of quarks. The accelerator is linear, rather than circular, to minimize the energy lost by bremsstrahlung, which is larger for lighter particles like the electron. The Stanford Positron Electron Accelerating Ring (SPEAR), an electron-positron collider completed in 1972, used SLAC as its injector. Both the ψ meson and τ lepton were discovered at SPEAR in 1974 and 1975.

strangeness quantum number – $\mathbb{S}$ The quantum number associated with the third quark s (or Λ_0). Like charm $\mathbb{C}$, it is an additive quantum number for hadrons conserved by the strong, electromagnetic, but not weak interactions. It is the negative of the difference between the number of strange and antistrange quarks in a system of hadrons. Strangeness was discovered by Nishijima in Japan in 1954, who called it η. It was popularized in the West in 1956 by Gell-Mann who renamed it strangeness.

strong interaction or force In contemporary particle physics the strong interaction, one of the four fundamental interactions, creates the force between quarks, which remains constant as the distance between them increases (when their separation is large). This force is created by the exchange of massless gluons as described by Quantum Chromodynamics (QCD). Originally the strong force referred to the interaction between hadrons mediated by other hadrons, initially through the exchange of the pion responsible for its long-range component. Unlike the strong force between quarks, it falls rapidly with increasing distance (slightly faster than exponentially). It is an emergent property of quarks, gluons, and their interactions.

SU(2) representations Hadrons very close in mass, with identical quantum numbers except for charge, form SU(2) representations (isospin multiplets). Particles in the same SU(2) representation respond identically to the strong interactions, and so are given a generic name, like the nucleon $\mathcal{N}$ for the proton and neutron, or the pion π for the π^+, π^0, and π^-.

tau – τ The heaviest of the three charged leptons, all having identical quantum numbers and interactions. With a mass of 1,777 MeV, it is about 3,500 times heavier than the electron, and almost twice as heavy as the proton.

TeV A unit of energy equal to 10^{12} electron volts.

vacuum The state of lowest energy; space-time without real particles. However, the vacuum is not empty; it contains all possible quantum fields with their virtual particle-antiparticle pairs that exist for time intervals inversely proportional to the pair's mass-energy. These transitory pairs have measurable consequences. For example, if particles in a pair are charged they can exchange virtual photons with real charged particles, changing the way these particles interact with an electric or magnetic field. The magnetic moments of electrons and muons are modified in this manner, and when measured, are consistent with their predicted values. This, and many other measurements, test our understanding of the vacuum, the electroweak theory, and quantum field theory more generally.

virtual particles Unlike real particles, virtual particles are not directly observable but are inferred from the effects their transitory existence produces on real particles and their interactions. Forces between particles are mediated by the exchange of virtual particles. Quantum fields in the vacuum create virtual particle-antiparticle pairs, only to have them disappear again before a violation of the conservation of energy can be detected. Such pairs are everywhere, sporadically appearing and disappearing. Virtual electron-positron

pairs inside a hydrogen atom modify the electron's motion, and so the frequency of light the atom emits. Such a shift in frequency was measured in 1947 (the Lamb shift).

wave function A complex-valued function, often designated by ψ, that encodes all information about a quantum system. The probabilities of all possible outcomes for measurements that can be made on the system, such as position, momentum, or angular momentum of all its particles, can be computed from it. The squared magnitude of the wave function, $|\psi|^2$, is related to the probability of finding the system in a particular state. For example, if $\psi(x, y, z, t)$ is the wave function of an electron in a hydrogen atom, $|\psi(x, y, z, t)|^2$ is the probability density of finding the electron at the point (x, y, z) at a specific time t, i.e., $|\psi(x, y, z, t)|^2 \Delta x \Delta y \Delta z$ approximates the probability that the electron will be found at time t within a small volume $\Delta x \Delta y \Delta z$ centered at (x, y, z).

wave-particle duality As explained by Louis de Broglie in his 1924 Ph.D. thesis, under certain circumstances a particle exhibits the properties of a wave with wavelength $\lambda = h/p$, where p is the particle's momentum and h is Planck's constant. Conversely, waves in quantum field theories sometimes act like particles. The wave-particle duality provides one explanation for why electrons in an atom have discrete Bohr orbits. Traveling around the nucleus, electron waves must return in phase. Electron waves fit orbits, like sound waves fit organ pipes. In both cases standing waves are created.

weak interaction or force One of the four fundamental interactions, along with the gravitational, electromagnetic, and strong (gluonic) interactions. The weak interaction is mediated by the exchange of three heavy elementary virtual spin-1 particles called intermediate vector bosons — the $W^\pm$ and Z^0. The weak interaction has been united with the electromagnetic interaction to form the electroweak theory which, together with the Higgs field and QCD, forms the Standard Model. It is the only interaction that can change the flavor of quarks (with the $W^\pm$). It is not symmetric under parity (P) or charge conjugation (C) transformations, but is almost symmetric when they are combined (*CP* transformations).

X-rays Photons produced primarily by bremsstrahlung when an energetic beam of electrons are suddenly decelerated by shooting them into a metal with a high nuclear charge. X-ray energies lie between those of ultraviolet radiation and gamma rays created by nuclear decay.

Zweig's rule When a hadron decays through the strong interactions its constituents (aces) flow into its decay products, which also receive contributions from one or more virtual ace-antiace pairs that arise from the vacuum [Figures 2.6 and 6.4, pp. 25 and 102]. Applied to ϕ decay, Zweig's rule forbids the reaction $\phi \to \rho + \pi$. The rule also specifies the strength of allowed decays, and more generally, the coupling constants for all strong interactions.

Zweig's rule, and the graphical calculus on which it is based, can also be used to calculate the color interactions between quarks and gluons, because both hadronic and strong (gluonic) interactions have an SU(3) symmetry.

Acknowledgments

Organizing, integrating, and expanding material written at different times for different occasions, without destroying their original spirit, has been a daunting task. They have been forged into a description of modern fundamental physics, and its participants. Understanding what readers coming from different backgrounds would find accessible was even more challenging. Family and friends were extremely helpful. I thank Ravi Chander, Adolf Cusmariu, George Wei-Shu Hou, Jeffrey Mandula, Joe Mitchell, and my brother Michael Zweig for their many suggestions and help in improving the text.

Shane Haas has read and reread much of the book, forcing me to simplify. Marek Karliner, with his understanding of ideas in their historical setting, was very helpful in bringing the manuscript up to date. Jonathan Rosner, one of the book's reviewers, found innumerable problems. His insights and remarkable attention to detail are much appreciated.

Special thanks are also due to my cousin Michael Frohlich, a botanist, who provided a myriad of comments, corrections, and musings. As stated in the Preface, Max Delbrück's ideal reader is "infinitely ignorant, but infinitely smart." Michael fails miserably in meeting Max's first metric, but more than makes up for it in the second.

Finally, I thank Erica Jen, my wife and editor, who not only found problems, but suggested ways to fix them. Her intellect continues to amaze me.

Chapter 4: Richard Feynman

I thank Cormac McCarthy for his editorial input and banishment of the "historical present," and Thomas Kiley for his perceptive remarks.[1] I am also grateful to Finn Ravndal for historical information and catching a thinko, to Gaetana Spedalieri, a biologist, who needed to know more about constituent quarks, and to Pan Yue (Alexa), a student at NTU in Singapore in 2018, who reminded me that it is worth remembering.

Appendix A: A Primer on Quarks

The Primer is dedicated to Francis Bello, an Associate Editor at *Scientific American*, with whom the Primer was started more than 50 years ago. I appreciate the time we spent together, and what he taught me about writing science articles for "grandmothers in Kansas." Frank died in 1987 at the age of 69.

Cambridge University Press

Vince Higgs' patience and guidance in bringing this book to fruition are much appreciated. Rachael K. Lazenby designed the remarkable playing cards that decorate the front cover, while Vince imaged how they might be arranged, by printing a deck, and scattering their cards into place.

---

[1] However, in his recent novel *The Passenger*, Cormac fuses fact with fiction when his protagonist, an ex-physicist, who is describing the discovery of quarks declares: "And he [Murray] swept the field and won the Nobel Prize and George went into therapy." Not true. I didn't go into therapy, and Murray didn't get the Nobel Prize, for quarks. A Nobel prize for the discovery of quarks has not been awarded, only two for the confirmation of that discovery.

Further Reading

Laurie M. Brown, Max Dresden, and, Lillian Hoddeson (eds.), *The Birth of Particle Physics*, Cambridge University Press, Cambridge, UK (1983).

Laurie M. Brown, Max Dresden, and Lillian Hoddeson (eds.), *Pions to Quarks: A Collection of Articles on the Development of Particle Physics between 1947 and 1963*, Cambridge University Press, Cambridge, UK (1989).

Robert P. Crease and Charles C. Mann, *The Second Creation*, revised ed., Rutgers University Press, New Brunswick, NJ (1996).

Richard Feynman, *The Character of Physical Law*, The MIT Press, Cambridge, MA (1965).

Richard Feynman and Steven Weinberg, *Elementary Particles and the Laws of Physics*, Cambridge University Press, Cambridge, UK (1987).

Sheldon Glashow, "Quarks With Color and Flavor," *Scientific American,*" **233** (4), 38 (October 1975).

George Gamow, *Thirty Years That Shook Physics*, Doubleday & Co. Inc. NY, 1966; reprinted by Dover Publications, Mineola, NY (1985).

James Gleick, *Genius*, Vintage Books, New York, NY (1999).

George Johnson, *Strange Beauty*, Vintage Books, New York, NY (1993).

Lawrence M. Krauss, *Quantum Man*, W.W. Norton & Company, New York, NY (2011).

Thomas S. Kuhn, *The Structure of Scientific Revolutions*, 3rd ed., University of Chicago Press, Chicago, IL (1996).

Rod Nave, *Hyperphysics: Index*, hyperphysics.phy-astr.gsu.edu/hbase/index.html; *Quantum Physics*, hyperphysics.phy-astr.gsu.edu/hbase/quacon.html# quacon; *Quarks*, hyperphysics.phy-astr.gsu.edu/hbase/Particles/quark.html.

Wolfgang Pauli, Charles P. Enz and Karl von Meyenn (eds.), *Writings on Physics and Philosophy*, Springer-Verlag, Berlin (1994).

Andrew Pickering, *Constructing Quarks*, University of Chicago Press, Chicago, IL (1999).

Ben Still, *Particle Physics Brick by Brick*, Firefly Books, Buffalo, NY (2018).

Chiang Tsai-Chien, *Madame Wu Chien-Shiung, The First Lady of Physics Research*, World Scientific, Singapore (2014).

B. L. Van der Waerden (ed.), *Sources of Quantum Mechanics*, North-Holland Publishing, Amsterdam, 1967; reprinted by Dover Publications, Mineola, NY (1967).

Steven Weinberg, *Third Thoughts* The Belknap Press of Harvard University, Cambridge, MA (2008).

Up-to-date Summaries, Reviews, Tables, and Plots of particle physics data may be found at pdg.lbl.gov.

Bibliography

[1] R. Aaij et al., *Phys. Rev. Lett.* **122**, 222001 (2019).

[2] E. Abers, F. Zachariasen, and C. Zemach, *Phys. Rev.* **132**, 1831 (1963).

[3] H. Abreu, et al. (FASER Collaboration), *Phys. Rev. D* **104**, L091101-1 (2021).

[4] N. Agafonova, et al. (OPERA Collaboration), *Phys. Rev. Lett.* **120**, 211801-1 (2018).

[5] S. L. Adler, *Phys. Rev. Lett.* **14**, 1051 (1965).

[6] H. L. Anderson, E. Fermi, E. A. Long, R. Martin, and D. E. Nagle, *Phys. Rev.* **85**, 934 (1952).

[7] H. L. Anderson, E. Fermi, E. A. Long, and D. E. Nagle, *Phys. Rev.* **85**, 936 (1952).

[8] H. L. Anderson, E. Fermi, R. Martin, and D. E. Nagle, *Phys. Rev.* **91**, 155 (1953).

[9] J. Baggott, "Forward," *Higgs: The Invention and Discovery of the "God Particle,"* p. XX, Oxford University Press, Oxford, UK (2012).

[10] P. Ball, "Honey, I Shrank the Motor," *The Guardian* – Science Section (June 9, 2004).

[11] H. Becquerel, *Comptes Rendus* **122**, 420 (1896).

[12] H. Becquerel, *Comptes Rendus* **122**, 501 (1896).

[13] R. E. Behrends, J. Dreitlein, C. Fronsdal, and W. Lee, *Rev. Mod. Phys.* **34**, 1 (1962).

[14] L. Bertanza et al., *Phys. Rev. Lett.* **9**, 180 (1962).

[15] H. Bethe, *Phys. Rev.* **72**, 339 (1947).

[16] J. D. Bjorken, and S. L. Glashow *Phys. Lett.* **11**, 255 (1964).

[17] J. D. Bjorken, *Phys. Rev.* **148**, 1467 (1966); erratum **160**, 1582 (1967).

[18] J. D. Bjorken, *Phys. Rev.* **163**, 1767 (1967).

[19] J. D. Bjorken and E. A. Paschos, *Phys. Rev.* **185**, 1975 (1969).

[20] R. Bjorklund, W. E. Crandall, B. J. Moyer, and H. F. York, *Phys. Rev.* **77**, 213 (1950).

[21] E. D. Bloom, et al., *Phys. Rev. Lett.* **23**, 930 (1969).

[22] N. Bohr, *Journal of the Chemical Society*, 349 (1932).

[23] A. Borrelli, "The Great Yogurt Project," *Model and Mathematics: From the 19th to the 21st Century*, p. 221, M. Friedman and K. Krauthausen (eds.), Birkhäuser, Cham Switzerland (2022).

[24] N. Brambilla, "The XYZ States: Experimental and Theoretical Status and Perspectives," arXiv:1907.07583v2 (2020).

[25] M. Breidenbach, et al., *Phys. Rev. Lett.* **23**, 935 (1969).

[26] G. Breit, E. U. Condon, and R. D. Present, *Phys. Rev.* **50**, 825 (1936).

[27] S. J. Brodsky, P. Hoyer, C. Peterson, and N. Sakai, *Phys. Lett.* **93B**, 451 (1980).

[28] V. D. Burkert and C. D. Roberts, *Rev. Mod. Phys.* **91**, 011003-1 (2019).

[29] N. Cabibbo, *Phys. Rev. Lett.* **10**, 531 (1963).

[30] J. Chadwick and E. S. Bieler, *Phil. Mag.* Series 6 **42**, 923 (1921).

[31] J. Chadwick, *Proc. Royal Soc. A* **136**, 692 (1932).

[32] C. T. Chase, *Phys. Rev.* **36**, 1060 (1930).

[33] G. F. Chew and S. Frautschi, *Phys. Rev. Lett.* **7**, 394 (1961).

[34] G. F. Chew, *S-matrix Theory of Strong Interactions*, UCRL-9701 (May 15, 1961).

[35] G. F. Chew, M. Gell-Mann, and A. H. Rosenfeld, *Scientific American* **210** (2), 74 (1964).

[36] W. Chinowsky and J. Steinberger, *Phys. Rev.* **95**, 1561 (1954).

[37] S. K. Choi, et al. (Belle Collaboration), *Phys. Rev. Lett.* **91**, 262001 (2003).

[38] J. H. Christenson, J. W. Cronin, V. L. Fitch, and R. Turlay, *Phys. Rev. Lett.* **13**, 138 (1964).

[39] S. Coleman and S. L. Glashow, *Phys. Rev. Lett.* **6**, 423 (1961).

[40] S. Coleman and J. Mandula, *Phys. Rev.* **159**, 1251 (1967).

[41] P. L. Connolly, et al., *Phys. Rev. Lett.* **10**, 371 (1963).

[42] R. T. Cox, C. G. McIlwraith, and B. Kurrelmeyer, *Proc. Nat. Acad. Sci.* **14**, 544 (1928).

[43] R. P. Crease and C. C. Mann, *The Second Creation*, p. 281, Rutgers University Press, New Brunswick, NJ (1986, revised ed., 1996).

[44] R. P. Crease, *Phys. Perspect.*, **7**, 404 (2005).

[45] R. H. Dalitz, *Phys. Rev.* **94**, 1046 (1954).

[46] B. d'Espagnat and J. Prentki, *Phys. Rev.* **99**, 328 (1955).

[47] B. d'Espagnat and J. Prentki, *Phys. Rev.* **102**, 1684 (1956).

[48] B. d'Espagnat and J. Prentki, *Nuc. Phys.* **1**, 33 (1956).

[49] B. d'Espagnat and J. Prentki, *Suppl. del Nuovo Cimento*, Serie X No. 2, **4**, 639 (1956).

[50] P. A. M. Dirac, *Physikalische Zeitschrift der Sowjetunion* **3**, 64 (1933).

[51] S. Drell, D. J. Levy, and T. M. Yan, *Phys. Rev.* **187**, 2159 (1969).

[52] R. S. Edgar, R. P. Feynman, S. Klein, I. Lielausis, and C. M. Steinberg, *Genetics* **47**, 179 (1962).

[53] P. C. England, P. Molnar, and F. M. Richter, *American Scientist* **95**, 342 (2007).

[54] V. Fanti et al, *Eur. Phys. J. C* **12**, 69 (2000).

[55] T. Fazzini, G. Fidecaro, A. W. Merrison, H. Paul, and A. V. Tollestrup, *Phys. Rev. Lett.* **1**, 247 (1958).

[56] E. Fermi, *Zeit. für Physik* **88**, 161 (1934).

[57] E. Fermi and C. N. Yang, *Phys. Rev.* **76**, 1739 (1949).

[58] R. P. Feynman, *Physics Today* **1** (2), 8 (1948).

[59] R. P. Feynman, *Phys. Rev.* **76**, 769 (1949).

[60] R. P. Feynman, *Phys. Rev.* **97**, 660 (1955).

[61] R. P. Feynman, *Rev. Mod. Phys.* **29**, 205 (1957).

[62] R. P. Feynman and M. Gell-Mann, *Phys. Rev.* **109**, 193 (1958).

[63] R. P. Feynman, *Engineering & Science* **XXIII** (5), 22 (February 1960).

[64] R. P. Feynman, M. Gell-Mann, and M. Lévy, OSTI.GOV_4185945 *Technical Note No. 17* (January 1960).

[65] R. P. Feynman, M. Gell-Mann, and G. Zweig, *Phys. Rev. Lett.* **13**, 678 (1964).

[66] Richard Feynman, *The Character of Physical Law*, The MIT Press, Cambridge, MA (1965)

[67] R. P. Feynman, "Consequences of SU_3 Symmetry in Weak Interactions," *Symmetries in Elementary Particle Physics*, pp. 111–174, A. Zichichi (ed)., Academic Press, New York, NY (1965).

[68] R. P. Feynman, "Closing Lecture," *Symmetries in Elementary Particle Physics*, pp. 400–418, A. Zichichi (ed.), Academic Press, New York, NY (1965).

[69] R. P. Feynman and A. Hibbs, *Quantum Mechanics and Path Integrals*, McGraw-Hill Companies, New York, NY (1965).

[70] R. P. Feynman, *Oral History Interviews*, Session IV, American Institute of Physics, College Park, MD (1966).

[71] R. P. Feynman, *Oral History Interviews*, Session V. American Institute of Physics, College Park, MD(1966).

[72] R. P. Feynman, *Phys. Rev. Lett.* **23**, 1415 (1969).

[73] R. P. Feynman, M. Kislinger, and F. Ravndal, *Phys. Rev. D* **3**, 2706 (1971).

[74] R. P. Feynman, *Nuclear Phys. B* **188**, 479 (1981).

[75] R. P. Feynman, *Surely You're Joking, Mr. Feynman! The Adventures of a Curious Character as told to Ralph Leighton*, p. 286, Centenary Edition, W.W. Norton & Co., New York, NY (1985).

[76] R. P. Feynman, F. B. Morinigo, and W. G. Wagner, *Feynman Lectures on Gravitation*, B. Hatfield (ed.), Addison-Wesley Publishing Co., Reading MA (1995).

[77] A. Franklin, *Experiment, Right or Wrong*, Chapter 5, Cambridge University Press, Cambridge, UK (1990).

[78] H. Fritzsch, M. Gell-Mann, H. Leutwyler, *Phys. Lett.* **47B**, 365 (1973).

[79] S. Fukui and S. Miyamoto, *Il Nuovo Cimento* **XI**, 113 (1959).

[80] M. Gaillard, B. Lee, and J. Rosner, *Rev. Mod. Phys.*, p. 277 (1975).

[81] F. Galton, English Men Of Science: Their Nature and Nurture, Macmillan & Co., London (1874).

[82] H. Geiger and E. Marsden, *Proc. R. Soc. Lond. A* **82** (557), 495 (1909).

[83] M. Gell-Mann, *Phys. Rev.* **92**, 833 (1953).

[84] M. Gell-Mann and A. Pais, "Theoretical Views on the New Particles" in *Proceedings of The 1954 Glasgow Conference on Nuclear & Meson Physics*, pp. 342–352, E. H. Bellamy and R.G. Moorhouse (eds.), Pergamon Press, London & New York, NY (1955).

[85] M. Gell-Mann, *Suppl. del Nuovo Cimento*, Serie X No. 2, **4**, 848 (1956).

[86] M. Gell-Mann, and A. H. Rosenfeld, *Annu. Rev. Nucl. Sci.* **7**, 407 (1957).

[87] M. Gell-Mann and E. P. Rosenbaum, *Scientific American* **197** (1), 72 (July 1957).

[88] M. Gell-Mann, *Phys. Rev.* **106**, 1296 (1957).

[89] M. Gell-Mann, and M. Lévy, *Il Nuovo Cimento* **16**, 705 (February 1960).

[90] M. Gell-Mann, "The Eightfold Way: A Theory of Strong Interaction Symmetry" *Caltech Synchrotron Report CTSL-20* (March 15, 1961).

[91] M. Gell-Mann, *Phys. Rev.* **125**, 1067 (1962).

[92] M. Gell-Mann, D. Sharp, and W. G. Wagner, *Phys. Rev. Lett.* **8**, 261 (1962).

[93] M. Gell-Mann, *Phys. Lett.* **8**, 214 (1964).

[94] M. Gell-Mann, *Physics* **1**, 63 (1964).

[95] M. Gell-Mann, "Current Topics in Particle Physics," *Proceedings of the XIIIth International Conference on High Energy Physics*, pp. 3–9, M. Alston-Garnjost (ed.), University of California at Berkeley, Berkeley, CA (1966).

[96] M. Gell-Mann, *Proceedings of the 1971 International Conference on Duality and Symmetry in Hadron Physics*, Tel Aviv, Israel, April 5–7, p. 389, E. Gotsman (ed.) (1971).

[97] M. Gell-Mann, *Acta Physica Austriaca, Suppl. IX*, 733 (1972).

[98] M. Gell-Mann, *The Quark and the Jaguar*, St. Martin's Griffin, NY (1995).

[99] M. Gell-Mann, "Summary," *To Fulfill a Vision: Jerusalem Einstein Centennial Symposium on Gauge Theories and Unification of Physical Forces*, p. 257, Y. Ne'eman (ed.), Israel Academy of Sciences, Jerusalem (1981).

[100] M. Gell-Mann and Y. Ne'eman (eds.), *The Eightfold Way*, Perseus Publishing, Cambridge, MA (1964, 2000).

[101] M. Gell-Mann, *Murray Gell-Mann, Selected Papers*, pp. 35–37, Harald Fritzsch ed., World Scientific, Singapore (2010).

[102] S. L. Glashow and J. J. Sakurai, *Il Nuovo Cimento* **12**, 337 (1962).

[103] S. L. Glashow, *J. Phys. Cong. Ser.* 196, 011003 (2009).

[104] J. Gleick, *Genius*, p. 433, Vintage Books, New York, NY (1992).

[105] H. Goldberg and Y. Ne'eman, *Il Nuovo Cimento* **XXVII** (1), 1 (1963).

[106] R. Gomez, H. Kobrak, A. Moline, J. Mullins, C. Orth, J. Van Putten, and G. Zweig, *Phys. Rev. Lett.* **18**, 1002 (1967).

[107] F. Goodall, *The Reminiscences of Frederick Goodall R.A.*, p. 131, The Walter Scott Publishing Co., Ltd., London (1902).

[108] D. L. Goodstein and J. R. Goodstein, *Engineering & Science* **59** (3), 15 (1996).

[109] J. Greensite, *Prog. Part. Nucl. Phys.* **51**, 1 (2003).

[110] L. Grodzins, *Proc. Natl. Acad. Sci. USA* **45**, 399 (1959).

[111] F. Gürsey and L. A. Radicati, *Phys. Rev. Lett.* **13**, 173 (1964).

[112] T. Hayashi, E. Kawai, M. Matsuda, S. Ogawa, and S. Shige-Eda, *Prog. Theor. Phys.* **47**, 280 (1972).

[113] U. Heinz and R. Snellings, *Annu. Rev. Nucl. Part. Sci.* **63**, 123 (2013).

[114] H. Heiselberg and M. Hjorth-Jensen, *Physics Reports* **328**, 237 (2000).

[115] W. Heisenberg, *Zeit. für Physik* **33**, 879 (1925).

[116] W. Heisenberg, *Zeit. für Physik* **43**, 172 (1927).

[117] W. Heisenberg, *Zeit. für Physik* **77**, 1 (1932).

[118] W. Heisenberg, *Zeit. für Physik* **120**, 513 (1943).

[119] W. Heisenberg, *Rev. Mod. Phys.* **29**, 269 (1957).

[120] W. Heisenberg, *Am. J. Physics* **43**, 389 (1975).

[121] W. Heisenberg, *Physics Today* **29** (3), 32 (1976).

[122] W. Heisenberg, *Glimpsing Reality: Ideas in Physics and the Link to Biology*, p. 15, P. Buckley and F. D. Peat (eds.), University of Toronto Press, Canada (1996).

[123] F. Henry, M. Glavin, and E. Jones, *IEEE Reviews in Biomedical Engineering* 16, 319 (2023).

[124] A. J. G. Hey, *Physics Today* 49 (9), 44 (1996).

[125] K. Higashijima, *Physics Today* 74 (11), 64 (2021).

[126] V. D. Hopper and S. Biswas, *Phys. Rev.* 80, 1099 (1950).

[127] J. Iizuka, *Progress of Theoretical Physics Supplement* 37, 21 (1966).

[128] M. Ikeda, S. Ogawa, and Y. Ohnuki, *Prog. Theor. Phys.* 11, 715 (1959).

[129] G. Impeduglia, R. Piano, A. Prodell, N. Samios, M. Schwartz, and J. Steinberger, *Phys. Rev. Lett.* 1, 249 (1958).

[130] W. Isaacson, *Einstein: His Life and Universe*, p. 65, Houghton Mifflin Company, New York, NY (2008).

[131] A. Janiak, *Newton: Philosophical Writings*, p. 102, Cambridge University Press, Cambridge, UK (2004).

[132] G. Johnson, *Strange Beauty*, Vintage Books, New York, NY (1999).

[133] S. Kegel, et al., *Phys. Rev. Lett.* 130, 152502 (2023).

[134] D. Kevles, *The Baltimore Case: A Trial of Politics, Science, and Character*, W.W. Norton & Co., Inc., New York, NY (1998).

[135] Lord Kelvin, *Phil. Mag.* Series 6 2, 1 (1901).

[136] D. Kennefick, *Physics Today* 58, 43 (September 2005).

[137] W. Kienzle, "Compiled Evidence for a Splitting of the A2 Meson," *Meson Spectroscopy*, p. 265, C. Baltay and A. H. Rosenfeld (eds.), W.A. Benjamin, New York, NY (1968).

[138] M. Kobayashi and T. Maskawa, *Prog. Theor. Phys.* 49, 652 (1973).

[139] K. Kodama, et al. (DONUT Collaboration), *Phys. Lett. B* 504, 218 (2001).

[140] C. M. G. Lattes, G. Occhialini, and C. Powell, *Nature* 160, 453 (1947).

[141] T. D. Lee and C. N. Yang, *Phys. Rev.* 104, 254 (1956).

[142] T. D. Lee and C. N. Yang, *Phys. Rev.* 105, 1671 (1957).

[143] T. D. Lee and C. N. Yang, *Phys. Rev.* 122, 1954 (1961).

[144] S. J. Lindenbaum, *Annu. Rev. Nucl. Sci.* 7, 317 (1957).

[145] H. J. Lipkin, "Quark Models and Quark Phenomenology," *The Rise of the Standard Model: Particle Physics in the 1960s and 1970s*, p. 542, L. Hoddeson, L. Brown, M. Riordan and M. Dresden (eds.), Cambridge University Press, UK (1997).

[146] H. J. Lipkin, *CERN Courier* 8, p. 234 (1969).

[147] H. J. Lipkin, "The Pre-QCD Quark Model," *Baryons '92*, p. 124, M. Gai (ed.), World Scientific, Singapore (1993).

[148] S. Lokanathan and J. Steinberger, *Il Nuovo Cimento* 2, 151 (1955).

[149] J. Mandula, J. Weyers, and G. Zweig, *Annu. Rev. Nucl. Sci.* 20, 289 (1970).

[150] R. E. Marshak, *Z. Naturforsch.* 52a, 3 (1997).

[151] R. W. McAllister and R. Hofstadter, *Phys. Rev.* 102, p. 851 (1956).

[152] J. Mehra, *The Beat of a Different Drum*, Clarendon Press, Oxford (1994).

[153] A. de Mille, *Martha, The Life and Work of Martha Graham*, p. 264, Random House, New York, NY (1991).

[154] H. Nagaoka, *Phil. Mag.* Series 6, 7, 445 (1904).

[155] T. Nakano and K. Nishijima, *Prog. Theor. Phys.* 10, 581 (1953).

[156] Y. Nambu, *Phys. Rev. Lett.* 4, 380 (1960).

[157] Y. Nambu and J. J. Sakurai, *Phys. Rev. Lett.* 11, 42 (1963).

[158] S. H. Neddermeyer and C. D. Anderson, *Phys. Rev.* 51, 884 (1937).

[159] Y. Ne'eman, *Nuclear Phys.* 26, 222 (1961).

[160] A. Niépce de Saint-Victor, *Comptes Rendus* 53, 33 (1861).

[161] K. Nishijima, *Prog. Theor. Phys.* 12, 107 (1954).

[162] K. Nishijima, *Prog. Theor. Phys.* 13, 285 (1955).

[163] K. Niu, E. Mikumo, and Y. Maeda *Prog. Theor. Phys.* 46, 1644 (1971).

[164] K. Niu, *Proc. Jpn. Acad., Ser. B* 84 1 (2008).

[165] S. Okubo, *Prog. Theor. Phys.* 27, 949 (1962).

[166] S. Okubo, *Phys. Letters* **5**, 165 (1963).

[167] M. Park, E. Leahey, and R. J. Funk, *Nature* **613**, 138 (2023).

[168] Particle Data Group, pdg.lbl.gov/2023/listings/contents_listings.html (2023).

[169] W. Pauli, Dissertation, Munich, 1921, *Ann. Physik* **68**, 177 (1922).

[170] L. Pauling, *College Chemistry*, 3rd ed., W. H. Freeman and Co., San Francisco, CA (1964).

[171] R. E. Peierls, "Wolfgang Ernst Pauli, 1900–1958," *Biogr. Mems Fell. R. Soc.* **5**, 174 (1960).

[172] D. H. Perkins, "Neutrino Interactions," *Proceedings of the XVI International Conference On High Energy Physics, Vol. 4* p. 189, J. D. Jackson and A. Roberts (eds.), Fermilab, Batavia, Il (1972).

[173] M. L. Perl, et al., *Phys. Rev. Lett.* **35**, 1489 (1975).

[174] A. Petermann, *Nuclear Phys.* **63**, 349 (1965).

[175] J. R. Pierce and G. Zweig, Electrical Stimulation of the Acoustic Nerve, Proposal to the HEW-Public Health Service (October 3, 1973).

[176] J. R. Pierce and G. Zweig, "Some Old Problems and New Challenges in Hearing and Communication," *Biosc. Commun.* **1**, 111 (1975).

[177] F. Ravndal, "How I Got to Work with Feynman on the Covariant Quark Model," *50 Years of Quarks*, pp. 127–148, F. Fritzsch and M. Gell-Mann (eds.), World Scientific, Singapore (2015).

[178] G. D. Rochester and C. C. Butler, *Nature* **160**, 855 (1947).

[179] M. Roos, *Rev. Mod. Phys.* **35**, 314 (1963).

[180] J. Rosner, *Phys. Lett. B* **385**, 293 (1996).

[181] E. Rutherford, *Phil. Mag.* Series 6 **21**, 669 (1911).

[182] E. Rutherford, *Phil. Mag.* Series 6 **27**, 488 (1914).

[183] E. Rutherford, *Phil. Mag.* Series 6 **37**, 537 (1919).

[184] E. Rutherford, *Phil. Mag.* Series 6 **37**, 581 (1919).

[185] E. Rutherford, *Proc. Royal Soc. A* **97**, 686 (1920).

[186] E. Rutherford and J. Chadwick, *Phil. Mag.* Series 7 **4** (22), 605 (1927).

[187] E. Rutherford, "The Development of the Theory of Atomic Structure," *Background to Modern Science: Ten Lectures at Cambridge Arranged by the History of Science Committee*, Reprint Edition, p. 68, J. Needham and W, Pagel (eds.), Cambridge University Press, UK, (2015).

[188] S. Sakata and T. Inoue, *Prog. Theor. Phys.* **1**, 143 (1946).

[189] S. Sakata, *Prog. Theor. Phys.* **16**, 686 (1956).

[190] J. J. Sakurai, *Phys. Rev. Lett.* **9**, 472 (1962).

[191] A. D. Sakharov, *Pisma Zh. Eksp. Teor. Fiz.* **5**, 32 (1967).

[192] P. Schübelin, *Physics Today* **23** (11), p. 32 (1970).

[193] J. Schwinger, *Phys. Rev.* **74**, 1439 (1948).

[194] H. K. Shepard, *Phys. Today* **42**, 100 (1989).

[195] J. R. Smyth and G. Zweig, "The Geochemical Classification of Fractionally Charged Particles," *LA-UR 87-2219*, Los Alamos National Laboratory (1987).

[196] W. R. Smythe, "Static and Dynamic Electricity," Edwards Brothers, Inc., Ann Arbor, MI. (1936), and McGraw-Hill, New York, NY (1950).

[197] D. F. Styer, M. S. Balkin, K. M. Becker, M. R. Burns, C. E. Dudley, et al., *Am. J. of Physics* **70**, 288 (2002).

[198] E. C. G. Sudarshan and R. E. Marshak, "The Nature of the Four Fermion Interaction," *Proc. Padua-Venice Conf. on Mesons and Newly Discovered Particles*, September 1957, reprinted in *Development of the Theory of Weak Interactions*, pp. 119–128, P. K. Kabir (ed.), Gordon and Breach, New York, NY (1963).

[199] E. C. G. Sudarshan and R. E. Marshak, *Phys. Rev.* **109**, 1860 (1958).

[200] I. Tamm, *Nature* **133**, 981 (1934).

[201] J. J. Thomson, *Phil. Mag.* Series 6 **7**, 237 (1904).

[202] M. A. Tuve, N. P. Heydenburg, and L. R. Hafstad, *Phys. Rev.* **50**, 806 (1936).

[203] S. Ulam, *Adventures of a Mathematician*, p. 70, University of California Press, Berkeley, CA (1976).

[204] R. L. Walker, D. C. Oakley, and A. V. Tollestrup, *Phys. Rev.* **89**, 1301 (1953).

[205] S. Weinberg, *Phys. Rev. Lett.* **27**, 1688 (1971).

[206] S. Weinberg, "The Search for Unity: Notes for a History of Quantum Field Theory," *Daedalus* **106** (4), p. 17, The MIT Press, Cambridge, MA (1977), www.jstor.org/stable/20024506.

[207] S. Weinberg, "Changing Attitudes and the Standard Model," *The Rise of the Standard Model: Particle Physics in the 1960s and 1970s*, p. 36, L. Hoddeson, L. Brown, M. Riordan, and M. Dresden (eds.), Cambridge University Press, UK (1997).

[208] S. Weinberg, *The Quantum Theory of Fields, Vol. II*, p. 231, Cambridge University Press, Cambridge, UK (2005).

[209] S. Weinberg, *Third Thoughts*, p. 108, Harvard University Press, Cambridge, MA (2018).

[210] S. Weinberg, *Foundations of Modern Physics*, p. 190, Cambridge University Press, Cambridge, UK (2021).

[211] W. I. Weisberger, *Phys. Rev. Lett.* **14**, 1047 (1965).

[212] J. A. Wheeler, *Phys. Rev.* **52**, 1107 (1937).

[213] K. Wilson, *Phys. Rev. D* **10**, 2445 (1974).

[214] F. L. Wright, *An Autobiography*, pp. 15 & 17, Faber & Faber Limited, London (1943).

[215] C. S. Wu, E. Ambler, R. W. Hayward, D. D. Hoppes, and R. P. Hudson *Phys. Rev.* **105**, 1413 (1957).

[216] Y. Yamaguchi, *Progr. Theoret. Phys. Suppl.* **11**, 1 (1959).

[217] H. Yukawa, *Proc. Phys. Math. Soc. Jpn.* **17**, 48 (1935).

[218] H. Yukawa, S. Sakata, M. Kobayasi, and M. Taketani, *Proc. Phys. Math. Soc. Jpn.* **20**, 319 (1938).

[219] F. Zachariasen, *Phys. Rev. Lett.* **7**, 112 (1961); erratum, ibid, 268 (1961).

[220] G. Zweig, "An SU_3 Model for Strong Interaction Symmetry and its Breaking," *CERN Report 8182/TH.401* (January 17, 1964); //cds web.cern.ch/record/352337?ln=en.

[221] G. Zweig, "An SU_3 Model for Strong Interaction Symmetry and its Breaking II," *CERN Report 8419/TH.412* (February 21, 1964). Also in *Developments in the Quark Theory of Hadrons, A Reprint Collection, Volume I: 1964–1978*, pp. 22–101, D. B. Lichtenberg and S. P. Rosen (eds.), Hadronic Press, Inc., Nonantum, MA (1980); cds.cern.ch/record/570209?ln=en.

[222] G. Zweig, *Il Nuovo Cimento* **XXXII**, No. 5, 689 (1964).

[223] G. Zweig, "Fractionally Charged Particles and SU_6," August 27, 1964, *Symmetries in Elementary Particle Physics*, pp.192–234, A. Zichichi (ed.), Academic Press, New York, NY (1965); www.researchgate .net/search.Search.html?type=publication&query=Fractionally%20Charged%20Particles%20and%20SU6.

[224] G. Zweig, "Meson Classification in the Quark Model," Caltech Report CALT-68-162 and *Meson Spectroscopy*, C. Baltay and A. H. Rosenfeld (eds.), W.A. Benjamin, New York, NY (1968).

[225] G. Zweig, *Science* **291**, 973 (1978).

[226] G. Zweig, "Origins of the Quark Model," *Baryon 80: Proceedings of the Fourth International Conference on Baryon Resonances*, pp. 439–479, N. Isgur (ed.), University of Toronto, Canada (1980), www-hep2.fzu.cz/~chyla/talks/others/zweig80.pdf.

[227] G. Zweig, "Memories of Murray and the Quark Model," *Proceedings of the Conference in Honour of Murray Gell-Mann's 80th Birthday*, pp. 7–20, H. Fritzsch and M. Gell-Mann (eds.), World Scientific, Singapore (2011). Also in *Int. J. Mod. Phys. A* **25** (20), 3863 (2010).

[228] G. Zweig, "Concrete Quarks," *50 Years of Quarks*, pp. 25–55, H. Fritzsch and M. Gell-Mann (eds.), World Scientific, Singapore (2015). Also in *Int. J. Mod. Phys. A*, **30** (1), 1430073 (2015).

[229] G. Zweig, *J. Acoust. Soc. Am.* **138**, 1102 (2015).

[230] G. Zweig, *J. Acoust. Soc. Am.* **139**, 2561 (2016).

Index

For EU product safety concerns, contact us at Calle de José Abascal, 56–1°,
28003 Madrid, Spain or eugpsr@cambridge.org.